"Peter (Zak) Zakrzewski and David Tamés go beyond the hype to deliver a long-overdue critical and robust methodology for creating a 'better human+computer ecosystem.' Comprehensive, well-researched, and drawn from practical experience, this is not only a 'how to' make but a 'how to think about' book that helps practitioners to deeply leverage the full range of expressive affordances offered by immersive tech."

Celia Pearce, *Professor of Game Design, XR and game designer and author of* PLAYFRAMES *and* Communities of Play

Mediating Presence

Mediating Presence: Immersive Experience Design Workbook for UX Designers, Filmmakers, Artists, and Content Creators provides the foundation for a broader understanding of the impact of XR as the next medium. The methods presented draw from a wide range of disciplines and professional practice to provide a practical guide to design methods, production techniques, best practices, and terminology that provides an effective road map for developing immersive experiences using augmented reality, virtual reality, augmented virtuality, and other emerging immersive media forms.

In this book, media designers Peter (Zak) Zakrzewski and David Tamés present a comprehensive framework for XR media experience design (XRXD) that will allow UX designers, filmmakers, artists, and content creators to connect their audiences to rich, multimodal, immersive experiences for entertainment, learning, creating, and healing. This book guides the reader through a 12+1-step design-based model for creating XR experiences. Each step is accompanied by specific media design methods expressly created for or adapted to XR content creation. The theoretical concepts and methods covered address the specific context and meaning aspects of the XR media environments being created.

Peter (Zak) Zakrzewski is a design strategist, researcher, and design and innovation coach who advocates for applying transformative design thinking and participatory design methods to the creation of media, experiences, and social change. He is an assistant professor of communication and visual arts at Thompson Rivers University. He is a former creative director of the emerging technologies conference EmTech at MIT. He is a recipient of the MIT President's Citation Award for his research, design, and strategy work for the MIT Enterprise Forum.

David Tamés is a media maker working in traditional and immersive media forms. His creative practice focuses on media co-creation and immersive media production. David recently completed an MIT Open Documentary Lab Fellowship, conducting research for this book. He is a teaching professor in the Department of Art + Design at Northeastern University with a joint appointment in communication studies. David is on the board of directors of Filmmaker's Collaborative and serves on the advisory board of MedVR. David has produced award-winning short films and recently co-directed and co-produced *Reality's Fugue: Genesis*, a virtual reality experience that provides a new take on the Genesis origin story to suggest how we create a shared world.

Mediating Presence

Immersive Experience Design Workbook for UX Designers, Filmmakers, Artists, and Content Creators

Peter (Zak) Zakrzewski and David Tamés

Routledge
Taylor & Francis Group
NEW YORK AND LONDON

First published 2025
by Routledge
605 Third Avenue, New York, NY 10158

and by Routledge
4 Park Square, Milton Park, Abingdon, Oxon, OX14 4RN

Routledge is an imprint of the Taylor & Francis Group, an informa business

ISBN: 978-1-032-58084-5 (hbk)
ISBN: 978-1-032-58083-8 (pbk)
ISBN: 978-1-003-44246-2 (ebk)

DOI: 10.4324/9781003442462

Typeset in Times New Roman
by Apex CoVantage, LLC

Access the Support Materials: https://xrxdesign.com

Peter's Dedication
I dedicate this book to the students in my immersive media class at Thompson Rivers University. You are the future of extended reality, and extended reality is your future.

David's Dedication
I dedicate this book to Jan Kubasiewicz and Gunta Kaza, designers who taught me to think of design as experience and to embrace theory as illuminating what is implicit in creative practice.

Research for this book was conducted during an MIT Open Doc Lab Fellowship 2021–22 and 2024; special thanks to the fellows, faculty, and staff for their support and encouragement, especially William Uricchio.

Contents

Acknowledgments

We thank Daniel Kershaw, Danielle Zarfati, and Alyss Barraza at Focal Press for their support during the production of this book. We express our gratitude to Kimberly Husband and Kate Fornadel for their assistance with the copyediting of the final manuscript. We would like to thank all of the people in the XR community who generously shared their experiences with us to expand our understanding of this new and rapidly evolving medium: Zev Berman, Christopher Boulton, Danger Donaghy, Bernard Francois, Rus Gant, Sofie Hodara, Jeffrey Jacobson, Oscar Keyes, Eileen McGivney, Vittorio Netti, Patrick Nugent, Steven Max Patterson, Celia Pearce, Nicolas Robbe, Mandy Rose, Mark Sivak, Graham Roberts, J. J. Ruscella, and Jamal Thorne.

Introduction

Engulfed in Flames: The Most Persuasive Medium in History

It was a cool and windy but otherwise inauspicious day in 1990 when David walked into the San Francisco Museum of Modern Art (SFMoMA) to see a new media exhibition.[1] Little did he know at the time that the show he was about to experience would be remembered as a landmark event that symbolically marked the birth of an imperative for a new-media art movement. Perhaps more importantly to our story here, it also became a formative experience that shaped David's fascination with the transformative potential of immersive media.[2] Strolling through the exhibition with no specific idea of what to expect, David came upon Jim Campbell's installation titled *Hallucination*. He was caught in the gaze of a black-and-white surveillance camera and saw himself on a large rear-projection video monitor. Just as he was about to turn to a friend to wonder out loud what this was all about, he witnessed himself being engulfed by flames. David watched the electronic reflection of his digital self burn up in vivid color. The accompanying sound of a crackling fire heightened the eerie sensation. This was no ordinary museum experience. The immersive out-of-body quality was accompanied by a distinct awareness of the artifice of what was happening. As David stared at the monitor, a woman appeared in the video reflection. Sometimes, she observed the environment passively. Other times, she would interact with it. Seeing another person seemingly enter the exhibit, David turned, expecting to see a real person behind him, only to realize that figure was an apparition in the virtual world. This uncanny sensation triggered a mild sense of anxiety followed by a philosophical inquiry: What was going on here regarding observable reality? This was 1990, a world in which e-mail was the killer app. The closest even the most adventurous techno geek got to the magic of the world of shared information was the disconnected text-based terminal screens.[3] Personal computers of the day only provided a tiny window onto the data stored on their own hard drives. They were devices used primarily for word processing, spreadsheets, e-mail, or simple graphical applications for those lucky enough to own a Macintosh or a Unix workstation. In this prehistoric context, *Hallucination* dramatically signaled how far computational media could and had to go and how it would expand our understanding of what it meant to participate in computationally mediated experiences.

DOI: 10.4324/9781003442462-1

The exhibit curator, Robert Riley, wrote that his "measure of art" was "seduction" manifested when the audience became "absolutely convinced by the authority of a work" to believe that "this is the way things are." Riley suggested that profoundly moving immersive content could become "the ultimate exchange of humanity, wisdom, and intellect that we have."[4] A work like *Hallucination* stood out because it moved beyond the confines of the installation and media artworks seen before and into a realm that dissolved the boundary between a participant's subjective experience and a computational environment. Campbell's piece achieved a high level of engagement by embedding the participant in the work as both an actor and an observer interconnected in the production process. Presence was enabled by a novel computational component that performed the image processing required to composite the flames onto the moving bodies of the spectators. The piece went beyond the mechanical interface expected from museum exhibits. Gallery visitors who wandered into the space became immersed in the virtual world created by the artist. In a moment of epiphany, David saw how this new art form would use software as a medium and data as material. In the fire of *Hallucination*, he saw the possible future of immersive computational media. Notions of the computer as a medium, our connection with computational objects, and new-media art, as both substrate and form, came to David in that single moment of combustion. The reverberations of experiencing *Hallucination* revealed to him that it would be just a matter of time before personal computers would collectively transport us into what we know today as extended reality (XR). In the mid-2020s, we acknowledge that the technology is not quite ready for mass adoption of XR beyond VR gaming and specialized applications; but we also argue that the time is now for content creators to not only get ready for those soon-to-come technological developments but to start co-creating a richer and more humane immersive media. What a time it is to be alive for this massive media revolution not just as a witness but as an active participant.

Dreaming of Better Human+Computer Ecology

Peter's immersive conversion occurred during his doctoral research, eventually culminating in the book *Designing XR: A Rhetorical Design Perspective for the Ecology of Human+Computer Systems*. Over the course of five years, Peter had an opportunity to observe the new domain slowly but steadily emerging from experimentation in the areas of ubiquitous computing, wearable devices, artificial intelligence (AI), the Internet of Things IoT (IoT), and dedicated augmented reality (AR) and virtual reality (VR) technologies. Since the early days of computation, the interaction between humans and computers has become increasingly intimate. In the span of six decades or so, we have gone from room-size computers operated by engineers in white lab coats to everyday people pointing, clicking, and typing on their personal devices to kids walking around with smartphones, calling and texting their friends by touching responsive screens. With the rapid co-mingling of AI, XR, IoT, and robotics, the emergence of immersive human+computer (H+C) intelligence augmenting systems seemed like an inevitable conclusion.

When the *Spacewar!*-playing computational media pioneers like Stewart Brand dreamed of putting the "funky hands" of everyday people on the emerging computer technology to "do tricks for funky heads," they did not imagine that the gift of personal computing was going to come with a baggage of mental health and other wellness issues (Scott, 2015, p. 103). Personal computers have expanded our ability to perform creative work and accomplish cognitive tasks more efficiently. Mobile devices gave us the first taste of spatial computing, allowing us to be informed, entertained, and connected on the go. Yet spending hours in a head-forward posture with eyes fixed on our screens has created generations of people whose bodies are chronically tight and sore to the point of creating a pain epidemic. The more "natural" user interfaces of XR can solve many of the problems associated with the graphical user interface (GUI) paradigm by adapting to their human users, but they will come with their own challenges. Our embodied minds have a language of their own. Our nervous system sets the landscape for all of our experiences. The stories that define our lives by reminding us who we are and how we see the world we live in are built on the platform of our biology. No medium in the history of humanity ever had the potential to reach deeper into the human nervous system than the emerging medium of XR. As our symbiosis with computers grows more and more intimate, we need to make sure that our computational media can truly complement rather than hijack our neurology. We need the next medium to respect the embodied nature of the human experience in the physical and social worlds.

We believe that there is a need to make the next computational medium not just more human centered but more humane. The unique design affordances of immersive media make XR experience design (XRXD) a fundamental building block of the new chapter in our relationship with computers. The complex and the sensitive nature of building H+C systems requires XR system designers to engage in ecological thinking that can restore the power balance between human users and increasingly advanced and complex computer systems. The immersive experience design we describe in this book offers users an opportunity to engage in augmented imagination by means of immersive transportation that provides access to context-relevant information in the key areas of understanding the self, others, and the environment. It can lead to a truly immersive and multimodal access to advanced computation that supports and promotes physical and mental well-being.

XR: A New Media-Making Paradigm

Philosopher of science Thomas Kuhn in his influential book titled *The Structure of Scientific Revolutions* defined paradigms as "universally recognized scientific achievements that, for a time, provide model problems and solutions for a community of researchers" (Kuhn, 1996). Paradigms describe what phenomena are to be observed and scrutinized, what kinds of questions are supposed to be asked and probed, and how these questions are to be structured. A field's paradigm provides the members of a community of inquiry with a set of problems, a philosophical stance, and a set of rules on how to look at these problems and how to address them. The human–computer interaction (HCI) field is a paradigm dedicated to the problems associated with the rise of personal computing since the 1970s. HCI researchers made a number of contributions applicable to a generation of computers that were operated by human users interacting with a device via the device-centered windows, icons, menus, and pointers (WIMP) style of interaction that predominantly addressed the visual sensory channel. Graphical user interfaces using WIMP made great strides toward multimodality from the interaction with punch cards, line printers, command lines, and machine language programming. Personal computer manufacturers gradually dedicated more hardware to the auditory channel. Improved chips allowed for faster, higher-resolution video display that encouraged media scholars and makers to think of computation as a medium. Mobile computing added the basic sense of touch to the human–computer interaction paradigm with taps and swipes. Yet the HCI paradigm remains fixated on interaction between human users and computing devices rather than on the creation of conditions conducive to immersive presence.

People who grasped the true potential of human+computer systems considered the "desktop metaphor" to be only a transitory state of human–computer interaction. One of the greatest challenges in the transition to immersive media is our ability to rethink the fundamental media making paradigms that are rooted in audiences looking at flat 2D screens. As immersive experience designers, we need to collectively reimagine how to create interactions centered around the body actively navigating 3D environments that become an active part of the user interface. Muriel Cooper, the founder of the Visible Language Workshop at the MIT Media Lab, transcended the GUI desktop paradigm by dreaming of *Information Landscapes*. Working with her students throughout the 1980s, armed with a Silicon Graphics Reality Engine $250,000 computer, Cooper undertook a series of projects that explored such topics as interactive calendars, infinite information zooms, typography in space, multiperspective data presentation, and personalized information galaxies in which information was organized and filtered in a dialogue between an algorithm and the reader (Maudet, 2017). Cooper and her team found their design inspiration not in a traditional two-dimensional, page- or screen-restricted graphic design but in three-dimensional forms of theater, performance art, and Bauhaus happenings. Their explorations illuminated five unique features of the computer enabled information landscapes: (1) three-dimensionality of (2) multimodal media spaces, temporally navigated by (3) actively participating audience members within (4) nonlinear time frames through (5) multiple access points not directly controlled by the designers.[5]

One of the major premises behind this book is our argument that moving from HCI to immersive media making will require a major paradigm shift. The approaching wider adoption of XR as the newest communication medium will bring the society at large and XR media-making community into a new domain of artificially designed immersive sociotechnical systems that far surpass the complexity of the era of personal computing. This book describes why humane-centered XR experience design will matter a great deal to this brave new world. It outlines how the new paradigm is qualitatively different from HCI design. It also lays out a framework of how to go about creating meaningful immersive experiences. The still-emerging XR technologies have already substantially expanded the range and control over objective stimuli available to XR designers. XR content creators can engage in multimodal composition that lays the foundation for subjectively experienced cognitive and emotional states of immersive presence—a close coupling of advanced computation and the human nervous system. XR media offer media makers five unique areas of opportunities and challenges that make immersive experience design unique. (1) The XR continuum from augmented (AR) to mixed (MR) and virtual reality (VR) is the first type of media in history to allow its makers to engage in computational-based world-making that allows its audiences to enter synthetic experiences that appear real to the human nervous system. (2) XR content makers accomplish immersivity by practicing multimodal composition based on designing for the eight sensory channels.[6] (3) XR systems are moving us toward the world of human+computer immersion in which computers, unlike their predecessors, use sensors and artificial intelligence (AI) to adapt to the human users rather than forcing them to learn how to navigate the newest version of the underlying operating system. XR environments can offer users unprecedented bio and neuro feedback, hence giving design teams an opportunity to customize XR experiences and even elicit meaningful behavioral change. (4) Based on advances in computational social science and social physics, XR media systems can be expressly built for enhancing collective intelligence. (5) Immersive media systems can be designed for collective action and purposeful emergence within the ecologically balanced socio-technical systems that offer humane intelligence augmentation. We will expand on all of these points throughout the book.

The Future of Content Creation

Hollywood movies like *Minority Report*, *Ready Player One*, and *The Matrix* have introduced many of the concepts associated with extended reality (XR) into popular culture. XR technologies have steadily become cheaper, higher in quality, and more widely available to both enterprises and consumers. Billions of people can experience mobile AR on their phones to navigate physical places using the Google Maps app or to play *Pokémon GO*. Head-mounted displays such as Meta's Quest and HTC Vive allow gamers of all ages to immerse themselves in VR worlds of popular games such as *Beat Saber*. *XR* today is an umbrella term for three related but qualitatively different forms of media. Virtual reality applications require the audience to use headsets and other hardware to become fully immersed in computer-simulated worlds. VR hardware generates realistic stimuli that engage all five external senses of vision, audition, olfaction, somatosensation, and even gustation to create immersive virtual worlds. Augmented reality (AR) applications rely on devices such as mobile phones, tablets, or headsets to overlay digital data over real-world environments. Mobile AR apps are already familiar to those who might have used the IKEA Place app. Mixed reality falls somewhere in between VR and AR. MR hardware allows users to switch between and blend real and virtual environments where physical and digital elements can interact in real time. Some examples of MR headsets are Microsoft's HoloLens 2, Meta Quest Pro 3, and Apple Vision Pro. MR users can see their surroundings as in AR but mixed with interactive 3D objects as in VR.

Up to now, we have been consuming flat media, which we usually view from the third-person perspective. Immersive media will not just give us a front-row seat; it will place us in the middle of the action. XR is an invitation to climb into the world of data and become part of the immersive information system we refer to as the XRI world. In this new world, instead of watching digital content, we will become immersed in spatially presented experiences that surround us with a richness of virtual presence. When we eventually collectively enter the XRI world, we will immerse ourselves into an environment designed to act upon us as much or more than we can act upon it (Rosenberg, 2022). The fundamental shift predicated on the unique conditions of XR media will dramatically change how we access digital information and how it affects us. The first phase of the transition into XR will continue to be mobile friendly. VR experiences such as VRChat, *Fortnite*, and Roblox can be accessed on TV screens, tablets, web browsers, and mobile phones. But as Unity CEO John Riccitiello argues, the currently popular apps and games like Instagram and *Grand Theft Auto* that rely on touch screens and physical controllers will translate poorly to the multimodal world of XR (Wingfield, 2021). Immersive media will eventually compel media designers to reinvent the way their products are offered in the world of XR media in the same way that the invention of sound changed moviemaking. As the technologies mature, the ability to provide natural-language and gestural interfaces will define immersive experiences.

The youngest generation of media audiences regularly spends time in virtual spaces such as *Horizon Worlds*, *Fortnite*, *Minecraft*, and *Roblox*. Megabrands like Nike and Gucci are following this audience, staging events and product launches in these emerging virtual spaces. XR applications are already improving the effectiveness and efficiency of education and skills training, enhancing customer experience, and raising productivity in areas from product design and prototyping to assembly to equipment service, sales, and marketing. The wide rollout of 5G networks will enable the deep integration of XR technologies into every aspect of our lives, transforming how our society learns, shops, and makes critical decisions. Educators are using a VR platform, Engage, to allow their students to build virtual geothermal stations on the ocean floors. The architects who designed Hong Kong International Airport used the XR simulation engine Unity to create fully functional models of the facility, including changes in passenger volume and airplanes that responded to weather and air traffic conditions. Unity is currently being used by nongaming customers such as Volkswagen AG to develop virtual experiences to train employees. We believe that imaginative media makers working with XR technologies will open new frontiers that are hard to imagine today. Opening up immersive media to more creative minds and hands is the motivation behind creating this book. That said, a variety of XR applications are already quietly making social and economic impact. We believe that XR media will disrupt all areas of media making, but here are a few examples of changes already taking place in such areas as entertainment and gaming, marketing and advertising, e-commerce and retail, healthcare, education, and travel and tourism.

Entertainment and Gaming

The biggest promise of XR is the ability to deliver experiences that are not attainable in the physical world. In her book *Computers as Theatre* (1991), Brenda Laurel suggests that computational media experiences should be designed with the principles of theatrical storytelling, focusing on creating engaging, meaningful experiences for participants. Her framework has had a lasting influence on immersive experience design, especially in experiences in which narrative is crucial for participant engagement. This framework was elegantly realized in *Placeholder* (Laurel et al., 1994) created by Brenda Laurel, Rachel Strickland, and their team at Interval Research, which integrated virtual reality, spatial audio, and embodied interactivity to orchestrate a two-person immersive experience. Participants explored three distinct virtual environments—a cave, a waterfall, and hoodoo formations—based on real-world locations in Banff National Park as they embodied one of four archetypal animal avatars (Crow, Spider, Fish, and Snake) while moving objects and leaving "voice marks" that added narrative layers. Using "smart costumes," which altered participants' avatars, voices, and sensory perceptions based on their chosen animal, created unique embodied experiences (Laurel et al., 1994). *Placeholder* highlighted the potential of using immersive media's sensory and narrative affordances to orchestrate dynamic experiences emphasizing embodiment and interactivity over passive observation. *Placeholder* drew from anthropology, mythology, and psychology to demonstrate how virtual environments could evoke personal and collective stories, challenging conventional notions of storytelling. The use of spatial audio and symbolic avatars foreshadowed the current emphasis on mediated presence and sensory engagement in XR experience design and provides a reference point illustrating how immersive technologies can be applied to create emotionally resonant and intellectually stimulating virtual spaces focused on creating a sense of "being there," now recognized as central to XR experience media. Echoes of *Placeholder* can be felt in the multiuser interaction and asynchronous communication currently in *VRChat*. BeAnotherLab's *The Machine to Be Another* draws directly from *Placeholder's* emphasis on embodiment to explore empathy and learning with immersive media. Narrative VR experiences such as *Dear Angelica* and *The Under Presents* owe their nonlinear, exploratory storytelling structures to the foundational work demonstrated by *Placeholder*.

It comes as no surprise that gaming and entertainment currently constitute the number-one XR market segment worth billions of dollars. Gamers playing such games as *Half Life: Alyx*, *Horizon: Call of the Mountain*, and *Gran Turismo 7* have embraced XR technologies to immerse themselves in the virtual worlds that bring more visceral and exciting challenges, adventures, and exploration. Various flavors of *Fortnite* allow players to indulge their social side by cooperating to fight off zombie-like creatures, forming teams to defend castles, create new worlds, and compete to be the last person standing. When Anifa Mvuemba's Hanifa brand's New York Fashion Week runway show was canceled due to COVID-19, the designer transformed all of her garments into 3D images and fitted them onto avatars. The entire show became a virtual event streamed over Instagram

Live. Music and sports fans can now virtually experience live music and sporting events from the comfort of their VR and MR headsets. Electronic music pioneer Jean-Michel Jarre used the platform as a venue for his 2020 New Year's Eve event titled Welcome to the Other Side and for a concert titled Alone Together. Welcome to the Other Side was broadcast simultaneously across radio, television, and online platforms inside the virtual Notre Dame Cathedral. It attracted more than 75 million views.

Disney's 2019 *The Lion King* might be seen as a bridge between traditional and immersive forms of filmmaking. The movie was made possible by an experimental form of production that leveraged VR to shoot virtual sets along with traditional direction and analog camerawork. Filmmakers created a computer-generated virtual world and characters that could be walked through and filmed while wearing a VR headset. The process was akin to playing a video game aimed at creating a movie. During preproduction, *The Lion King*'s crew worked on master scenes using the industry-standard special effects Maya system. Using a custom-built asset management system, they translated the scenes into digital files compatible with the 3D game engine Unity. Key crew members, including director Jon Favreau, wore HTC Vive headsets to explore the virtual world as they walked through a physical set. Immersed in the VR environment, they could make creative decisions and discuss the camera placements, lenses, and locations of actors, sets, and props. The production team created 350 premade skies to make every shot uniquely fit the director's creative vision. Shooting in a 25-foot-square area called the Volume, Favreau and his crew moved a simulated camera around the stage to identify the best camera placements and movements. The chosen three-dimensional flight path of the camera was tracked inside the virtual world, allowing the team to record the master scenes in Unity. The virtual camera was designed to simulate the Arri Alexa 65 body and Panavision 70 Series cinema lenses to give the film a realistic documentary style reminiscent of an episode of the BBC's *Planet Earth* docuseries. The resulting 3D footage was delivered to the animators in London, who combined it with other assets and animations into the movie's final version. *The Lion King* will go down in cinema history as Hollywood's first, but most likely not the last, virtual production to show the creative potential of XR filmmaking (Giardina, 2019).

Less well-heeled content creators can conduct XR filmmaking experiments on platforms like VRChat, where their audiences can use sensors and haptic technologies to experience rather than just watch movies. A new breed of XR filmmakers is well exemplified by Joe Hunting, whose *We Met in Virtual Reality* documentary was shot entirely within the *VRChat* platform. The work examines immersive escapism during the days of the COVID-19 pandemic, showing a burgeoning cultural movement demonstrating the real power of immersive connection and digital intimacy. Hunting's subjects include such avatar performers as Jenny, who teaches American Sign Language (ASL) to classes of virtual students. Another avatar by the name DustBunny teaches virtual belly dancing. The film also follows two users named DragonHeart and IsYourBoi, who, despite living thousands of miles apart, have fallen in love on VRChat. Hunting's film is a testament to how VR platforms can offer users a chance to be free from social expectations associated with gender, sexual identity, or physical limitations by becoming animated entities or colorful anthropomorphic creatures (Daniels, 2022). Experimentation in immersive filmmaking has only just begun, so it is not wise to predict where it will take us. We know one thing, though. The next *Up* or *Star Wars* will not be told like the last one, with one exception: imaginatively presented stories which deeply resonate with audiences are not going away anytime soon.

Journalism

Nonny de la Peña created the new field of immersive journalism by integrating VR technologies with journalistic storytelling. Her work has demonstrated how immersive media can foster empathy and emotional engagement and contribute to a better understanding of social, political, and humanitarian issues. David remembers participating in de la Peña's *Project Syria* at the 2014 Sheffield Doc/Fest; at the time, it was his first encounter with VR journalism, and it left an indelible mark that is felt to this day. Participants are placed amid the Syrian civil war, bringing them directly into the realities of the conflict. *Project Syria* uses real audio and video footage combined with computer-generated imagery to recreate scenes of devastation and human suffering. Participants in VR headsets enter the room-scale VR experience, which begins in a bustling Syrian marketplace filled with the sounds and sights of vendors, children, and daily life. Suddenly, your walk through the marketplace is disrupted by the explosion of a bomb. You become a witness to the chaos, confusion, and cries for help. As the experience unfolds, you are then transported to a refugee camp to witness the struggles of displaced families, including children scavenging for resources and individuals coping with loss. Combining arresting visuals, spatial audio, and movement creates a beguiling sense of presence. Participants can look around, take in the devastation and human suffering, and form an emotional connection to the events portrayed in a manner that is not as easy to do with traditional media. *Project Syria* attempts to foster empathy and raise awareness of the refugee crisis by placing people amid harrowing scenes. The experience provides one of the most compelling examples of immersive journalism shifting participants from passive observers to active participants by proxy.

At the same time, the effectiveness of immersive media in communicating complex social issues remains an active area of debate and research. Regardless of the debates on whether virtual reality is or is not an empathy machine, Nonny de la Peña's work demonstrates that immersive media can be used for more than entertainment or escapism. XR experiences can invite people to engage with complex social realities and catalyze meaningful discussions. de la Peña's work, of which *Project Syria* is but one among many others, provides participants with a first-person perspective to enhance the emotional impact and has influenced subsequent immersive media projects focused on empathy-building and understanding complex social realities, such as *The Key* (Celine Tricart, 2019) and *We Live Here* (Rose Troche, 2020).

Media Art and Scientific Visualization

Osmose (1995) was the first virtual reality experience installed in a contemporary art museum. Char Davies, a co-founder of the 3D animation software company Softimage, created this work, deeply rooted in her interest in phenomenology and her background as a painter (McRobert, 2016). Participants fitted with a custom suit and headset navigated through a translucent, flowing environment populated by abstract forms and naturalistic imagery. They navigated the virtual space through a unique interface that responded to their breath and balance. This experimental form of natural interface encouraged participants to connect with their physical sensations in distinct contrast to the traditional computer interfaces available at the time. *Osmose* blurred the boundary between the participant's bodies and the virtual environment, inviting them to engage with immersive media as an extension of their inner and physical realities. Her work challenges conventional notions of space, perception, and interaction, highlighting the potential of immersive media as a vehicle for introspection, emotional resonance, and reembodiment, and her bio-responsive design influenced later developments in media art, natural user interfaces, and therapeutic uses of immersive media (McRobert, 2016). Char Davies's artistic vision provides the conceptual and experiential insights that prefigured and influenced the evolution of immersive media toward specialized and application-driven uses like scientific visualization like *BioVR*, bridging art and technology to expand how we engage with emotional and scientific complexities. While *Osmose* pioneered an artistic and meditative exploration of virtual environments, BioVR applies similar principles to scientific visualization, focusing on DNA, RNA, and protein structures. The ground broken by *Osmose* is evident in *BioVR's* embodied interaction, intuitive spatial exploration, and emphasis on natural input methods, allowing users to manipulate 3D molecular models with hand movements (Zhang et al., 2019).

Branding, Marketing, and Advertising

The XR world is already seeing a rise in a new breed of designers and architects focused on creating virtual places for immersive spatial experiences populated by AI bots, customized avatars donning digital fashion, objects, and accessories. Working for clients such as McDonald's, *Vogue*, and the Utah Jazz, design consultancies such as Polycount have been trailblazing this type of design since 2001. For the Qatar World Cup 2022, Polycount collaborated with the Swiss luxury watchmaker Hublot to transform a virtual stadium designed by MEIS Studio into a fully realized metaverse environment for soccer fans. Hublot's official Qatar World Cup 2022 watch became "the central design element for the virtual stadium" (Polycount, 2022). The virtual build included "15 Hublot brand ambassadors stationed around the pitch, each holding a specially designed ball and room for ninety-thousand virtual fans" (2022). The Hublot Loves Football Stadium was opened to the public on a social VR and live event platform, *Spatial*, which has more than 800,000 monthly visitors. The experience is available to fans via desktop, mobile, and Oculus. In another project, Polycount partnered with fashion designer Humberto Leon and IW Group to design a metaverse space celebrating the Lunar New Year. The Hall of Zodiacs exhibit, sponsored by McDonald's, was hosted on *Spatial* and *AltspaceVR*. It featured a spatial experience that allowed the visitors to explore the twelve Chinese zodiac animals. The animals were each placed in front of banners describing the years when their particular attributes became dominant. Visitors accessing the Hall of Zodiacs received personalized horoscope readings based on the year and month of their birth and the associated zodiac animal. The campaign was cross-promoted on Asian television news channels and through traditional gifts distributed at McDonald's stores.

E-Commerce and Retail

XR can contribute to both online and offline shopping experiences. Customers visiting brick-and-mortar stores can quickly and easily learn about products on display without seeking assistance from store staff. Mobile AR–enabled devices can recognize barcodes and text on the labels to provide shoppers with information about product features, ingredients, reviews, and promotions. Retailers can use the same features to assist their employees in inventory control. They can also offer personalized discounts to customers based on shopping pattern data. Product returns are estimated to cost online retailers hundreds of billions of dollars annually. In the realm of both e-commerce and traditional retailing, this can be reduced by letting shoppers virtually interact with the digital twins of the products before buying. XR technologies, rather than forcing consumers to imagine how the products will look and feel, can let the users experience the product firsthand. For brick-and-mortar giants like Macy's whose home goods and furniture product catalog features much bulkier offerings than Amazon, returns are a necessary but quite costly evil. To combat the problem, the company launched an experimental "try before you buy" program in 2019 using AR technology that allowed in-store shoppers to design and furnish their own virtual room mock-ups on iPad stations. Customers could step into virtual spaces to look around the room in 3D. A traditional furniture area in a retail store can take up thousands of square feet of retail space and cost about half a million dollars to build. The experimental space only took up 500 square feet and cost less than $50,000 to build (Boland, 2019), providing a glimpse of the potential for XR in retail sales.

Healthcare

There are multiple areas in which XR technologies combined with biosensors, machine learning, and predictive analytics can contribute to healthcare. These areas include mental well-being, physiotherapy, improved diagnostics, telesurgery, and education for both medical professionals and patients. Medical imaging technologies like magnetic resonance imaging (MRI) and CT scans can give medical staff 3D representations of the patients' bodies that offer a much greater level of detail than traditional 2D imaging. People like Jeffrey Jacobson, the XR Lead of

Immersive Design Systems at Boston Children's Hospital, are reenvisioning the future of medical training with XR tools like the HoloLens. Jacobson and his team developed an AR pharmacy training module using simulated needles and drug vials in addition to actual sterile cleaning supplies to assist in teaching the pharmacy personnel the fundamentals of aseptic technique for sterile compounding. The system allows up to four students to simultaneously train with the supervision of a senior pharmacy educator to decrease the amount of time needed to complete initial competency training from a full day to just 40 minutes. The system has significantly improved the pass module rates in competency tests to 90%.

Education

XR-based learning experiences are expected to grow in influence in education at all levels from elementary to high school to university and corporate and professional training. A study conducted by the National Training Laboratory found that VR-based learning modules have a retention rate of 75% compared to 5% for the lecture-style learning experiences and 10% for reading (Carlton, 2017). XR can also dramatically reduce the cost of training while providing educators with a lot of useful data on the learners' progress. Field trips have traditionally been powerful learning experiences for students of all ages, but tightening school budgets combined with the rising transportation costs and complexity of logistics have reduced these experiential learning opportunities. With a comparatively lower cost, schools can now use the spectrum of XR technologies to give their students immersive experiences that will positively impact their learning. AR technology allows students to get additional information about what they see. One of the main benefits of VR simulations is that they allow learners to make critical errors and experience real consequences within a safe environment. VR modules can enable students to explore complex scientific topics in unprecedented detail by manipulating 3D models of molecules. They can use multisensory engagement to experience what it would be like moving in outer space or performing a surgery on virtual patients. German car parts manufacturer Bosch uses the Oculus Rift headset to take its 10,000-strong army of service technicians on three-dimensional tours of the inner workings of various car parts to enhance the technicians' understanding of the functioning of direct-injection and braking technologies.

The Workplace

The gaming industry's success in popularizing VR technologies has now paved the way for the integration of XR in professional settings. Immersive media can effectively reconnect a hybrid workforce and remote employees in a more collaborative, immersive environment. Virtual workspaces enable remote and hybrid teams to collaborate, communicate, and increase productivity without a need for expensive and time-consuming travel. Immersive workspaces can also foster creativity by encouraging willingness to take risks and enabling inexpensive experimentation in a safe environment. Hyundai Motors supports its numerous teams engaging in multiple rounds of design feedback through its world's largest VR design evaluation facility, allowing 20 designers, engineers, and key stakeholders to collaborate simultaneously on a car design in a virtual environment. The setup combines high-performance Nvidia GPU-powered systems with VR headsets to let the teams modify the virtual car's colors, shapes, textures, and parts in real time. Design teams can view high-fidelity renderings of the car's interior to understand better what the car buyer may experience at the dealership. Designers can also place cars in real-world backgrounds to better assess every facet of the design.

XR brings unprecedented opportunities for on-the-job training in healthcare, manufacturing, technical support, and service industries to help employees learn hands-on and in the field. AR-based training can provide employees with practical experience in a realistic environment without the need for expensive physical equipment or facilities. Six Flags Amusement Park trains up to 40,000 seasonal employees using a custom-designed AR training module that allows trainees to interact with virtual customers. The immersive training module uses tablets rather than expensive cash registers to let the new employees test their point-of-sale (POS) system skills under simulated stress and to help them understand typical scenarios they will experience while on the job. The immersive program has improved employee training and scaled up training while cutting costs for the park.

Travel and Tourism

The COVID-19 pandemic had a devastating effect on tourism, but it also prompted creative travel operators to apply XR technologies to offer novel stay-at-home travel experiences. There are four ways in which XR can benefit both the consumers and the travel industry.

Virtual travel offers consumers the opportunity to travel the world safely while saving money and reducing environmental impact. Virtual travel allows visitors to experience destinations that are remote or need to be preserved from traffic. For instance, the *Patagonia VR* experience available from Meta's app store lets users explore Argentina's Monte Fitzroy mountain landmark and the neighboring remote glacial lake Laguna Sucia, both of which are hard to reach in real life. Virtual hotel tours let the operators of hotels and upscale resorts show off their stunning locations and world-class amenities to prospective visitors in immersive detail before they book. XR can also help travelers to immersively test drive different trips and excursions available at their chosen destinations. Mobile AR apps can help travelers with navigation. *Google Maps* now incorporates an AR feature called Live View for those who are navigating unfamiliar locations on foot. The feature is available on all ARCore and ARKit-enabled mobile devices for any locations where Google already has the Street View option. Live View overlays big arrows and easy-to-follow directions onto the user's screen over the street view, clearly showing them where to walk and turn (Marr, 2021).

Designing XR Media, and Why It Matters to You as a Media Maker

Media scholar Henry Jenkins summed up an insight concerning media disruption caused by networked computers by articulating a concept of media convergence. According to Jenkins, convergence goes beyond a mere technological change to create deeper social, cultural, and industrial transformation. The first stage of convergence brought all forms of content from print to photography, recorded music, and cinema into the world of ones and zeros. Web 2.0 added broadcasting and long-distance telephony to the mix of mediation available over the Internet. Mobile devices allowed audiences all over the world to access digital media on the go. This technological shift allowed media makers to integrate content and to widely disseminate it through multiple access points and platforms. Media audiences developed novel media habits and protocols, moving freely between platforms while consuming content, sharing content, and most importantly, developing an expectation of participation from likes, comments, and full-blown fan related content posted on social media channels. This cultural convergence linked media audiences, their communities, and media producers in ways that inspired new forms of creativity. Jenkins likened the product of convergence culture to traditional folk culture in which songs, gossip, and stories were transmitted freely throughout the community.

The trend of media convergence has not stopped. Extended reality might be the greatest hybrid of media technologies and protocols to date. XR is a blend of all forms of narrative media, gaming, apps, live events, and human–computer interaction. In our approach to immersive media, we are aligned with Laura Ermi and Frans Mäyrä's tripartite framework, which shows immersion as a product of three overlapping categories of sensory experiences, challenge-based (ludic) interactions, and imaginative (narrative) approaches. Within these three complementary domains, media makers can identify separate sources of rich opportunities to engage their diverse audiences. The multimodal composition familiar to user experience (UX) designers engages the audience on the deeply embodied sensory level. Immersive media projects also benefit from the lessons learned from the task-oriented software design and play-oriented game design traditions. XR applications can offer their users effective, efficient, engaging, error-tolerant, and easy-to-learn ways to accomplish life- and work-related tasks and challenging or relaxing play experiences. Finally, narrative immersion adds another dimension by drawing on the long and rich tradition of storytelling, books, and cinema.

The development of the early cinema offers a great parallel to immersive media. The medium started without a "language" of its own. Cinema pioneers such as Fritz Lang, Dziga Vertov, Elizaveta Svilova, Esfir Shub, Sergei Eisenstein, Alice Guy-Blaché, and F.W. Murnau had to invent new conventions that could help audiences to make meaning based on the series of flickering images projected on the walls of the makeshift movie theaters. The early masters who canonized this visual language told stories within the constraints of "silent" cinema. They had little choice but to engage in mostly visual storytelling.[7] Later filmmakers such as King Vidor, Alfred Hitchcock, Orson Welles, Akira Kurosawa, and Luchino Visconti used postsynchronous sound technology to immerse their audiences further by combining sight with sound. Since the 1960s, media makers have attempted to create a greater sense of immersion by experimenting with additional sensory channels and spatial dimensions. In the 1950s, Hollywood cinematographer Morton Heilig (1992) proposed that all the sensory intensity of life could be simulated with what he called "reality machines" (p. 240). Heilig saw what we now call virtual reality as a natural extension of cinema. He articulated a powerful vision of media makers controlling the multisensory stimulation that could provide audiences with an illusion and sensation of being present in virtual worlds. Heilig argued that "the cinema of the future would become the first art form" to present audiences with new worlds in "the full sensual vividness and dynamic vitality" of human consciousness (p. 239). He believed that the expanded form of media making would involve taste, touch, and smell in addition to sight and sound. Such multimodal cinema would be capable of dissolving the fourth wall of traditional movie theater performances to transport audiences into inhabitable virtual worlds he referred to as "experience theater" (p. 240). Heilig backed up his vision by inventing an arcade-style machine he appropriately named Sensorama. It was a mechanical device comprising a stereoscopic 3D color display, stereo-sound system, various fans, scent emitters, and a motional motorcycle chair. The immersive experience involved a motorcycle ride through Brooklyn. Participants experienced riding through the streets of a busy city, sitting on an imaginary motorcycle, while hearing the cars passing by, smelling petrol fumes and local pizza shops. Heilig's surrogate travel device did not find proper financial backing and did not become a popular success, but it influenced a generation of VR pioneers in the 1990s who sought to realize Heilig's vision of inhabitable movies using computers as a medium.

Another field advancing the multimodal language that has dominated media since the second part of the 19th century started with early magazine layouts for such publications as *Harper's* and *The Atlantic* and advertising for such iconic brands as Coca-Cola and Campbell's. The UX and UI canons behind such digital products as games, apps, and websites started from the tradition of communication design still frequently referred to as graphic design. Much of what has been taught in graphic design programs is based on the modernist canon established in Weimar Germany at such institutions as the Staatliches Bauhaus, known as the Bauhaus. The field further developed after WWII in Switzerland and the United States at such institutions as the Basel School of Design and the Institute of Design at the Illinois Institute of Technology, originally founded as the New Bauhaus. The graphic design field originally developed in an era when media technologies separated the senses to offer radio, film, and photography as distinctive forms of mediated communication. Various foundational figures in the field, such as Bruno Munari, stressed the visual aspect of communication to distinguish the predominantly print-based graphic design from discursive forms of communication such as literature. The emphasis on the visual aspect of communication design continued with the contribution designers made to the development of the graphical user interface (GUI). Since the days of the early personal computer, such as Apple's Macintosh, design has become the common denominator behind all computational media. More advanced forms of computational media offered expanded opportunities for the new generations of new media designers to evolve the field into more multisensory interaction and experience design. Today, UX/UI design is a well-developed field whose main focus on making advanced media technologies, such as mobile computing, more usable and accessible to the wider audience can greatly benefit the emerging multimodal XR medium.

Computer-as-a-medium has not only allowed media makers to think in terms of an added dimension of multimodal interaction, it has also allowed them to think in terms of systems. Two fields that simultaneously emerged from the personal computer era were digital games and software design. Both of these fields evolved around the concept of rule-based systems. The main task of both game and software design is to shape virtual play and work spaces by rules which allow the users maximum freedom while being easily internalized—two requirements that are essential to usability. Both gaming and UX/UI designers explicitly recognize the vital role users play in any type of interactive system. Like play, presence—the ultimate goal of immersive media design—cannot be coerced. Presence is a state that emerges from the audience members interacting with immersive systems. Traditional and digital games create powerful experiences because, like stories, they tap into some fundamental aspect of human nature. Play permeates all expressions of human culture. We grow up and learn to socialize by playing games. Digital games have captured the magic inherent in such activities separated by space and time as pre-Columbian ball playing and Chinese checkers. Gaming has evolved from consoles found in arcades, to Pong to Sega and Nintendo which were plugged into TV sets of suburban living rooms, to Sony PlayStation, PlayStation2, and the Microsoft X-Box. The medium has now expanded into highly social online and mobile gaming and VR games. Writing in an introduction to his 2001 book *Information Anxiety 2,* Richard Wurman predicted a "revolution" that will see gaming move from first-person shooters "to business gaming, problem solving and decision-making, vicarious travel, and ultimately to flying through information—perhaps all human knowledge" (Wurman, 2001, p. xxvii). He saw the potential of gaming to become a powerful tool for learning, strategic decision-making, and navigation of information. Whether gaming will become anything other than gaming, game designers have developed a language that cannot be ignored in any immersive media design. For this reason, the cross-platform game engine Unity has become one of the most popular authoring environments for AR and VR applications. From innovative user interfaces to experience design strategies, gaming has created a robust media design toolbox that is closer to Muriel Cooper's vision of spatial computing than GUI's flat desktop metaphor.

In the context of XR design, the convergence of media is much more than just the convergence of technologies; it is a convergence of methodologies. Bringing knowledge from a diverse set of immensely accomplished fields can shape immersive media experience design in ways that will only expand the media makers' creative potential. We see the merging of the methods, languages, and terminology from UX, game design, and narrative storytelling fields as the core challenge of developing a new immersive experience design (XRXD) paradigm. The tripartite cross-disciplinary approach combining the narrative, the challenge-based, and the multimodal sensory media-making traditions underpins the immersive media experience design framework we present in this book.

I Am the Camera: Immersion Is More Than Interaction

In contrast to the critique advanced by the opponents of rhetoric, we believe that in all forms of rhetorical communication, meaning is co-created between the media makers and the audience. Audiences always have to work hard to generate meaning out of a series of stimuli supplied by the medium. XR media makers, more than their filmmaking and design counterparts, need to embrace the fact that immersive content creation involves an act of co-creation. Each immersive media audience member becomes the co-director and the camera operator through their choice of gaze. The choices they make provide them with a unique perspective, path, and sequence through the immersive experience. This is why our approach to immersive experience design comes, first and foremost, not from the world of computer code but from neuroscience. It is based on the fundamental understanding that art or any aesthetic experience is only possible because the human brain actively constructs its own experiences. What our eyes, ears, and other sensory organs receive are not meaningful sights, sounds, and smells but "a barrage of light waves, chemicals, and changes in air pressure with no inherent significance" (Barrett, 2020, p. 65). Faced with an onslaught of ambiguous scraps of sense data, the brain, whose core task is to control the body to keep us alive and well, must figure out what to do next. The brain meets this challenge by drawing on a lifetime of past experiences, things that have happened to us personally and things that have happened to our ancestors. What we believe we see, hear, touch, taste, and smell is a combination of what is present in our physical environment and what is constructed by our brains (2020). When we recognize a human figure in a painting or on a screen, we understand what we see because of the memories of all human figures we have experienced that help our brains make sense of the represented abstract elements. Based on an intuitive grasp of this phenomenon, Marcel Duchamp argued that an artist is only responsible for 50% of the work necessary to create art. The remaining 50%, he suggested, was the domain of "the beholder's share" (2020, p. 69). Immersion, like all experience, is the domain of conscious human meaning-making, which emerges from the active, co-creative process of constructing our private sentient experiences.

Ultimately, the XR's success depends on us collectively answering the question: What sort of resonant experiences that are not possible today can immersive media make possible tomorrow? XRs future is not about the technological conditions of immersion but about what Sanchez-Vives and Mel Slater refer to as presence. They draw a clear distinction between presence and immersion, suggesting that immersion is a product of technology, while presence is an internal physiological and psychological state experienced by participants immersed in XR environments (Sanchez-Vives & Mel Slater, 2005). Presence is an emergent phenomenon, a product of immersive technology eliciting a subjective state in the user. The ability of

XR designers to create a sense of presence in the participant is the foundation of immersive transportation. A break-in-presence spells the breakdown of the virtual illusion. Glitches at the level of the physical and technological substrate, such as loss of tracking or bumping into objects in a physical environment, are detrimental to XR experiences.

One of the unique characteristics of XR is that it affects the human sensorimotor system in a way that prompts the brain to create internal psychological states that transport participants beyond the immediate physical environment and into the virtual world of digital media. Virtual reality researchers refer to the simulated sense of "being there" and "acting there" as presence. In the context of XR, presence can be understood as a form of visceral communication (Jerald, 2016). Virtualized presence can take many forms. We can experience mental and embodied presence or a social and emotional one, when virtual environments immersively mediate social interaction. Presence can be defined in at least two ways. One research perspective views presence as a subjective mental state in which a user "feels physically present within the computer-mediated environment" (Sanchez-Vives & Slater, 2005, p. 4). Researchers who consider reality to be formed through action rather than through mental filters define presence as the ability of the XR system to successfully support the participant's actions in the virtual environment (2005). Under this definition, the perceived reality of the user experience is relative to the functionality of digital objects and environments rather than their appearances (2005). While the focus of the experience-centered approach is on the creation of virtual environments that support the sense of "being there," the action-centered approach aims to create environments that support a targeted range of user actions and their agency to "act there."

Mel Slater argues that presence is based on two primary sensory illusions: the illusion of place and the illusion of plausibility. The place illusion addresses the question of location, while the plausibility illusion convinces the brain that whatever is happening is real. Technologically driven factors such as low-latency head tracking, accurate 1:1 body and eye tracking, and fast framerate lead the human perceptual system to believe that the person is present in a place other than their real physical location. A virtual world that accurately meets our expectations in the way it behaves and reacts to our actions makes us perceive the XR environment as being coherent in its construction (Bye, 2017).

The Wild West of XR Technology

There is no point hiding the fact that one of the greatest obstacles for media makers getting into XR is that the field resembles a Wild West when it comes to its technical standards. Developing XR experiences demands a solid understanding of the underlying immersive media technologies that can let the media makers develop a clear feasibility strategy to negotiate the fluidly changing landscape of a number of proprietary standards competing for market dominance. Currently, XR applications can be developed as mobile augmented reality (AR) apps for Android or iOS, VR/MR games and apps for Quest 3, Microsoft HoloLens, Apple Vision Pro, and other headsets. We can add to that a lineup of devices in the "smart glasses" category, which represents an AR-focused heads-up display (HUD) one can wear on one's face while walking on the street. These include Viture Pro XR, Rokid Max, Chamelo Music Shield, and Ray-Ban Meta Smart Glasses. There are also 360-degree video projection systems that record and play back in 360 degrees of azimuth and 180 degrees from nadir to zenith, viewable on flat displays or through spherical projections. There is simply no one way to immerse an audience in an XR experience at the moment. We cover the most important building blocks of immersive experience (XR) design to help media makers coming from outside of XR become more familiar and more comfortable with the most promising XR technologies.

The Rise of AI: Immersion Meets Our Digital Unconscious

XR lives in the shadow of another much more hyped technology. The rapid adoption rate of OpenAI's chat generative pretrained transformer (ChatGPT) app should convince even the most skeptical of us that the technological sea change we are living in the midst of is showing no signs of waning. ChatGPT reached its first one million users in only five days. It took such tech superstar start-ups as Instagram a whole month, Facebook ten months, and Netflix more than three years. General-purpose technologies such as the steam engine, the internet, and AI share a few common characteristics including continuous improvement, widespread proliferations, and complementary innovations. Watching the rapid advances in AI confirms that we are on the brink of another massive shift in human history. One thing is for certain: since the time of the Cognitive Revolution estimated to have occurred around 70,000 years ago, humanity has never faced an entity that can surpass our own powers of cognition. It is this artificial form of intelligence that XR will immerse its users in. The quietly-brewing-in-the-background extended reality revolution and AI are destined for each other. As they combine into a single domain, they will immerse the modern *Homo sapiens* in the world of hybrid human+computer (H+C) systems that go far beyond the old metaphor of human–computer interactions and into the domain of autonomous agents, chatbots, and virtual assistants. These intelligent conversational agents will be able to adapt to domain-specific scenarios and handle a variety of complex tasks through the application of open-source frameworks such as TaskWeaver. Natural language processing (NLP) and machine vision will become core technologies underpinning immersive experiences. From chatting bots to eye and hand tracking, AI will help shape immersive experiences, and XR will offer an immersive entrance point into the world of advanced computational media.

The combined world of XR, IoT, and robotics will be populated with all kinds of AI-based helpers capable of giving well-researched legal advice, translating between languages, offering mental health support, creating art, and writing code. With AI by its side, XR will take us from the age of information (Web 2.0) and into the age of knowledge. As good as this sounds, there is, as usual, when it comes to novel technology, one caveat. While knowledge might represent a higher level of consciousness, it still does not automatically rise to the level of wisdom, which is the ability to take purposeful and ecologically sound action in a complex and uncertain world.[8] The knowledge provided by AI systems might not hand us wisdom on a silver platter, but it can play a role in the process of making better decisions. Wearable AR devices will allow us access to mission-critical (yes, we mean shopping, but also things like navigating culturally rich urban environments and manufacturing-guided assembly) information while moving around the physical world with our heads up and our minds fully present. XRI world will replace typing commands with gesture and eye and voice commands fluently interpreted by AutoGen-powered agents as devoted to us as Mr. Carson was to the masters of Downton Abbey.

XRI immersion will be available at the shopping mall, the construction site, the street, and the classroom. The good news is that by using GPS-style systems, we will be able to navigate completely unfamiliar physical or cognitive environments. The downside of this cognitive offloading might be a gradual but pronounced change in the way that we process navigational information. People who follow GPS instructions do not form cognitive maps of their environments. This is poignantly illustrated by the case of the Ski-Doo–riding and GPS-dependent young generation of Inuit hunters in Northern Canada who are no longer able to follow oral directions using such references as wind direction, snowdrifts, memorably shaped rock, and the old Inuit place names that do not appear on maps. Traditional directions created by people traveling slowly and mindfully by dogsled while creating cognitive maps of the world around them mean nothing to the modern generation, which is used to moving fast by following clear-cut "turn left, turn right" GPS directions (Kemp, 2022). Finding the right balance between empowering users to explore unknown territories and depriving them of their natural ability to pay attention and form cognitive maps will be one of the many new challenges of XR immersion facing media designers.

AI technologies are also enabling creative professionals using AI to extend their skills and time to create magical 3D environments, objects, and code. Soon after the release of ChatGPT, Andrew Sibert, a lead product designer at gaming and mobile developer Backbone announced on X (Twitter) that he had been able to use the AI app to build a fully functional notes app in under an hour. The Swift-based app created by the ChatGPT's NLP interface only required minor debugging. OpenAI's generative image-making model, DALL-E, can create realistic art from a description given in natural language. OpenAI's Sam Altman suggests that the image-making app has crossed the critical threshold to produce photorealistic images that suggest a creative intelligence. DALL-E 2 makes an appearance of understanding meaningful cultural concepts well enough to combine images that elicit emotional responses from people viewing them. Altman described working with DALL-E to be akin to "talking to a colleague who's a graphic artist" (Heaven, 2022). The AI tools will not eliminate the role of designers and content creators from the process. Still, they will lower the price of entry and offer new efficiency and exploratory capabilities. Gaming, perhaps more than any other media so far, has taken advantage of the technologies forming the foundation of immersivity. AI is taking the nonplayer characters (NPCs) in digital games to the next level, allowing them to advance the narrative and enhance gameplay in surprising and rewarding ways.[9] NPCs can inspire all AI-powered digital objects that can go far beyond simple chatbots. Designers can express digital objects' infinite action potential by turning them into instigators of action, helpers, tutors, and advisors. We cover the application of Julien Greimas's Actantial Model to inspire the creative search for imaginative AI-based actants.

The Most Rhetorical Medium of All

Rhetoric has acquired a somewhat checkered reputation since the time of Plato. Even today, rhetoric can still be associated with duplicity, political blustering, and overdramatic posturing. Plato, who saw Sophists, the major proponents of rhetoric, as mortal enemies of his beloved teacher, Socrates, posed one of the earliest and most aggressive critiques of rhetoric. Brian Vickers (1989) argues that Plato's hostility to any form of oratory was motivated by his mistrust of the power it gave the sophisticated rhetors operating within the Athenian democracy whose social structure and the legal system jointly condemned Socrates to death. Once delegated to the study of style, rhetoric was resurrected in the 20th century by such communication scholars as I. A. Richards, Kenneth Burke, Chaim Perelman, Marshall McLuhan, and James Crosswhite to represent the communicative dimension that forms the basis of human coexistence. All forms of mediation and media making contain at least some element of persuasion. Media makers who narratively transport us into the imagined worlds of superheroes and supervillains implicitly persuade us to embrace the logic of the world not as it is but as it could be.

Jerome Bruner distinguished between two modes of ordering our experience and of structuring reality. Good stories and well-formed arguments can both be used as means for convincing others of our worldview, yet they convince in fundamentally different ways. Arguments use procedures for establishing formal and empirical truth. Stories draw their persuasive power by establishing not truth but verisimilitude or lifelikeness (Bruner, 1986, p. 11). Immersivity like narrative transports the audience into the world made by the media makers. One of the key premises of this book is that extended reality is a highly rhetorical medium, very likely more rhetorically potent than any other medium in history. This does not mean that XR must necessarily be manipulative. On the contrary, we who are collectively co-creating this new medium must make sure that it is not. Coercion and persuasion are two entirely different phenomena. As Peter argued in his previous book, ensuring the user's freedom within the context of immersive systems involves two components: "the existence of a set of genuine alternatives, and informed choice, which refers to the user's ability to choose their course of action" (Zakrzewski, 2022, p. 187). Rhetorical XR design is based on the willing participation of conscious human actors who are empowered "to make informed choices by selecting from the possible options of future actions" (p. 187). Immersive systems aiming to be persuasive rather than coercive must be designed to provide audiences with a wide array of choices for participation. User agency is a critical dimension of ecologically designed immersive systems. Despite the persuasive power of immersion, presence is not something that designers can create directly. While designers can manipulate the objective immersive stimuli to elicit target cognitive and emotional states, presence is a private and subjective state generated by the audience based on their life context and goals.

Co-Creating Humane XR: Your Mission, Should You Accept It

American historians Melvin Kranzberg and Carroll W. Pursell articulated a set of insights about technology. They saw technology not as a mysterious, alien entity but as a very human activity. They humorously suggested that invention was the mother of necessity. In other words, once people saw a technology they liked, they would increasingly find more uses for it until usage became a regular habit. They pointed out that while technology was deeply implicated in many social issues, the nontechnical factors involved in creating and adopting novel technologies were far more consequential than the technical ones (Kranzberg & Pursell, 1967). Media technologies do not develop in a vacuum. The human-made substrate of our material realities develops over time within specific social, cultural, and ideological contexts. Frequently, it has been the storytellers imagining the future who articulated the most powerfully resonant cultural imperatives our technologies eventually deliver to broader audiences. Science fiction writer Isaac Asimov gave us the "Three Laws of Robotics" in his 1942 short story *Runaround.* The laws were conceived as the organizing principle for his fictional robotic-based universe. Asimov also introduced us to the concept of autonomous vehicles with his "robot-brain cars." Philip K. Dick's *Minority Report* described, admittedly in quite unsettling ways, the idea of predictive analytics. The process of prescient storytelling, which Neal Stephenson refers to as "making shit up," has also given us the early vision of the metaverse (Robinson, 2017). Stephenson's *Snow Crash* not only painted a vivid picture of virtual reality but imbued it with such devices as avatars and digital currency. Enchanted objects have always played a unique role in fictional worlds, empowering characters to act in supernatural ways. Long before the cryptocurrency boom and blockchain technology, Stephenson's virtual currency represented a resonant vision of freedom and virtual value exchange in cyberspace. Adopting the Hindu concept of the "avatar," Stephenson used it to describe the digital twins of the metaverse and captured the possibility of virtual representation as a way toward flexible exploration of identity.

One of Kranzberg and Pursell's most astute observations was that any technology in and of itself was neither good nor bad, nor was it neutral. Technologies do not, by themselves, create dystopias or utopias. The human purpose and particular applications of technology make technology a critical component of social phenomena. At the moment when Web 2.0 and mobile app innovations have run their course, the future of XR is emerging from the present, powered up by the rich imaginative experiments conducted by immersive content creators who are not limited by yesterday's engineering boundaries. Immersive media innovators are leveraging XR's basic building blocks to envision novel ways to address core human needs, including expression of identity, socialization, exchange of value, storytelling, and co-experiencing the world of human creativity. While mainstream media commentators obsess over the vicissitudes of the corporately owned version of the metaverse, many other immersive experiments are bubbling up from the collective imaginations of XR content creators. For anyone looking closely, these emergent signals are coming from many directions. This new world is being built bottom up in a way Neri Oxman might refer to as a growth-not-assembly mode.

XR technologies offer immersive media makers unprecedented opportunities to push the boundaries of creativity in the new and unexplored area of designing for presence. Suggesting that XR will alter how humans relate to one another and the world around us is an understatement. XR applied to education can help to place students into navigable experiences that explore otherwise daunting abstract concepts. In medicine, it can assist medical professionals in visualizing patient data. AR can overlay digital information onto our physical spaces such as tourist destinations and retail stores to provide just-in-time information for all kinds of daily decision-making. But for all its incredible potential, by further blurring the physical and virtual boundaries, XR media can present a higher level of the potential risk to the users' well-being than its Web 2.0-era digital predecessor. Extending our symbiosis with advanced computation will face its fair share of potential hazards, including issues involving evidence-based reality, trust, power asymmetry, and mental health. In a report titled *Waking Up to a New Reality*, researchers at strategic consultancy Accenture outline three areas of risk associated with the wide adoption of XR: manipulation of data profoundly connected with personal identity, intimate behaviors, and thoughts; direct connections to not-yet-fully-understood human mental faculties and their perceptions of reality; and the irreversibility, opaqueness, and complexity of the distributed and decentralized advanced digital tools (Accenture Research, 2019).

We are only starting to grasp the implications and possibilities of manipulating the codes that underlie our societies' fundamental structures. Media scholar Marshall McLuhan poetically suggested that we effortlessly step into the media we consume as we step into a warm bath. Media environments are seductive by design. This is why they just feel right to the point that we do not think critically about the medium being the message. McLuhan asked us to think of media as complex and dynamically evolving systems. When Sherry Turkle cautioned us about computational media, arguing that the price for lowering the barrier of entry into computing for the masses by way of usability would lead to simulation without understanding, she addressed the issue of deep psychological effects media technologies can have on the users. She suggested that a societywide acceptance of taking things offered by computer systems at "(inter)face value" could have unintended consequences for the well-being of individuals and society as a whole (Turkle, 1997, p. 78). Social media examples we cover in what follows confirm the prescience of this prediction. As the computing medium grew more complex, the increasingly advanced software engineering was masked by the magic of UX design, making its simulations more and more opaque. A growing cadre of UX designers made the digital illusion appear real by blending the mastery of visual communication with key insights from cognitive science and neuroscience. Over the years, Silicon

Valley's digital disruptors have become increasingly skilled at seducing users to seamlessly adopt increasingly advanced computation into their everyday lives. The two parallel developments of the opaque complexity and the persuasive power of UX design have combined forces to spur the growing symbiosis of humans and computers, with the balance tipping toward those shaping and controlling the technology (Zakrzewski, 2022). We wrote this book based on the belief that the highly consequential task of creating the new immersive medium will be best accomplished through the participation of many creative individuals coming from diverse fields and perspectives. Our collective task as immersive experience designers is to make sure that our co-creation ends up benefiting all of its users.

The Structure and the Magic: Trusting the Process

While many of the specific software and hardware tools that are currently used in the XR field will undergo extensive transformation over the next few years of rapid development, the principles, process, and methods we discuss will remain valid for many generations of XR technologies. As VR pioneer Jaron Lanier argues, the most important question to ask about any novel technology is how it affects people. While we cover the technological feasibility in this book, we focus on the human experience first. Every novel technology offers media designers certain unique affordances that can shape the user experience in creative ways. This was true for film, radio, television, personal computers, and mobile phones. This is why this workbook is concerned primarily with form rather than the specific content. We leave the creative challenge of sharing meaningfully immersive experiences to the reader. What we present here is the comprehensive process that includes a suite of principles, methods, tools, and building blocks uniquely suited to creating XR content. We hope that the guidelines we propose, like grammatical syntax, while grounding the form of expression based on the realistic affordances of XR, will allow content creators near-infinite communication creativity in this new and exciting medium. The disciplined framework we propose will offer XR media makers a solid foundation of a clear and explicit strategy for their work. The magic in media design comes from the empathic co-creation between the designers and their audiences. The structure we propose aims to support the kind of magic that happens at the subjective moment of presence. By focusing on conceptual tools and methods enabling immediate exploration, we hope to encourage broader participation in the co-creation of the emerging XR medium. Our focus on immersive experience design highlights the critical importance of ensuring that the new XR medium is, above all, a humane one.

Seminal design theorist Herbert Simon suggested that every problem-solving effort must begin with creating an appropriate problem representation. With that in mind, Simon introduced the concept of a problem space within which the search for solutions can take place. The XRXD logic model we present in Section II outlines the 15 steps within the five areas of the XR media makers problem space. It is meant to help XR media makers to act strategically and purposefully by being able to focus on the particular features of immersive system design that are relevant to the context of their unique immersive experience. The model addresses the logical structure within the problem space by constructing a symbolic representation for particular classes of immersive system design problems. The outline of this book follows the phases and the steps of the model filling details at each progressive step. The theoretical concepts and methods covered in the following chapters address the specific context and meaning aspects of the XR media environments being created. The theater of the mind created by the human nervous system is the magical place within which all content creators operate. Before we enter the world of this immersive theater, let us discuss a structure for designing immersive experiences that fully explore all kinds of the magic that this wonderous new medium will allow.

Chapters at a Glance

Before moving to the specific theoretical underpinnings, process, and methods of XR media design, we lay the foundation for a broader understanding of the impact of XR as the next medium. We introduce the importance of immersivity as the next phase in human+computer symbiosis and our interpretation of the importance of this new medium in Section I: "What Is Immersive Media?" We introduce the concept of immersive experience design (XRXD) and discuss the design model in detail in Section II, titled "What Is XRXD?"

Section III, "Discovery and Pre-Suasion," starts to delve into the details of the XRXD model. The general concept of applying logic models to transformative design has its theoretical basis in the research on policy planning and evaluation in nongovernmental organizations and social enterprises. Logic models are conceptual tools that help to explicitly describe the proposed theory of change used to activate the intervention components such as program inputs, processes, and outcomes necessary to accomplish social change. The early version of the XRXD logic model was first introduced by Peter in his book *Designing XR*. We build and expand on the model and use it as a guiding set of simple rules to help designers identify key steps toward creating meaningful immersive experiences. The model comprises five major phases: (I) Discovery and Pre-Suasion, (II) Ideation and Inscription. (III) Planning and Building, (IV) Launching and Learning (From Play), and (V) Evaluating and Iterating.

Section III covering Phase I (Discovery and Pre-Suasion) of the model includes three chapters. In Chapter 1: "Vision and Purpose," we cover the why behind the project and outline some steps toward developing a strong project vision. This step in the model is focused on answering the following question: What do we want to accomplish with this project? What is our vision for the outcomes of this project? How will the audience be transformed by this XR media experience? In Chapter 2: "Understanding Your Audience," we highlight the importance of clarifying and developing a strategically empathic understanding of the intended audience. The chapter guides the reader through the process of discovering the micro insights about audience interests, goals, needs, and motivations. It also explores the fundamental desirability (people and culture) building block of persuasive media design.

This step is focused on answering the following questions: Who is the audience? What do we know about our audience and their media consumption needs? What themes might resonate with our audience? Chapter 3, titled "Developing Your Value Hypothesis," continues the audience identification theme with guidelines for the development of a well-defined value proposition for the core audience. It guides the reader through the process of answering the question: What value are we providing for our audience?

Section IV covers Phase II (Ideation and Inscription) of the model and includes three chapters describing the next three steps of the immersive experience creation process. Chapter 4, titled "Creating the Context Scenario," describes the process of leveraging narrative as a design tool. In this chapter, we help the reader answer such questions as: What will the immersive experience be like? How will the audience interact with the system? How can the XR system best respond to the user to support their goal (effectiveness) efficiently while being engaging, error tolerant, and easy to use? In Chapter 5: "Engineering the Magic: The Feasibility of Immersive Experiences," we introduce the XR technologies as the core building blocks of multimodal media composition. This chapter is aimed at helping the reader to estimate core competencies needed for the project, including internal and external resources, skills and processes, hardware, software, timelines, and budgets. The methods presented in the chapter guide media makers through a series of questions that test their leap-of-faith hypotheses about the project including (a) the availability of technologies to support the desired XR experience and (b) the team's core competence regarding the essential activities and processes needed. Chapter 6: "Building the Proof of Concept" describes the importance of testing the key leap-of-faith hypotheses covering the design teams' assumptions regarding the value provided to the audience, the feasibility/usability of the experience, and the scalability potential of the proposed project.

Section V covers Phase III (Planning and Building) of the model and includes three chapters describing the next three steps of the immersive experience creation process. Chapter 7: "Designing the System Prototype" delves into systems thinking theory and the process of applying these principles into the design of digital products. The chapter discusses systems design principles as applied in gaming and software design to give users a more rewarding experience. In Chapter 8, titled "Building the Functional System Prototype," we cover the basics of creating functional prototypes for immersive experience products. XR applications pose some of the most technologically complex challenges, and navigating them requires both a disciplined process and applied knowhow. In Chapter 9: "Playtesting the Prototype," we discuss the important stage of bringing the audience into the system. We also share some of the approaches to the evaluation of the initial user feedback.

Section VI (Launching and Learning) covers Phase IV of the model and includes three chapters describing the last three steps of the immersive experience creation process. Chapter 10: "Attracting the Audience" covers the various strategies for inviting the audience into the designer-inscribed XR system. Like any other digital product or media experience, XR applications need to attract core audiences to scale. We share some of the practices from such fields as gaming, software, and media marketing to outline the possibilities available to XR media makers in a way of attracting participants to their offering. In Chapter 11: "Immersing the Audience," we cover the important step of managing the audience entry into the immersive experience. We discuss the importance of incentives and shaping strategies. We also cover the relationship between such media design concepts as "System as Designed," "System as Encountered," and "System as Understood" by the participants. In Chapter 12: "Extending Play," we talk about building durable transmedia franchises by extending play beyond the original XR application. The focus of Chapter 12 is based on the research done by media scholar Henry Jenkins and his industry collaborators, Joshua Green and Sam Ford, who coined the term "spreadability." Extending play involves a process of successfully blending the key attributes of the XR media experience to appeal to motivations of particular user communities in a way that inspires them to embrace the XR media franchise and share content with others in their network.

In the final Section VII, "Iterating Immersive Systems," we discuss the truly unique opportunities afforded to XR media designers for auditing and evolving XR systems toward their intended function. The section returns to such areas of system design as designing the ecology of sociotechnical systems and optimizing XR systems for collective intelligence. The chapter also outlines Donella Meadows's concept of leverage points and her counterintuitive hierarchy of their respective effectiveness in transforming sociotechnical systems.

The proposed framework is offered not as a specific linear process but as a flexible, iterative set of simple yet essential principles that explicitly outline key activities involved in designing meaningful immersive experiences. The deliberately iterative approach of dealing with the uncertainty of proposing innovative solutions means that designers may cycle back and forth through all of the stages, pivoting back to earlier ones if necessary. The built-in hypothesis testing cycles allow immersive media design teams to move between thinking-to-build and building-to-think phases to pragmatically deal with the unavoidable challenge of facing uncertainty necessarily present in transformative design practice. We refer to our toolkit as the model-plus-tools approach to reflect the practical nature of complementing all of the steps of the logic model framework with appropriate conceptual tools and methods to guide the applied work of immersive media making. Now, let us explore immersive media experience design together.

Notes

1 The exhibition, curated by Robert Riley, included work by Bill Fontana, Doug Hall, Paul Kos, Tony Labat, Lynn Hershman, Chip Lord, Mickey McGowan, and Alan Rath. Lynn Hershman's *Deep Contact* also stood out, inviting me to touch an image of the guide who appeared on a touch sensitive computer screen, but since it involved actual touch, I found it less immersive. For some reason, touching the computer monitor made me all too aware of the surface tension between myself as participant and the guide as a virtual presence. Reference: Bay Area Media, exhibition, San Francisco Museum of Modern Art, March 15 to May 13, 1990.

2 Jim Campbell created the experience with a large-screen rear-projection video monitor, a video camera, custom image processing, custom electronics, two videodisk players,

and a Macintosh computer. He used *HyperCard* and a Macintosh simply as a sequencer; the heavy lifting had to be accomplished with custom electronics and image processing code. Reference: "Hallucination," JimCampbell.tv, https://new.jimcampbell.tv/portfolio/hallucination/ (accessed October 24, 2022).

3 Between 1989 and 1991, Tim Berners-Lee developed the first web browser at CERN, see Wikipedia contributors, "Tim Berners-Lee," *Wikipedia*, https://en.wikipedia.org/wiki/Tim_Berners-Lee (accessed October 24, 2022).

4 Robert Riley, "Voices Lecture Series," Gallery 400, program notes, University of Illinois at Chicago, January 14, 2002, www.uic.edu/aa/college/gallery400/02_spring2002/02_spring2002-v01.htm (accessed October 1, 2010, no longer available online).

5 All of these features will be explained in detail and elaborated on throughout the book.

6 We will discuss all of the channels and their submodalities in detail later in the book.

7 Silent cinema was never really silent. Since the early days of cinema exhibition, movies were accompanied by some form of music, typically a piano, and sometimes an entire orchestra in large movie houses. Dziga Vertov developed an elaborate score for his seminal *Man With a Movie Camera*. See "Music for Man with a Movie Camera," Silent Film Sound & Music Archive, www.sfsma.org/music-for-man-with-a-movie-camera/

8 For an excellent review of emerging unintended consequences of the application of AI to media and communication, see Melissa Heikkilä, "How AI-generated text is poisoning the internet," *MIT Technology Review*, December 20, 2022, www.technologyreview.com/2022/12/20/1065667/how-ai-generated-text-is-poisoning-the-internet/

9 Nonplayer characters (NPCs) are digital game characters not controlled by a player. The term originated in traditional tabletop role-playing games, such as Dungeons & Dragons, which applied it to characters controlled by the gamemaster. NPCs have a predetermined set of behaviors that can impact the narrative and gameplay. Early forms of NPCs were not necessarily a product of fully fledged AI.

References

Accenture Research. (2019). *Waking up to a new reality: Building a responsible future for immersive technologies*. Accenture.

Barrett, L. F. (2020). *Seven and a half lessons about the brain*. Houghton Mifflin Harcourt.

Boland, M. (2019, October 21). Macy's reduces product returns to <2% with VR. *AR Insider*. https://arinsider.co/2019/10/21/macys-reduces-product-returns-to/

Bruner, J. (1986). *Actual minds, possible words*. Harvard University Press.

Bye, K. (2017, July 14). *Mel Slater's theory of VR presence vs an elemental theory of presence* [Podcast]. https://voicesofvr.com/556-mel-slaters-theory-of-vr-presence-vs-an-elemental-theory-of-presence/

Campbell, J. (1990). Hallucination [Portfolio]. *Jim Campbell*. www.new.jimcampbell.tv/portfolio/hallucination

Carlton, B. (2017). Virtual reality: Opportunities for learning? *The Masie Center*. https://bighorngov.com/wp-content/uploads/2020/12/MASIE-VR_Learn_Report-2017.pdf

Daniels, R. (2022, July 27). Capturing the truth of the space: Joe Hunting on we met in virtual reality. *RogerEbert.Com*. www.rogerebert.com/interviews/capturing-the-truth-of-the-space-joe-hunting-on-we-met-in-virtual-reality

Giardina, C. (2019, July 19). 'The Lion king' "virtual production" could be a game-changer for filmmaking. *The Hollywood Reporter*. www.hollywoodreporter.com/movies/movie-news/lion-king-virtual-production-could-be-a-game-changer-filmmaking-1224990/

Heaven, W. D. (2022, December 16). Sam Altman: This is what I learned from DALL-E 2. *MIT Technology Review*. www.technologyreview.com/2022/12/16/1065255/sam-altman-openai-lessons-from-dall-e-2/

Heilig, M. L. (1992). EL Cine del Futuro: The cinema of the future (Uri Feldman, Trans.). *Presence: Teleoperators and Virtual Environments*, 1(3), 279–294. https://doi.org/10.1162/pres.1992.1.3.279

Jerald, J. (2016). *The VR book: Human-centered design for virtual reality*. Association for Computing Machinery; Morgan & Claypool.
Kemp, C. (2022). *Dark and magical places: The neuroscience of navigation*. W.W. Norton & Company.
Kranzberg, M., & Pursell, C. W. (1967). *Technology in Western civilization*. Oxford University Press.
Kuhn, T. S. (1996). *The structure of scientific revolutions*. University of Chicago Press.
Laurel, B. (1991). *Computers as theatre*. Addison-Wesley.
Laurel, B., Strickland, R., & Tow, R. (1994). Placeholder: Landscape and narrative in virtual environments. *ACM Siggraph Computer Graphics*, 28(2), 118–126. https://doi.org/10.1145/178951.178967.
Marr, B. (2021, June 11). Extended reality in tourism: 4 ways VR and AR can enhance the travel experience. *Forbes*.
Maudet, N. (2017). Muriel Cooper, information landscapes. *Back Office*, 1. www.revue-backoffice.com/en/issues/01-making-do-making-with/nolwenn-maudet-muriel-cooper-information-landscapes
McRobert, L. (2016). *Char Davies's immersive virtual art and the essence of spatiality*. University of Toronto Press.
Polycount. (2022). Hublot loves football virtual stadium for Qatar World Cup 2022. *Polycount—Case Studies*. www.polycount.io/case-studies
Riley, R. (2002, January 14). Voices lecture series, gallery 400, program notes. *University of Illinois at Chicago*. www.uic.edu/aa/college/gallery400/02_spring2002/02_spring2002-v01.htm
Robinson, J. (2017, June 23). Neal Stephenson explains Silicon Valley's latest obsession. *Vanity Fair*. www.vanityfair.com/news/2017/06/neal-stephenson-metaverse-snow-crash-silicon-valley-virtual-reality
Rosenberg, L. (2022). Regulation of the metaverse: A roadmap: The risks and regulatory solutions for largescale consumer platforms. *2022 the 6th international conference on virtual and augmented reality simulations*, 21–26. https://doi.org/10.1145/3546607.3546611
Sanchez-Vives, M. V., & Slater, M. (2005). From presence to consciousness through virtual reality. *Nature Reviews Neuroscience*, 6(4), 332–339. https://doi.org/10.1038/nrn1651
Scott, F. D. (2015). Networks and apparatuses, circa 1971: Or, Hippies meet computers. In A. Blauvelt (Ed.), *Hippie modernism: The struggle for utopia*. Walker Art Center.
Turkle, S. (1997). Seeing through computers. *The American Prospect*, March/April.
Vickers, B. (1989). *In defence of rhetoric*. Oxford University Press. https://doi.org/10.1093/acprof:oso/9780198117919.001.0001
Wingfield, N. (2021, June 21). *The big interview: Unity CEO predicts AR-VR headsets will be as common as game consoles by 2030*. www.theinformation.com/articles/unity-ceo-predicts-ar-vr-headsets-will-be-as-common-as-game-consoles-by-2030
Wurman, R. S. (2001). *Information anxiety 2* (2nd ed., expanded and updated). Que.
Zakrzewski, P. (2022). *Designing XR: A rhetorical design perspective for the ecology of human+computer systems*. Emerald Publishing Limited. http://doi.org/10.1108/9781802623659
Zhang, J. F., Paciorkowski, A. R., Craig, P. A., & Cui, F. (2019). BioVR: A platform for virtual reality assisted biological data integration and visualization. *BMC Bioinformatics*, 20, 78, http://doi.org/10.1186/s12859-019-2666-z

Section I

What Is Immersive Media?

eXtended Intelligence: It Is Not AI But IA That Matters

One would have to hide under a rock deep in the Amazon (the real one) to avoid the onslaught of AI-related news since the record-breaking adoption of generative artificial intelligence (AI) tools such as ChatGPT. AI has been the topic of prediction, speculation, and much hype to rival any technology from nuclear fission to the internet. The reality of AI entering all aspects of our everyday lives might be gaining so much traction because it fits into both the dreams and the nightmares we humans have about our technologies. To properly understand and assess AI's future impact on human affairs, it is essential to focus on its close relative. Intelligence augmentation (IA) is a field concerned with amplifying existing human abilities by augmenting them with AI systems, Internet of Things (IoT) objects, robotics, and other advanced tools. The ability to experience information and to seamlessly communicate with our advanced tools is one of the most significant ways to empower people to be more creative and effective at work, play, and decision-making. Walterio Mayol-Cuevas, who studies collaborative interaction between humans and robotics, argues that timely delivery and accuracy are the two most essential factors in displaying task-sensitive information. People concerned with humane and ecologically sound cognitive augmentation might add a third critical factor—alignment. In the AI field, alignment refers to the match between the outcomes of sociotechnical systems and the individual and our shared human purpose and ethical principles. Our ability to effectively interact with our technologically advanced tools will profoundly impact the outcomes we co-create.

Within the field of information theory, information—the resource at the center of action-taking and decision-making—can be abstractly thought of as the resolution of uncertainty. This insight seems intuitively obvious when considering the real-life context of solving problems and navigating the world. We know that having proper information available at the right time will have a profound impact on the decisions we make. To illustrate this further, let us assume that we want to entertain ourselves while being stuck in a tent camping on a rainy day (yes, we should have estimated the probability of rain better) by playing games for small stakes to make things more interesting. We could place small bets on a probability game, such as heads or tails. A simple game of flipping a coin has two equally

DOI: 10.4324/9781003442462-2

likely outcomes: heads or tails. Now, hypothetically, consider if you would place your bets differently if you knew the outcome of each flip in advance with one hundred percent certainty. Would you not bet the maximum allowable stake on the coin coming up heads if you knew for certain that the coin would come up heads? Reducing the uncertainty to zero would put you in a position of having everything to win and nothing to lose. It would most likely cause you to make very different decisions from facing a fifty/fifty probability.

Real life is a bit more complicated than that. Still, whether the task is to estimate the probability of rain while planning a camping trip (our original problem) or playing games of chance, we humans are predictably irrational in our decision-making. In the famous "Linda" experiment, Kahneman and Tversky gave their experimental subjects personal information about an individual named Linda. These included Linda being 31 years old, single, outspoken, and very bright. Linda majored in philosophy and, as a student, was deeply concerned with social justice issues. Subjects were shown a number of hypotheses about Linda. These included: (A) "Linda was a bank teller," (B) "Linda was active in the feminist movement," and (A and B) "Linda was a bank teller and was also active in the feminist movement." The subjects were asked to judge whether (A) or (A and B) options were more probable. In Kahneman and Tversky's studies, as many as 80% to 90% of participants concluded, based on the information provided, that it was more probable that Linda was both a bank teller and an active feminist than that Linda was just a bank teller. The answer violates the actual probability of A and B being more likely to occur than A or B alone since the mathematical probability of a conjoint hypothesis of A and B cannot possibly exceed that of either of its constituents (Tversky & Kahneman, 1993).

In terms of information theory, flipping a coin or attempting to identify two equally likely outcomes offers lower entropy and less uncertainty than specifying the outcome of a roll of a die with six equally likely outcomes. What makes things more complicated in real life is that everyday decision-making poses more complex challenges than playing dice. It seems life might be more like a random-variable game based on an equivalent of rolling multiple dice with unknown dot patterns. There is a matter of distinguishing between uncertainty and risk. In the case of risk, the outcome might be unknown, but because of the availability of past data, the probability distribution governing that outcome can be calculated.

Uncertainty, on the other hand, is a phenomenon associated with doing something new. That means that uncertainty is characterized by both an unknown outcome and an unknown probability distribution (De Groot & Thurik, 2018). This, by the way, is the reason we recommend throughout this book the iterative (trial-and-error) design approach that allows media makers to pragmatically manage the innovation process. The most important problems, from placing our bets on a career path to planning vacations to investing for retirement, require access to the essential resources of information, appropriate cognitive strategies, and computational processing that an average human does not possess individually. Our human ability to solve problems is limited by the three constraints articulated by Herbert Simon in his argument for the concept of bounded rationality.

In our problem-solving and decision-making, we are all limited by our cognitive system, which has not evolved to deal with the complex challenges of the modern

Anthropocene age. As *Homo deus*, we face the unprecedented challenge of dealing with the global consequences of our collective local actions, such as stock market bubbles, climate change, and the economy, which completely boggle our minds. We are also faced with a mystery of scientific problems that would have seemed like science fiction or fantasy only a century ago. In our grand and prosaic everyday challenges, such as loading an IKEA loveseat into our Toyota Prius, we are all constrained by imperfect information and our limited capacity for information storage.

Most of all, we struggle with the cognitive processing limitations of our otherwise very evolutionarily adaptive brains. Despite being the most powerful such instrument on planet Earth, the size of our brain and its architecture severely limit the amount of information we can store and the speed at which we can process it. The interactions between an average of 86 billion neurons in the human brain allow for massively parallel operation; however, the neurons themselves are the weak link because of their slow firing capacity and reliance on chemical processing. A typical neuron can fire up to 200 times per second. Compare that to the ever-expanding speed of computers, which can currently operate at the rate of trillions of times per second. Additionally, most neural connections depend on chemical messengers known as neurotransmitters, which perform three critical functions: excitation, inhibition, and modulation.

The mixing of chemical and electrical signaling is not necessarily a flaw. It allows for great flexibility in timing and the development of diverse neural pathways. However, one obvious flaw concerns our ability to meet the absolute standards of rationality according to the norms necessary for advanced deductive logic, probability, and utility theory. As Dan Ariely argued in his famous book, we are predictably irrational at our own peril.

With this in mind (cheap pun intended), let us now imagine a common task, such as driving around an unfamiliar city with a physical map on your lap. The good news is that the information we need to find our destination is available. What is not so good is the fact that shifting our gaze from navigating the traffic to the map and back again is turning us into distracted drivers. Diverting our attention can be downright hazardous. The audio commands of most GPS navigation systems solve this problem rather well. They not only serve us the right information at the right time, but they also serve it in a way that makes the interaction, our decision-making, and task accomplishment easier. Even at this early stage of immersive intellect and sensory augmentation, visually impaired tourists equipped with AI-enhanced OrCam MyEye devices can walk around unfamiliar cities and simply point them at various objects in the environment to get accurate information that goes beyond what a normally sighted person could infer.

In another task domain, Figure, a robotics company, has created an AI-based solution allowing older workers to direct autonomous humanoid agents, exceeding the limits of physical human strength, using natural language. Shoppers with dietary restrictions can use Scandit smart data capture–powered mobile AR apps on their smartphones to compare ingredients, allergens, and other features among multiple options on store shelves. Retail employees can check product expiry dates or delivery dates for missing stock and validate prices with a single AR scan of a store shelf. Personal shoppers working for grocery delivery services such as

Instacart, picking customer orders in a store, can quickly find the exact product ordered among many similar products on the shelf. Field service staff can review plans and diagrams and seek advice from experienced colleagues as they work to reduce the downtime of the equipment being serviced. Surgeons at the Imperial College London performing reconstructive surgery on patients with severe leg injuries, including broken bones, can now precisely locate the blood vessels surrounding the break while wearing the HoloLens mixed reality (MR) headset, which overlays the required imaging data right over the patient's body in real time during the procedure. Previously, the information containing computed tomography angiography (CTA) 3D visualization of the blood vessels, bones, and soft tissue in the leg had to be analyzed and memorized before the surgery. The new generation of XR heads-up displays (HUD), such as the HoloLens, improve decision-making by allowing all kinds of wearers to access task-relevant information to be combined with real-life situations without having to shift attention.

But our definition of extended reality (XR) does not rely on a specific pair of goggles or wearable AI devices; it concerns the extension of the human nervous system to augment our naturally bounded cognitive resources with the power of advanced computation that allows all of us to perform everyday tasks, from learning to work to socializing, shopping, and entertainment, with expanded capacity to access information, to remember, to reason, to make better decisions, to solve problems, to create, and to imagine new possibilities. The power of XR lies in its ability to connect users to all kinds of advanced computation and information in real time and in a more multimodal manner than any medium that had come before. To get a fuller picture of the potential of XR, we can extend the augmented intelligence scenarios to DIY plumbers getting professional-grade real-time instructions, shoppers effortlessly hunting through big box stores, tourists wandering foreign destinations without maps or guides, students exploring the depth of the ocean, and analysts examining data visualizations by walking through 3D models.

Our definition of extended reality (XR) does not rely on a pair of goggles or wearable AI devices. It concerns extending the human nervous system to augment our naturally bounded cognitive resources with the power of advanced computation. XR systems aim to allow all of us to perform everyday tasks, from learning to working, socializing, shopping, and entertainment, with expanded capacity to access information, reason, remember, make better decisions, solve problems, create, and imagine new possibilities.

Information Landscapes: Multimodal Access to Augmented Intelligence

In our definition of XR, we borrow an argument from media scholar Marshall McLuhan, who suggested that the first era of human co-evolution with technology had been marked by mechanical extensions such as hammers extending the power of the hand and the car extending the locomotive power of our feet and of our horses. During these mechanical ages, our technology allowed us to extend our physical bodies in space. Between the early 1960s and the time of his passing in 1980, McLuhan observed that the electronic and the emerging computational media began extending the human central nervous system itself in a kind of "global embrace" that gave us the illusion of abolishing space and time. McLuhan considered the technological simulation of consciousness—the creative process of knowing being collectively and corporately extended to the whole of human society—the next phase in the long history of technological extension of human potential (McLuhan, 1994).

Back in 1957, when film and television were considered new media, McLuhan's colleague, anthropologist Edmund Carpenter, suggested that broadcast media represented entirely new languages that shaped their audiences' ways of thinking and behaving. McLuhan himself went further to refer to electronic media as psychic environments. When Carpenter wrote about television, its grammar and vocabulary were still developing. Looking back on a more established medium of writing, he argued that written language did not merely record oral language; it introduced an entirely new means of communicating that rewired the brains of its users to think and behave differently (Carpenter, 1956) and later argued that each medium, if its bias is properly exploited, has the potential to reveal and communicate a unique aspect of reality.

Each medium offers us a different perspective, a different way of seeing ourselves and the world. (Carpenter, 1960). While, as George Lakoff and Mark Johnson (2003) argued, the semantic component of spoken language was rooted in the embodied nature of human cognition and our sensory and motor systems, writing emphasized abstract logic and syntax. It privileged and encouraged lineal argumentation, linear communication, and analytical thinking. Written communication made the subject distinct from the verb and the adjective from the noun. It separated the actor from the action and essence from form by ordering words in ways that conformed with externally imposed rules. Writing made natural language, a once richly creative but ambiguous medium of expression, both more reliable and a lot more restrictive.

Rule-based grammatical notation also made language computable. It also linked it to theoretical computer science and programming language theory. Written arguments were procedurally developed from start to finish, sentence to sentence, paragraph to paragraph, and chapter to chapter. The printing press technology made this process more durable by creating an efficient way to mass-produce uniform black-and-white pages that were read silently, subvocally, and alone (Carpenter, 1956).

The supremacy of the printed page was first challenged in the second half of the 19th century with the invention of photography, the gramophone, and moving pictures. Early cinema initially replaced words with images then added sound and color but essentially followed the linearity of the written text. Pioneers of the cinema invented a new media language in the realm of vision, motion, and sound but invited audiences to lean back and enjoy the movement of narratives scripted in time and fixed in a unique perspective.

Before the turn of the millennium, the long tradition of linearity could eventually be disrupted by personal computers running graphical user interfaces (GUIs) and hypertext. The personal computer age launched with users communicating with their devices through strings of typed commands. Microsoft's Disk Operating System (MS-DOS), developed by Microsoft and licensed by IBM during the 1980s, had its roots firmly planted in the rules of the written language. Hyperlinks and the graphical user interface (GUI), first introduced in the 1970s at Xerox PARC, offered another much more compelling vision of computational medium.

The Xerox Star (Xerox 8010) workstation, introduced on April 27, 1981, opened the door to freedom of navigation, with users able to actively interconnect chunks of information via mouse clicks. Xerox 8010 features, including a bitmapped display and two-button mouse combo, a window-based graphical user interface of icons and folders, Ethernet networking, print servers, and email, have since formed the standard interaction paradigm for personal computers thanks in no small part to the success of Apple's 1984 Macintosh.

GUI ushered in the human–computer interaction (HCI) paradigm that had users interact with their personal computers via windows, icons, menus, and pointing devices such as a mouse or a trackpad. The tangible user interface (TUI) proponents later referred to the entire scheme as WIMP (windows, icons, menus, pointer). Hyperlinked applications of the 1980s became precursors to web browsers, which sped up the mass adoption of the internet. Graphical browsers such as NCSA Mosaic and Netscape Navigator made connected multimedia content accessible to non-technical users. The success of the Internet led the HCI community to be concerned with usability, which spawned the fields of user interface (UI) and user experience (UX) design.

People who grasped the true potential of human+computer (H+C) systems considered the desktop metaphor to be only a transitory state of human–computer interaction and argued for expanded metaphors because the desktop metaphor "blinded user interface designers and researchers to other interaction styles" (Green & Jacob, 1991). Muriel Cooper, the founder of the Visible Language Workshop at the MIT Media Lab, transcended the GUI desktop paradigm by dreaming of information landscapes. Working with her students throughout the 1980s, armed with a Silicon Graphics Reality Engine $250,000 computer, Cooper undertook a series of projects that explored such topics as interactive calendars, infinite information zooms, typography in space, multiperspective data presentation, and personalized information galaxies in which information was organized and filtered in a dialogue between an algorithm and the reader (Maudet, 2017). Cooper and her team found their design inspiration not in a traditional two-dimensional, page- or screen-restricted graphic design but in three-dimensional forms of theater, performance art, and Bauhaus happenings. Their explorations illuminated five unique features of the computer-enabled information landscapes: (1) three-dimensionality of (2) multimodal media spaces, temporally navigated by (3) an actively participating audience within (4) nonlinear time frames, through (5) multiple access points controlled by the users.

Many of the features of immersive computation envisioned by Cooper's team were advanced and popularized by the medium that developed parallel to personal computing. Writing in an introduction to his 2001 book, *Information Anxiety 2*, Richard Wurman predicted a "revolution" that would see gaming move from first-person shooters "to business gaming, problem-solving and decision-making, vicarious travel, and ultimately to flying through information—perhaps all human knowledge" (Wurman, 2001, p. xxvii). Gaming has evolved from the 1940s experiments in interactive electronics to home video game consoles and arcade games like *Pong* in the 1970s to more advanced consoles like Sega and Nintendo to Sony PlayStation and the Microsoft Xbox. The medium has expanded into highly social online, mobile gaming, and XR games.

Whether gaming will become anything other than gaming, game designers have developed a powerful multimodal interaction language that cannot be ignored in immersive media design. The potential for new forms of human–computer interaction Wurman saw emerging from digital gaming offers a clear vision of how immersive forms of computational media can become powerful tools for navigation of information, learning, and strategic decision-making. From innovative user interfaces to experience design strategies, gaming has created a robust media design toolbox that is closer to Muriel Cooper's vision of spatial computing than GUI's flat desktop metaphor. The immersive quality of gaming experiences might be why the cross-platform game engine Unity has become one of the most popular authoring environments for AR and VR applications.

Apple's introduction of the half-phone, half-computer chimera enabled the convergence of all the previous media into a single device that made calls, took photos and videos, played movies, acted as a video game console, and added mobility. In 2007, mobile phone users lucky enough to hold the new iPhone could access media anytime and anywhere for the first time in human history. With the explosion of the app ecosystem, smartphone users could effortlessly communicate with any other smartphone user on the planet, entertain themselves, or augment their decision-making by overlaying media content over the real world while playing *Pokémon Go* or using Google Maps. Smartphones became media devices that, on the one hand, looked backward by converging the previous media, but on the other, also looked forward by bringing humanity to the doorstep of immersion into information landscapes.

It is no coincidence that the company that was arguably responsible for the widespread adoption of personal computers and introduced the world to smartphones referred to its next big product launch metaphor as spatial computing. But Apple did not invent the concept of spatial computing. Getting people away from having to sit in front of desktop workstations, punching keyboards, has been a cultural imperative among computer scientists such as Alan Kay, Mark Weiser, Pattie Maes, Paul Milgram, Fumio Kishino, Hiroshi Ishii, Mark Weiser, and others for as long as GUI-based personal computers have existed. The goal of spatial computing is to bring the virtual and physical worlds together in a seamless way, enabling computation to augment the physical world with vital information.

Drawing the Line: The Information-Rich World of Ubiquitous Computing

The demarcation line, which conceptually separates the robust world of immersion from the more narrowly focused human–computer interaction (HCI) paradigm of computational media, was initially drawn by researchers working on separate aspects of cognitive augmentation. Mark Weiser's seminal 1991 article "The Computer for the 21st Century" outlined a vision for what we now commonly refer to as the Internet of Things (IoT). Weiser, who was the chief technology officer at Xerox's Palo Alto Research Center (PARC), drew the picture of embodied virtuality—a world of wirelessly connected devices that could seamlessly automate helpful actions—he called ubiquitous computing. In the article, Weiser asserted that "The most profound technologies are those that disappear" (Weiser, 1991, p. 1). Writing later in the 1993 special issue of the *Communications of the Association for Computing Machinery* (ACM) titled *Back to the Real World*, he advocated an augmentation framework that would enrich the real world with digital functionality rather than force users to enter into virtual worlds. Weiser envisioned computer technology weaving into "the fabric of everyday life" until it was "indistinguishable from it" (Weiser, 1991, p. 1). According to this vision, computers embedded within the physical environments could make interactions with computers as refreshing as "taking a walk in the woods" (p. 8).

Weiser is considered to be one of the most significant visionaries of ubiquitous computing to advocate for turning the entire world into one seamless interface with physical artifacts smartened up by "intelligent" data networks. This new ubiquitous immersion would fully embrace the richness of human sensory perception and intuitive skills of navigating the physical world. Weiser's Ubicomp vision looked past the mobile revolution to argue that even the state-of-the-art connected mobile computers still forced users to focus their attention on a single device. His team's experimental immersive environments anticipated further advancements in AI to have smart objects placed within the Ubicomp environment. Within the Ubicomp world, doors would open automatically to the right badge wearers, rooms would greet people by name upon entry, and appointment diaries would quietly write themselves.

Weiser's proposal anticipated information-rich environments enabled through networks of cheap chips and wireless sensors that could enable novel forms of user interfaces. He argued that cheap, small, and efficient embedded chips could give all sorts of smart objects their own source of computing and communicative power (p. 2). In a proposal blending ubiquitous computing with augmentation, he envisioned thermostats, stereos, and ovens interconnected within computer networks and embedded in physical environments. About two decades later, David Rose, working within the tangible user interface (TUI) paradigm, referred to such devices as "enchanted objects."

Whatever the term used, the enchantment of the enchanted objects is aligned with a new paradigm of decentralized information architecture known as edge computing. The new architecture envisions moving data storage and processing closer to the source of data to lower the dependence on cloud computing. Small computing devices ranging from the inexpensive Raspberry Pi to Apple's latest smart watches and iPhones will play an increasingly pivotal role in the immersive transformation. This updated domain of the immersive intelligence augmentation is now commonly referred to as the Internet of Things (IoT).

In 2011, Holz et al. proposed a unified framework for classification, design, and construction of intelligent agents straddling the virtual and corporeal worlds. They referred to it as mixed reality agents (MiRAs) and defined them as agents embodied within the spectrum of mixed reality environments (Holz et al., 2011). The accompanying taxonomy classifies MiRAs along three vital axes of agency, corporeal–virtual presence, and the interactive capacity within the AR spectrum. Holz et al. visualized their framework as a MiRA Cube with agency represented as Y, corporeal presence represented as X, and interactive capacity represented as the Z dimension. Agency, which we will discuss later, is conceptualized based on the weak and strong notions outlined by Wooldridge and Jennings in their paper, "Intelligent Agents: Theory and Practice" (1995). Corporeal presence describes the representation within the virtual-physical continuum. An agent's interactive capacity is characterized by its ability to sense and act within both the virtual and physical environments. Extended Ubicomp environments as envisioned by Holz et al. consist of both virtual and smart physical objects imbued with agent-like behavioral capabilities and possessing a degree of actantial or interactive properties in both the virtual and physical dimensions.

The Milgram-Kishino Continuum and Beyond

Shortly after Weiser's ubiquitous computing proposals became public, Paul Milgram and Fumio Kishino, in their article titled "A Taxonomy of Mixed Reality Visual Displays," introduced the now famous reality–virtuality (RV) continuum. The article also introduced the now commonly used term "mixed reality" (MR) (Milgram & Kishino, 1994). Milgram and Kishino anchored one end of their continuum with a purely physical environment consisting solely of real objects. They anchored the other with a purely virtual environment populated solely by virtual objects. They proposed that the space between the real and virtual realities be considered a mixed reality (MR). In mixed reality, the real world would be augmented with virtual content called augmented reality (AR). Further up the continuum, they envisioned a world of augmented virtuality (AV) where most of the content would be virtual but with some awareness or inclusion of real-world objects. Like Weiser before them, Milgram and Kishino's vision was bounded by the screen-based paradigm to be primarily concerned with visual displays and display hardware. Reflecting the reality of the 1990s technological landscape, the users within the continuum were not connecting to the augmented world, as they would to the natural one, through eight sensory channels.

In the succeeding decades, Milgram and Kishino's proposal has been cited, expanded on, and critiqued by many researchers in the field. Richard Skarbez, Missie Smith, and Mary Whitton quite reasonably argue that the fully virtual environment endpoint of the original continuum, within which the users interacted with a completely synthetic world, was "ill-defined" (Skarbez et al., 2021, p. 3). They likened it to a popular sci-fi conception of the Matrix, in which the participant's nervous system would be disconnected from the natural world and directly stimulated by technology. We wholeheartedly agree that the physical substrate of all immersive experiences cannot be overlooked. Not only are all of the sensory display devices real physical objects, they are also situated in real physical environments. In the case of virtual reality, walking off stages and bumping into walls would be a couple of really sure ways to create breaks in presence, with perhaps some further unintended consequences to the participants' physical well-being. In the cases of AR, the entire secret lies in the accurate mapping of the augmented and the physical worlds. Skarbez et al. argue that for most intents, all XR experiences are, in some ways, mixed reality experiences.

Returning to the possibility of the existence of the extreme of pure virtuality, the concept of the direct brain–computer interface (BCI), sometimes called a brain–machine interface (BMI), involves the elimination of the intermediary component of the exteroceptive sensory channels has been a topic of serious research since the 1970s. BCI researchers who straddle the worlds of neuroscience and computation explore the direct communication pathways between the brain's electrical activity and external devices, such as computers or robotic limbs, that augment or repair human cognitive or sensory-motor functions. Neuralink might be one of the best-known neurotechnology companies that is developing implantable brain–computer interfaces.

A body of neuroscientific research shows the feasibility of direct brain stimulation. For instance, patients suffering from Parkinson's disease have been treated with the aid of a procedure known as deep brain stimulation involving surgically implanted electrodes in the subthalamic nuclei area of the brain. Stimulating the targeted brain regions for brief periods, a painless procedure carried out with the patient fully conscious produces dramatic effects. Yves Agid at the Pitié-Salpêtrière in Paris found that minute variation in the position of the electrode could cause a patient to experience states that are wholly fused with "an automatic recruitment of cognitive, emotional, and motor aspects of being" and experienced at "the highest phenomenological level" (McGilchrist, 2019, p. 226). One patient, in a span of minutes, underwent a dramatic change in personality from an impassive, immobile Parkinsonian state to actively expressing feelings of sadness, guilt, and uselessness. Eventually, for five minutes, less than 90 seconds after stimulation was stopped, the patient entered a mildly hypomanic state, laughing and joking with the researcher while playfully pulling his tie. The hypomanic state could be induced in the patient repeatedly by minutely moving the placement of the electrode.

Iain McGilchrist argues that this research points to the existence of "experiential wholes that are completely coherent" across all layers of consciousness. He suggests that these wholes and not just the act of moving the electrode are already present well below consciousness while affecting the patients at the conscious level. The observed emergence of the complex mix of motor restrictions of Parkinson's disease and manic signs of depression has to involve the need for the highest levels of cortical function and integration (p. 226).

Since the brain is a living physical object, the question of whether direct brain stimulation qualifies as pure virtuality is deeply philosophical. While being a distinctly feasible possibility, implantable brain–computer interfaces are not the topic of this book, but in the light of this argument, we feel that it would be fair to argue that the kind of immersion we discuss in the book can reasonably be considered as lying within the mixed reality (MR) spectrum.

The Milgram-Weiser Continuum: The Brave New XRI World of Augmented Intelligence

Weiser singled out virtual reality as an anathema of his dream of "embodied virtuality" (p. 2). He expressed particular scorn for interfaces involving such VR staples as gloves, bodysuits, and goggles that projected an "artificial" scene onto users' bodies (p. 2). To him, the notion of virtual reality meant an attempt to create a world inside the computer rather than bringing computing into the world people lived in. He reserved the application for virtual immersion to a limited in use exploration of otherwise inaccessible realms, such as the micro worlds of cells or the information webs of complex databases, not as an approach to general computing. Weiser wanted computing environments to include rather than exclude everyday objects and extend to physical environments.

A group of researchers led by Joseph Newman suggests that even though Weiser placed Ubicomp in direct opposition to virtual reality, Ubicomp's and VR's positions within the immersive paradigm are not opposite to one another but rather orthogonal. They argue that the Ubicomp framework can be seen as another continuum complementary to that of the Milgram-Kishino's. The proposed "Weiser's Continuum" shows ubiquitous computing as one extreme, with monolithic mainframe-based terminals at the other (Newman et al., 2007).

Consequently, Newman et al. build on both of the earlier frameworks by cross-referencing Weiser's visualization of ubiquitous computing ranging from terminal-based interaction to the ubiquitous immersion in computationally rich environments, with Milgram and Kishino's continuum of mediated of possible "realities" between the real environments, augmented reality (AR), augmented virtuality (AV), and finally virtual reality (VR). They argue that the resulting matrix defined by the real-to-virtual and monolithic-to-Ubicomp axes identifies a vast untapped potential for research in novel immersive applications. They suggest that while the virtuality axis demonstrates the varying degrees in which virtuality and physical reality could be mixed in the XR paradigm, the ubiquity axis adds the dimension of the level of information access being dependent or independent from being in a fixed place (2007). The combined visualization laid the foundation for an XR research program aimed at defining the future of human+computer immersion.

The big picture of intelligence augmentation comes into focus as a hybrid of XR immersion and the distributed computing of the Internet of Things (IoT). A growing number of researchers agree about the inevitability of a fruitful creative collision between the two paradigms that will produce hybrid XR-IoT (XRI) systems allowing people equipped with edge devices to roam freely between two worlds, one populated by information-rich IoT agents embedded in MR environments and the other that of IoT networks accessible through advanced immersive interfaces. Mixing the worlds of XR and IoT is an outcome of a steady path forward marked by an evolution away from computing devices actively operated by users cognitively situated within the physical realities of their homes and offices and toward MR spaces that immerse users in the information. The combined visions of Milgram, Kishino, and Weiser are leading media makers outside of the tight boundaries of HCI and into a new and unknown territory of the corporeal world of smart objects seamlessly blended with the virtual world of digital twins.

Human in the Loop: Human+Computer Emergence

One of the critiques of the original Milgram and Kishino paper offered by Skarbez et al. argues that the authors did not account for the possibility of an XR system being interactive. Interactions within an XR system can take at least three forms: the machine to machine or code to code, the machine to human, and the human to human. Bringing humans into the loop makes immersive systems infinitely more complex, more challenging to manage, and a lot less predictable. Much of this book is dedicated to dealing with this particular nontechnical challenge of immersive system design.

In his earlier work, Peter addressed the importance of two fundamental forces behind the sociotechnical systems: the designers' rhetorical power and the users' agency and choice. Consequently, he adapted Newman et al.'s Milgram-Weiser Continuum to include the third axis, describing the critical components of the emergence of immersive system states with participation on one end and control on the other. The inclusion of the emergence of the state of presence at the individual level of each participant and the emergence of system function at the level of system behavior adds the crucial third dimension besides ubiquity and virtuality to enrich the picture of the critical issues facing immersive human+computer (H+C) system design (Zakrzewski, 2022, pp. 83–84). The resulting framework, which we show in Figure I.1, is based on the three axes mentioned earlier: Virtuality, Ubiquity, and Emergence (Figure I.1). The framework anticipates and accommodates user participation at all levels of immersive system design, which we will continue to discuss in the following chapters.

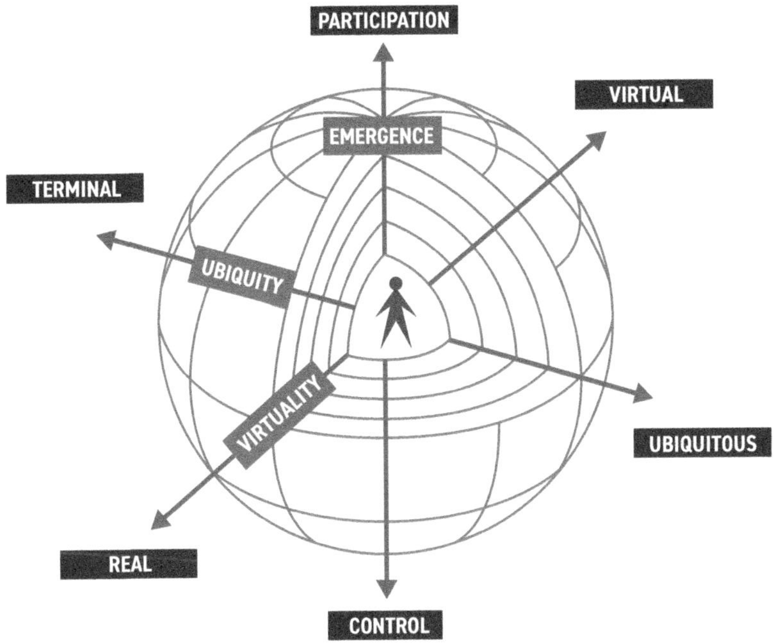

Figure I.1 Crossing the immersive line

Source: Illustration by © 2024 Peter (Zak) Zakrzewski, based on Dieter Schmalstieg & Tobias Höllerer, AR Mapping.

The Three Dimensions of Immersively Augmented Intelligence

It is important to put Milgram and Kishino's seminal work in the proper context. In the mid-1990s, at the time when their original work was published, the custom-made immersive systems were better suited to research labs than to real-world applications. Head-worn display and computer hardware needed for immersive simulation were uncomfortably bulky and prohibitively expensive. Three decades on, things are quite different. Many XR experiences can be implemented using off-the-shelf hardware components or even off-the-shelf display devices such as mobile phones, tablets, and head-worn AR and VR displays and standardized modern general-purpose, high-level programming languages such as Microsoft's C# programming language used by Unity or declarative programming environments such as Apple's Swift. Maybe most importantly, modern sensing technologies combined with the world of the IoT devices, originally imagined by Weiser as ubiquitous computing, can now deliver much richer interactions and access to information about the real world that would have been difficult to imagine in the 1990s.

Milgram and Kishino introduced a three-dimensional taxonomy to categorize various extended reality technologies in their paper. The extent of world knowledge

(EWK) described the level of modeling of the real world that included the identification of objects and the locations of objects showing where they were located within the XR environment. The reproduction fidelity (RF) addressed the display technology's capacity to deliver the sensory stimuli. The extent-of-presence metaphor (EPM) accounted for the level of "naturalness" of user interactions the technology was capable of supporting.

Skarbez et al. offered a revised taxonomy that accounted for three decades of technological and conceptual advances to focus on the three core aspects of immersive computation. Immersive experiences are based on three unique dimensions: (1) the extent of world (environmental) knowledge (EWK), (2) the set of valid actions supported by that system they refer to as immersion (IM), and (3) coherence (CO). To avoid confusion, we will refer to the set of valid actions supported by the XR system simply as interaction (I). Within this framework, the EWK dimension captures the level to which the system is connected to its real-world environment and its ability to respond to any changes in those surroundings.

While Milgram described EWK as a combination of the "what" and the "where" of the known objects within the immersive space, Skarbez et al. argue that a "perfect" EWK could add to this knowledge by taking advantage of the more advanced

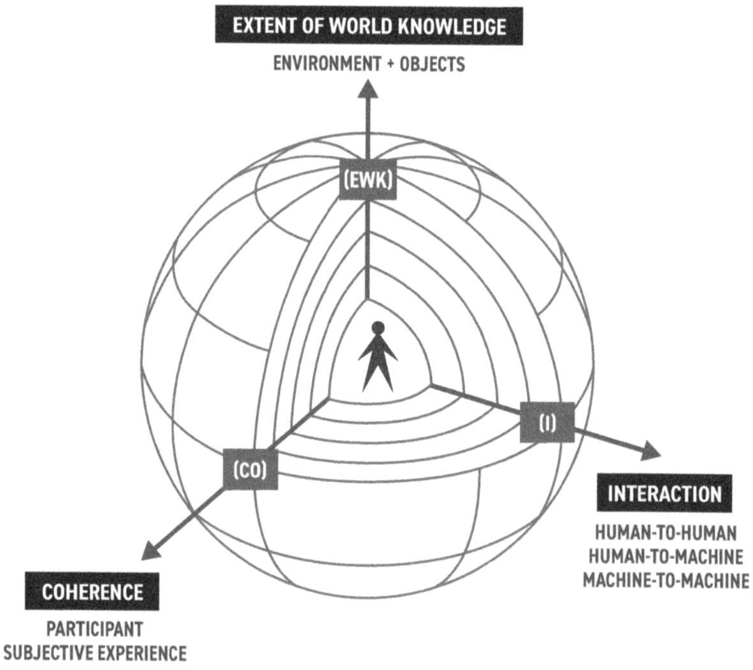

Figure I.2 The three dimensions of immersively augmented intelligence

Source: Concept and illustration by © 2024 Peter (Zak) Zakrzewski, based on Skarbez et al. Taxonomy of MR Experiences.

sensing capabilities to let the user know how things work and when certain actions might happen (Skarbez et al., 2021). The XR system's full range of interaction (I) is defined by the set of valid actions supported by that system. Slater suggests that immersive interactions comprises two types of valid actions: sensorimotor valid actions that result in changes to a user's perception of the environment and effectual valid actions that affect changes to the environment itself (Skarbez et al., 2021). Finally, coherence of the experience (CO) within an XR system addresses how the sensory outputs of the immersive system support a unified experience for the participants. This dimension has been previously discussed as "fidelity" by Alexander et al. (2005) and as "authenticity" by Gilbert (2016).

With their multidimensional taxonomy, Skarbez et al. point to the defining aspects of immersive mediation. The knowledge about the world addresses the environment and the objects within it. Thinking in terms of interactions allows media makers to strategically design the range of possible actions between the human users of the system and the objects and the AI-based agents contained within the system. Coherence refers to the subjective experience the audience members have with the system as predicated on the immersive quality of the stimuli provided by the XR system.

Within all three dimensions, XR substantially expands the repertoire of design affordances from the past HCI and GUI paradigms. XR media is defined by the possibility of creating entire virtual environments populated by AI agents and the ability to map real environments and overlay them with information. The sensor technology mixed with multimodal inputs and outputs and paired with advanced AI algorithms significantly broadens the range of possible interactions over the previous computational media. The multimodality of XR media allows for the levels of coherence that bring us ever closer to the goal of immersive presence or near-complete simulation of conscious sensory experience of the world. We will expand on these fundamental dimensions of XR media and the interrelationships between them when we discuss the alignment between the XR technologies and the subjective immersive experiences they can foster.

Designing the Artificial: Computational World-Making and the Five Unique Features of Immersive Media Making

Computational world-making might be the aspect of XR that is most appealing to media makers exploring the design of immersive experiences. The uniquely human ability to function in what Yuval Noah Harari calls "dual reality" makes our cultural and media environments function as invisible psychological grids that shape our perceptions, habits, and behaviors (Harari, 2015, p. 34). The birth of representation marks a dramatic turn in the history of humanity. The cognitive revolution has turned our fortunes around from surviving as the "naked apes" hunted by stronger species to becoming the rulers of the planet mapping out the course of our own Anthropocene age. We have accomplished this feat by representing complex ideas in the abstract form of mathematics, language, and art and by mediating the flow of our ideas to think and dream together.

We can see the early forms of human-made media environments in the 30,000+-year-old paintings of the Pettakere Cave found 36 kilometers from Makassar City in Indonesia and the Lascaux cave paintings in the Vezere Valley in France. Researchers believe that these cave-painted environments played an important communal role in trance-dancing rituals. The caves were the magic spaces created for transmitting information and experiencing alternative realities. Werner Herzog's film *Cave of Forgotten Dreams* offers us a unique glimpse into these magical worlds full of contrasting color and the illusion of three dimensions. The cave painters used a form of perspective in their nonfigurative images, and their depictions of the night sky correlated with constellations.

Like its prehistoric predecessors, the Sistine Chapel is a virtual space where reality is suspended in a metaphorical container. It was created for spiritual journeys during which participants could engage with the out-of-reach sacred dimension, referencing Christian mystical practice and mythology. Walking through places like the Sistine Chapel activates our holistic premodern patterns of consciousness, leading to alternative perceptual experiences. It is harkening to this ability to create magical worlds that makes immersive media so consequential.

While computational world-making, or the ability to design the artificial, tops the list of what makes XR media different from all other previous media, throughout this book, we discuss and expand on Peter's list of four other unique areas of opportunities and challenges that make immersive experience design fundamentally different from the design of the HCI paradigm. In the next section, outlining the XR experience design framework, we discuss one of the principal tools of immersive world-making, the ability to engage in multimodal composition, or XR's capacity to deliver experiences designed for the eight sensory channels. No medium in history allowed a near-complete mediation of all of the sensory channels. The painters of Pettakere could decorate the existing cave in any way they wanted, but they could not create an illusion of an entirely artificial one from scratch.

Besides engaging a full spectrum of sensory inputs, XR media makers now have the opportunity to design and iterate ecologically sound sociotechnical systems optimized for purposeful emergence. Immersively combining humans and computers opens up a possibility of coordinating our actions toward well-defined system outcomes. We will discuss a systems thinking approach to XR media design in Chapter 7. Purposeful emergence is predicated on the next defining characteristic of immersive systems—the leveraging of our understanding of computational social science, sometimes referred to as social physics, to give XR system designers the ability to design systems that support collective intelligence.

Closely linked to social physics is another unique feature of immersive systems. XR's capacity to generate bio- and neurofeedback offers XR media designers an opportunity to create persuasive environments capable of eliciting behavioral change. We address these groundbreaking features of XR systems in the last section, which covers the topic of system auditing and iteration.

Why XR Must Be Designed and Not Just Invented

With all due respect to the ingenuity and creativity involved in inventing new things and the essential contribution of technology, we can best explain the difference between design and invention by adapting Janet Murray's description of playfulness. Like playfulness, pre-commercialized invention is an exploratory search motivated by curiosity and active personal or group engagement, marked by "desire to try things out and see what will happen" (Murray, 2011, p. 380). Inventors seek "surprisingly pleasurable" and "emotionally charged" but also technically feasible results that can provide them with a sense of agency, discovery, and mastery (p. 380).

In contrast to invention, design can be described as a purposeful, transdisciplinary, problem-solving activity aimed at changing undesirable status quos into preferred states (Simon, 1996). Design-minded media makers are guided by two guiding sources of action: the clarity of strategic purpose and the very next step toward it. Design methods, such as empathic discovery, vision articulation, iteration, hypothesis testing, and rapid prototyping, allow designers to pursue possible futures with clear direction without completely defining the solutions from the start. There is little doubt that the success of XR as a medium depends on the invention and improvement of its core technologies.

Despite the promise of wearable computing, there is still a large chasm between the usability of light and highly wearable mechanical objects such as jewelry and glasses and computing devices people can comfortably wear on their bodies, let alone on their faces. Some technical challenges, such as battery life and weight, heat dissipation, and increasing complexity in processor design, push hard against the known engineering methods and even the laws of physics. With that being said, no technology is ever widely adopted based on its technical brilliance alone, nor can technological brilliance ever assure the alignment between its application and our deeply held human values and goals.

To illustrate the difference between inventive and designerly ways of creating products, we need to look no further than the two-wheeled, self-balancing, electric battery-powered, individual transporter called Segway, created by the brilliant engineer Dean Kamen. The device enabled users to command the vehicle by simply shifting their weight. It detected any change in its center of mass to adjust both speed and direction through the application of gyroscopic and fluid-based leveling sensors. Silicon Valley heavyweights John Doerr and Steven Jobs were so taken by Segway's engineering brilliance that they mistook its technical feasibility for desirability. John Doerr pronounced the Segway to be a major transportation breakthrough and a potentially more important invention than the Internet. Segway's forecasted sales at its optimistic launch in 2001 were 40,000 units a month. The actual sales of this not-quite-a-bike and not-quite-a-car vehicle amounted to 6,000 units in 2 years. Segway makers did not lack engineering brilliance or money, and yet their product failed to be widely adopted by the general public because it did not address a well-defined audience and a resonant need people could relate to.

Like Segway, Apple's Vision Pro is an acknowledged engineering marvel, yet consumers are not sleeping outside Apple stores to line up to get one as they did with the iPhone. The founder of the social media giant Meta Platforms has spent billions of dollars pursuing the dream of the metaverse. Still, the metaverse has far to go in terms of mass adoption. Since at least the 1950s, inventors such as Morton Heilig and Ivan Sutherland have dreamt of building virtual worlds.

However, a widely adopted medium is more than just a cultural imperative of a select group of technologists or a set of disruptive technologies cobbled together into a device. Strictly speaking, a successful medium is a technology that facilitates communication, but it is also a set of associated social and cultural protocols and practices that emerge alongside the technology. Above all, a new medium is a new way to experience the world. Designers behind significant innovations are, first and foremost, advocates for the user. XR media, like all successful innovations, will be built out of the three building blocks of the desirability of compelling experiences addressing core human needs, the feasibility of usable technology that can support the experience, and the viability of the business model that puts real value in the hands of well-defined audiences. We firmly believe that XR will become the dominant medium, but only when immersive media makers design experiences that perform meaningful jobs in the lives of its users and when those experiences start enriching the lives of their audience.

Summary

We introduce extended reality (XR), emphasizing the role of immersive media in amplifying human capabilities through technology. We contrast artificial intelligence (AI) with intelligence augmentation (IA) tools that enhance human decision-making and creativity. Key XR technologies, like heads-up displays and augmented reality (AR) interfaces, illustrate the seamless blending of digital information with physical environments, as seen in applications ranging from medical surgery with HoloLens to retail enhancements and real-time information for visually impaired users. The Milgram-Kishino continuum, which defines a spectrum from real to virtual environments, highlights the potential of XR to enrich human perception and interaction. We emphasize the shift from traditional human–computer interaction (HCI) to immersive, spatial computing, where ubiquitous computing—pioneered by Mark Weiser—enables technology to invisibly augment the physical world without disrupting it. The convergence of XR with the Internet of Things (IoT) is suggested as the future direction, creating "information landscapes" that integrate XR and IoT for rich, interactive, real-world data experiences. We argue that the potential of XR lies in creating purposeful, socially impactful designs that align with human needs, cognition, and values, moving beyond technical innovation to design experiences that users find meaningful and enriching. This shift underscores XR's emerging role as a technological advancement and a transformative medium reshaping human interaction with both computational and physical worlds.

References

Alexander, A. L., Brunyé, T., Sidman, J., & Weil, S. A. (2005). From gaming to training: A review of studies on fidelity, immersion, presence, and buy-in and their effects on transfer in pc-based simulations and games. *DARWARS Training Impact Group*, 5, 1–14.

Carpenter, E. (1960). The new languages. In E. Carpenter & M. McLuhan (Eds.), *Explorations in communication: An anthology*. Beacon Press.

Carpenter, E., & McLuhan, M. (1956). The new languages. *Chicago Review*, 10(1), 46–52. www.jstor.org/stable/25293194

De Groot, K., & Thurik, R. (2018). Disentangling risk and uncertainty: When risk-taking measures are not about risk. *Frontiers in Psychology*, 9, 2194. https://doi.org/10.3389/fpsyg.2018.02194

Gilbert, S. B. (2016). Perceived realism of virtual environments depends on authenticity. *PRESENCE: Teleoperators and Virtual Environments*, 24, 322–324. https://doi.org/10.1162/PRES_a_00276

Green, M., & Jacob, R. (1991, July). SIGGRAPH '90 Workshop report: Software architectures and metaphors for non-WIMP user interfaces. *SIGGRAPH '90*. SIGGRAPH, Dallas.

Harari, Y. N. (2015). *Sapiens: A brief history of humankind* (H. Watzman & J. Purcell, Trans.). HarperCollins Publishers.

Holz, T., Campbell, A. G., O'Hare, G. M. P., Stafford, J. W., Martin, A., & Dragone, M. (2011). MiRA—Mixed reality agents. *International Journal of Human-Computer Studies*, 69(4), 251–268. https://doi.org/10.1016/j.ijhcs.2010.10.001

Lakoff, G., & Johnson, M. (2003). *Metaphors we live by*. University of Chicago Press.

Maudet, N. (2017). Muriel Cooper, information landscapes. *Back Office*, 1. www.revue-backoffice.com/en/issues/01-making-do-making-with/nolwenn-maudet-muriel-cooper-information-landscapes

McGilchrist, I. (2019). *The master and his emissary: The divided brain and the making of the Western world* (New expanded ed.). Yale University Press.

McLuhan, M. (1994). *Understanding media: The extensions of man* (1st MIT Press ed.). MIT Press.

Milgram, P., & Kishino, F. (1994). A taxonomy of mixed reality visual-displays. *IEICE Transactions on Information Systems*, E77D(12), 1321–1329. https://doi.org/10.1117/12.197321

Murray, J. H. (2011). *Inventing the medium: Principles of interaction design as a cultural practice*. The MIT Press.

Newman, J., Bornik, A., Pustka, D., Echtler, F., Huber, M., Schmalstieg, D., & Klinker, G. (2007, March). Tracking for distributed mixed reality environments. *Workshop on trends and issues in tracking for virtual environments, IEEE Virtual Reality Conference (VR'07)*.

Simon, H. (1996). *The sciences of the artificial*. The MIT Press.

Skarbez, R., Smith, M., & Whitton, M. C. (2021). Revisiting Milgram and Kishino's reality-virtuality continuum. *Frontiers in Virtual Reality*, 2, 647997. https://doi.org/10.3389/frvir.2021.647997

Tversky, A., & Kahneman, D. (1993). Probabilistic reasoning. In A. I. Goldman (Ed.), *Readings in philosophy and cognitive science*. The MIT Press. https://doi.org/10.7551/mitpress/5782.003.0006

Weiser, M. (1991, September). The computer for the 21st century. *Scientific American*, 265(3), 94–104. https://doi.org/10.1038/scientificamerican0991-94

Wooldridge, M., & Jennings, N. R. (1995). Intelligent agents: Theory and practice. *Knowledge Engineering Review*, 10(2), 115–152.

Wurman, R. S. (2001). *Information anxiety 2* (2nd ed., expanded and updated). Que.

Zakrzewski, P. (2022). *Designing XR: A rhetorical design perspective for the ecology of human+computer systems*. Emerald Publishing Limited. http://doi.org/10.1108/9781802623659

Section II

What Is XRXD?

Designing Immersive Experiences

Immersion Is More than Interaction: Transitioning From UX to XR Experience Design

When the developers of the VR version of the best-selling digital game *Minecraft* tested their prototypes with volunteers, they found that the immersive I-person view offered by the game was so intense and emotionally taxing that the testers had to take a break after every hour of play (Kelly, 2016). Holding a pick and chopping the cartoon pixel blocks inside the virtual-reality version of *Minecraft* seemed so immediate and real that the player's own presence within the virtual world was dramatically amplified over the flat-screen first-person version played on PCs, tablets, and phones. The impact of immersion was so profound that the players' sense of being markedly shifted inward. This shift had little to do with wearing the VR gear. Playing the virtual flat-screen version of *Minecraft* in the traditional 2D first-person view on a virtual monitor gave the testers so much affective and cognitive rest that they could return to playing the fully immersive mode for another short period of time. Performing the same role from the immersive I-person perspective feels dramatically more intimate than it does in first person on a flatscreen, because immersion puts us in the position of feeling rather than the position of observing (2016).

Even the traditional forms of media, such as books, can mentally transport readers to places and situations not related to their current position in the natural world by engaging their imagination. The embodied theory of language comprehension supported by neuroimaging studies suggests readers represent situations described in language by constructing images from existing basic sensory and motor representations. Stories we read dynamically activate parts of our brain associated with specific visual, motor, and conceptual features of similar experienced activities. Viewed with neuroimaging, brain regions involved in processing such goal-directed human activities as navigating spatial environments and manually manipulating objects show an increase in activation at points when analogous aspects of the narrated situation change in the story (Speer et al., 2009). A research project conducted by Speer et al. found that imaging of readers processing descriptions of story characters interacting with objects showed increased activation in precentral and parietal areas of the brain associated with motor planning and execution, such as grasping hand movements. Still, the media we have widely adopted so far, from

DOI: 10.4324/9781003442462-3

books to film, television, and personal and mobile computing, affect only a limited number of sensory channels and layers of consciousness. It seems that some unspoken cultural imperative spurs us forward toward richer forms of mediation that allow us to fully experience the worlds of our imagination.

Virtual reality has been not just a distant dream or science fiction but a widely embraced cultural imperative among immersive media pioneers since at least the 1950s. Between 1957 and 1962, filmmaker and VR pioneer Morton Heilig built and patented a big and bulky video arcade game–shaped viewing cabinet called Sensorama, which gave the audience a simulated experience of riding a motorcycle on the streets of Brooklyn. The immersive experience included the 3D view of the streets, the feeling of the wind on one's face, the vibration of the motorcycle seat underneath, and even the smells of the Big Apple. A few years later, computer scientist Ivan Sutherland called for the creation of "The Ultimate Display" in a now-famous essay by the same title and built the first proof-of-concept prototype of a head-mounted display showing virtual computer-generated graphics. Sutherland wrote, "There is no reason why the objects displayed by a computer have to follow the ordinary rules of physical reality. . . . The ultimate display would, of course, be a room within which the computer can control the existence of matter" (Sutherland, 1965, p. 508). A chair displayed in Sutherland's magic room "would be good enough to sit in" (p. 508). Sutherland's VR prototype relied on an elaborate ceiling suspension mechanism that allowed head tracking and see-through optics. The device was so cumbersome that it was dubbed the "Sword of Damocles."

After decades of experimentation, we have arrived at immersive XR (AR, VR, and MR) technologies, which, while still not at the stage of broader consumer adoption, would absolutely astound and vindicate Heilig and Sutherland. Today's XR hardware and software have already given us a usable medium capable of delivering the most sensory-rich and truly mind-bending experiences any medium has ever done. What gives XR its unique status among all media is that it affects the human sensorimotor system in a way that can trick the brain into creating internal psychological states that transport participants beyond the immediate physical environment and into the virtual worlds. The ultimate goal of immersive experience design—an XR-mediated presence—is a form of visceral communication that has never been available to media audiences before. With its core promise of immersing participants in virtual worlds that seem real enough to make them feel as if they were physically present in a given computationally simulated space, XR takes mediation a significant step further from the electronic and computational media that were disrupting human societies between the 1950s and the early 2000s. For the forest time in human history, by directly affecting how the nervous system receives messages about the inner and outer environments, XR designers will profoundly affect their audience's ability to take appropriate action and make motor and behavioral responses.

Today, immersive technologies can not only augment our experience of the physical world, but they can also present us with entirely new worlds of games, virtual communities, workplaces, explorations of outer space, and simulated 3D data environments. In XR, we can roam imagined worlds as self-designed avatars. We can explore real but inaccessible environments located at the depth of the oceans. We can view our living rooms as if a yet-unpurchased piece of furniture had already furnished them. We can be guided, in real-time, to repair complex equipment we have never seen before. We could even create experiences that would allow biology students to see the world through the "eyes" of a butterfly. Pre-XR media makers might have been able to mediate and even integrate the visual and auditory channels into media experiences, but they could not have offered their audiences simulated smells and the tactile feel of a battlefield or a tropical beach to enhance the interactive experience. In addition to the five visual, auditory, tactile, gustatory, and olfactory senses, the XR experience design will gradually include all eight channels of human sensory perception.

Our embodied experience in the physical world is predicated on three additional extrinsic sensory systems. The vestibular system regulates our sense of balance and orientation in physical space. Proprioception is responsible for sensing the position of our muscles and joints while we move to give us the ability to coordinate the body parts as we interact with our environment. The interoceptive system is responsible for addressing the self-awareness of the body's internal physiological and physical states. In addition to the eight senses, our present experience of the world is built on deeply integrated layers of consciousness.

No medium in history has ever been able to rhetorically address the entire human nervous system and the deepest layers of consciousness as comprehensively as XR promises to do. Even though the mediation of the gustatory, olfactory, and tactile channels is still in its infancy, the basic experimental work has already given researchers access to those communication channels. Immersive neuro- and biofeedback systems have opened new avenues of therapy for such conditions as post-traumatic stress disorder (PTSD) and sensory processing disorder (SPD). Extended reality experience design (XRXD) will affect more than just the narrative transportation of movie audiences and gamers. Common daily activities, such as shopping, working, cooking, staying healthy, or traveling, will undergo a disruptive transformation to become immersively augmented.

While immersive experience design builds on human–computer interaction and traditional UX design, XR offers system designers several unique creative media-making affordances. XR designers have an unprecedented opportunity to create entire virtual universes that immerse users in the world of computer code by reintegrating the eight senses and the three layers of consciousness fragmented by the previous iterations of media technology. Not since early humans interpreted social and environmental signals in the natural world could all of the senses and all layers of consciousness come together in a single communication environment.

The fact that XR spaces are sensorially experienced environments which are a product of intentional human activity makes the premises of the XR research problem and design opportunity quite different from the concerns of HCI. Interaction designers did not have nor could they ever address the entire nervous system that connects us to the outside world. As we will explain later in the book, the partial mediation offered by traditional and computational media could not go far beyond the extended layer of consciousness. The aesthetic rhetoric afforded to

XR experience designers draws its unique power from the unconscious, nonverbal processing of its appeals that tap into our evolved predispositions such as the preference for well-lit places or the like of sweet tastes. With XR, media makers can now explore the previously inaccessible, vast, and mysterious world of human consciousness in depth that was unimaginable to the previous generations of media makers.

The Mystery of Conscious Experience

The title of this book comes from the concept of fully mediated experience known in XR research as presence. We will define the concept in detail as discussed in the literature by such researchers as Mel Slater. But before talking about mediated conscious experience, we feel it is imperative to discuss unmediated presence—the everyday embodied experience of being alive in the physical world we all know well but understand very little.

Conscious presence is still a scientific and philosophical mystery at the heart of the human condition. We take consciousness for granted as a core part of our everyday world and forget that it is one of the most elusive-to-explain phenomena in the natural world. Despite there being an active field of science of consciousness, with many researchers coming from such disciplines as cognitive psychology, neuroscience, psychiatry, and philosophy, the empirical knowledge about consciousness is so fragmentary and incomplete that what one believes about this phenomenon might be considered to be more a matter of philosophical stance and no more than a handful of contested and competing theories or even a disciplined speculation.

On an intuitive level, we have very clear expectations for conscious organisms. We see ourselves and the people around us as conscious agents capable of making independent decisions. We have clear expectations of our coworkers and family members. We get upset at drivers who do not follow the traffic rules. Yet we cannot clearly define what it is that gives us our agency.

Well, if we cannot conclusively define consciousness, we can perhaps best understand it aside from formal theoretical frameworks through its absence. Anesthesia, a state of controlled, temporary loss of sensation and awareness induced for medical purposes, is one such instance. It not only meets the intuitive standard but also the more rigorous materialist scientific one for measuring conscious activity. Neuroscientist and professor of cognitive and computational neuroscience Anil Seth describes general anesthesia as "the art of turning people into objects" (Seth, 2021, p. 1). Anesthesia creates a state unlike any other—awake or asleep—it almost completely quiets the brain's electrical activity to the point that consciousness seems totally absent. When we wake from a deep sleep, we still have an impression of time having passed by, but patients emerging from general anesthesia have no idea whether they have been absent for five minutes, five hours, or five years (2021). Those who study consciousness presume that it has evolved due to processes "taking place within the causally integrated spatiotemporal system we think of as Nature" to offer some evolutionary benefits (Kriegel, 2020, p. 1).

The phenomenon of conscious experience makes the appearance of a single and unified reality possible (Metzinger, 2009). It enables organisms to plan and execute complex and evolutionarily beneficial actions. Lisa Feldman Barrett traces the origin of the brain to the Cambrian period over 500 million years ago. During that time, something very significant appeared on the evolutionary scene. It was not thinking but hunting. Somehow, creatures became able to sense the presence of other creatures, which enabled them to eat those other creatures deliberately (Barrett, 2020). Hunting did not require a complex brain, but the evolution of the brain made hunting possible and purposeful.

The hunting hypothesis about the possible origin of consciousness addresses the essential functional aspects of behavior. However, conscious experience cannot be reduced to the product of the complex computation of neural firings in the brain alone. Metzinger argues that "What sets human consciousness apart from other biologically evolved phenomena" is the philosophically significant insight about consciousness making "reality appear within itself" (Metzinger, 2009, p. 15). From this perspective, the products of consciousness are inextricably bound with reflection and reflexivity. The life process within a living organism can not only act purposefully, it can also become aware of itself.

The third and most interesting immersive experience design perspective on consciousness is that of phenomenology. The phenomenon of a first person is necessary for a first-person point of view. There could be no subjective experience without what philosophers refer to as phenomenology. For philosophers such as Edmund Husserl, Martin Heidegger, and Maurice Merleau-Ponty, all knowledge and aesthetic judgments we form about the world have their source in the ebb and flow of the lived world. Our subjectively lived experience of objective reality is the ultimate source of meaning and value. A conscious organism must engage in some kind of phenomenology to experience such commonly taken-for-granted qualia as the subjective experience of seeing "red" (Seth, 2021).

Evolutionary biologist E. O. Wilson identified the critical interplay between our sensory perception and mental modeling by arguing that what we think of as color does not actually exist in nature. The visible spectrum, which predicates our vision, is formed by a continuous wave of light varying in frequency between 400 and 800 nm. What we refer to as color vision is a product of a complex visual signal-processing system comprised of the photosensitive cone cells located in the retina, mediating ganglion cells, the optic chiasma, the lateral geniculate nuclei of the thalamus, and finally, the cells in the primary visual cortex.[1] The response to the resulting signal depends on other kinds of information received by the nervous system combined with stored memories they might potentially trigger. Transparent to us, neural patterns generated in the brain may cause us to think of cultural units referring to colors such as red, blue, and white or to culturally relevant patterns such as the American flag (Wilson, 1999).

Much of what we focus on in this book concerns the mediation of the phenomenology of subjective experience by immersive technology. It is easy to measure the long-wavelength end of the visible light spectrum classified as the color red. It is a different matter to assess how an audience member will experience and interpret that particular wavelength when it is projected within the XR goggles. Our individual color perception depends not only on our unique genetic makeup but also on our life experience and cultural background. Some people viewing the red, blue, and white might think of the French flag. Simultaneity of color contrast also suggests that adjacent colors influence each other to change how we view them. Color perception is usually a part of a more complex process of perceiving objects, or surface properties of objects, to be more exact.

Color perception is directly related to only one of five of our external sensory channels, but it gives us a good understanding of the relationship between objectively measurable stimuli and subjectively experienced phenomena. Perceiving colors involves discernment of hue, value, and saturation. These properties can be objectively controlled by the design team. Within the additive RGB (red, green, and blue) color system used to project digital images, each of the three colors can be dialed up or down and mixed to form other secondary and tertiary hues. The experience of redness, on the other hand, is a property of a subjective conscious experience that emerges spontaneously within a sentient organism. This experience might include associations with memories of our favorite red sweater, our mother's umbrella, or any number of objects in our lived experience.

We media makers can be most impactful when, as conscious beings, we factor in our own phenomenological experience of the world as a basis to communicate with others. The best chefs can give explicit cooking instructions, but they do not cook by numbers; they make their judgments by tasting the food. Musicians whose performances move us emotionally listen to the chords they are playing. One of Miles Davis's telling responses about the ballads he played so beautifully was that they sometimes felt too beautiful. Addressing the phenomenology perspective of conscious experience leads us to an essential insight that immersive media, despite its advanced technology, must be both art and science in equal measure.

The Sense of Self: Are You the Same Person Today You Were a Year Ago, and Would You Want to Be?

Various forms of media have always shaped our sense of self, but what we know so far about the nature of conscious selfhood, mixed with the ability of immersive media to simulate virtual forms of presence, strongly suggests the unprecedented identity-shaping power XR experiences can exert on their audiences. What gives us the feeling of experiencing the world from a single first-person perspective is our deeply felt sense of self. Our unique first-person identity is the foundation of how we see the world, what we believe in, and how we act.

In a book dedicated to the interplay between identity formation and material culture, anthropologist Grant McCracken quotes a telling passage from Marilyn Monroe's (Norma Jeane Mortenson's) diary in which she confesses to falling in love with herself, not as she was but the way she could be (McCracken, 2008). Norma Jeane did not just dream of being Marilyn Monroe. She first convinced herself, and then the entire world, that she was Marilyn Monroe. And so she became

Marilyn Monroe, and millions of people believed her. In the process, Norma Jeane co-created a branded mythical identity we still collectively consume in our own quest to find out who we want to be. These mythical archetypes presented to us by the media include powerful, single, successful, little-black-dress-wearing women looking for a prince charming (*Sex and the City*), tough, ragged independent men (The Marlboro Man), secret agents (James Bond), the brave, courageous, and beautiful heroes (Barbie), and superheroes (Black Panther).

Shaping and reshaping the unique sense of self is a major project throughout our lives. Immersive role-playing games (RPG) like *Kingdom Come Deliverance* and social VR platforms like VRChat allow our avatars to explore hundreds of VR worlds and, more importantly, try on a person we might want to be. But how are such explorations and transformations even possible when we are absolutely convinced that our one-in-a-seven- or eight-billion self is a well-defined, solid, stable entity looking out on the world of other well-defined, solid, stable entities and objects? Here again, neuroscience offers many fascinating and counterintuitive insights.

Our selfhood might be a lot more malleable and fragile than we imagine. One of the best-known narrative examples of this fragility of the single self has been captured by the Scottish author Robert Louis Stevenson in his Gothic horror novella titled *Strange Case of Dr Jekyll and Mr Hyde*. Neuroscientists are advancing a hypothesis that the self is a special kind of bundle of perceptions, yet another controlled hallucination created by the brain. The sense of personal identity and the experience of being a body are our brain's evolutionarily evolved best guesses designed to keep us alive (Seth, 2021, p. 153).

Consciousness researcher Anil Seth uses a compelling example to prove this point. In the summer of 2014, Seth's mother, who was admitted to the surgical emergency unit of the John Radcliffe Hospital in Oxford for a scheduled surgery, suddenly slipped into a vegetative state caused by an undiagnosed disease of the brain known as encephalopathy. Two days after admission, Seth's mother developed intense hallucinations and delusions, which included seeing people crawling up the walls of her hospital room. Her grip on reality had faded to the point that she could not remember where she was or why she was there. She was convinced that she was subject to a cruel experiment being performed on her, which included her mad scientist son orchestrating the entire evil plot as a ringmaster. A usually kind, cooperative, and charming woman, she barked angrily at the hospital staff, demanding that her son be removed, and tried to escape numerous times. Witnessing his mother in a state of delirium, Seth could not help but reflect that the woman who, for all intents and purposes, looked like his mother was not his mother, at least not according to the construct he had of her based on decades of shared life experiences (2022).

What happened to the neuroscientist's "real" mother under the temporary state of delirium? One consequence of our intuition of self is the assumption that there can only be one indivisible, immutable self. The diverse elements of selfhood like having a body, a first-person perspective, and a sense of agency are all bound up together, subsumed within an overarching unified familiar experience we think of as being "me." The unified character of the self-experience seems so natural that we take it for granted (pp. 159–160). The experience of being me is irresistible.

This deeply felt intuition seems to reveal to us a genuine property of the actual self rather than being a special collection of perceptions. We all experience our selfhood as a sense of having a first-person perspective, a subjective viewpoint that seems to reside within our bodies, usually inside our heads. We also feel a sense of volition and agency that strengthens with every object we pick up at will and every action we take, even though many of our actions are based on habit and executed without much conscious control. Conditions such as schizophrenia and alien hand syndrome can rob individuals entirely of the volitional self. We think of our unique personalities, we think of our emotional makeup as the bedrock layer of selfhood. This intuition, again, does not square with neuroscience research. The deeper neuroscientists dig into the mechanisms behind emotions and moods, the more they discover formless feelings without clearly definable spatial extent or specific content, which we share at the core layer of consciousness with most living organisms.

Our indivisible, immutable self is based on a life story that can be told and retold to ourselves, our friends, and our family. It is created by an active brain processing and coding events in time and space, known as episodic memory. Its particular focus is the memory of events involving ourselves known as autobiographical memory. But even this aspect of subjective stability of the self is not as rock solid as we might imagine.

Consider the case of British choirmaster and scholar of Renaissance music Clive Wearing. At the height of his career, Wearing suffered a devastating brain infection that left him with massive damage to his hippocampus, producing one of the most profound amnesias ever documented in neuroscience. Wearing could not recall old memories and, even more dramatically, was unable to lay down new ones. As a result, he seemed to exist in a permanent present that lasted between 7 and 30 seconds, allowed by his working memory. After the onset of his dramatic illness, Wearing experienced life as a series of mini awakenings as if he had kept emerging from an intermittent coma. Without a firm autobiographical self, Wearing had no stable "I" around which to organize the flow of perceptions of himself, people in his life, or the world around him. In a book recalling their experience, Wearing's wife, Deborah, reflected that with no evidence present in his own mind of ever being awake before the short moments of conscious awareness, her husband was so dislocated that he would question whether he was, in fact, alive (Seth, 2021, p. 171).

Wearing's case is one of many curious neurological conditions that convince neuroscientists that all of the hallmarks of selfhood can mutate over time, erode, or even disappear entirely. Our experiences of embodied selfhood relate directly to our bodies. We believe in a unique sense of ownership we apply to our physical bodies but seemingly not to other objects in the world. This is actually not quite what happens to expert skiers and race car drivers who extend their sense of bodily ownership to their skis and cars.

There is no better proof of the changeable and precariously assembled nature of identifying with particular objects in the world as our own bodies than the laboratory experiment known as the rubber hand illusion. In the experiment, a volunteer places their real hand out of sight on one side of a table divided by a cardboard partition. The experimenters place a rubber hand in front of the subject in the location and orientation where their real hand would generally be. The experimenter gently strokes the real and the rubber hands with two paintbrushes. When the hands are stroked synchronously, the subjects develop the uncanny feeling that the rubber hand is somehow part of their physical body, even though they know it is not the case (Seth, 2021, p. 161). The brain assembles all available sensory information to conclude that the actions performed on the visually perceived object combined with the sensation of being stroked are connected to form an impression of the rubber hand belonging to the body.

Other experiments show that it is not only body parts that we can experience as our own. This phenomenon extends to the whole body. Experimenters used VR technology to generate out-of-body-like experiences in their subjects. In one study conducted by a group led by Olaf Blanke, subjects wearing head-mounted VR displays were shown an image representing the back of their own body from a distance that appeared to be about 2 meters. From that perspective, they then witnessed their virtual body being stroked with a paintbrush, synchronously and asynchronously, with strokes applied to their bodies. As in the case of the rubber hand experiment, when the stroking was synchronous, most participants reported feeling that the virtual body was their own. Seth argues that these experimental findings hint at the moment-to-moment flexibility of our body ownership (Seth, 2021, pp. 162–165).

One of the most fascinating cases of flexibility of body ownership for XR experience designers is Seth's own experience with BeAnotherLab, the brainchild of Mel Slater. The lab's system, known as *The Machine to Be Another*, allows participants to perceive the world from within the virtual body of a co-volunteer. The experience involves two participants putting on VR headsets and first looking down at their laps to see their partner's body instead of their own. The two partners then follow detailed instructions to engage in a series of coordinated movements to observe their new body appearing to respond to their commands. After some time, mirrors are held up to them, and they each see the mirror image of the other as if seeing their mirror image. Next, the experimenters remove the curtain separating the two individuals, allowing them to look at themselves from within the other's body. In the final act of this carefully choreographed experience, the two partners approach each other and hug "themselves." Seth found the experience of feeling himself inhabiting someone else's body and seeing himself from another's perspective unexpectedly compelling and powerfully moving (Seth, 2021, p. 167).

As with any technology, when it comes to XR media's impact on identity, there is good news and bad news. Immersive media can help us to explore different identities and social scenarios. It has been used by therapists to help heal deep emotional scars of trauma. It can offer us rich simulated learning environments to make better real-life decisions. It can transport us to the currently inaccessible parts of the world. All of these incredible benefits are contingent on XR media makers' willingness to use this powerful medium with humane care that respects the psychological impact of simulation of reality. Our newfound powers as immersive media makers must be matched with humility and care for our fellow sentient beings who make up our audiences.

The Tree of Consciousness: Damasio's Three Layers of Conscious Selfhood

One neuroscientific model of consciousness that can help immersive experience designers navigate the space between the objectively created immersive sensory stimuli and the subjective experience of presence was proposed by Antonio Damasio in his 1999 book titled *The Feeling of What Happens*. Damasio argued that consciousness was a dynamic, emergent state that arose from complex interactions between the brain, the body, and the environment. He proposed a model that conceptually separated consciousness into three hierarchically stacked layers, each of which emerged from the activity of the previous one. Damasio's model of consciousness relies on three fundamental concepts: emotion, feeling, and feeling a feeling. At the bottom of the pyramid, we have emotions, which are internal reactions to external stimuli that cause observable changes in the organism.

Emotions are a collection of unconscious neural responses that give rise to feelings that arise when the organism becomes aware of the changes experienced as a result of received stimuli. The emotions we feel like happiness, sadness, or joy are patterned collections of chemical and neural responses produced by the brain when it detects the presence of emotionally competent stimuli such as meaningful objects or challenging social situations (Damasio, 2001). This emotional processing of external stimuli is generated automatically. Our emotional responses are evolutionarily shaped to keep us safe by quickly responding to certain classes of objects and events with well-matched repertoires of actions.

Emotions allow all living organisms to successfully cope with encountered objects and situations that can be potentially dangerous or advantageous. The main area of the emotional responses resides in the viscera and the musculoskeletal system of the body and the monoaminergic nuclei in the brainstem tegmentum. The result of the body-targeting emotional responses is the emergence of an emotional state that affects the organism's homeostatic balance, leading to an enactment of such well known behaviors as freezing, fight-or-flight, or the production of what in social science is referred to as honest signaling in the form of particular facial expressions. As the term honest signaling suggests, emotional responses are directly observable with the naked eye. Their neurobiology can be investigated with psychophysiological and neurophysiological measurement methods (2001).

Feelings are mental representations that result from the physiological changes produced by emotions. They can be probed, but they require different methods to investigate. One of the evolutionary benefits of feelings is their ability to amplify the impact of a given situation to enhance learning and to prepare an organism to respond to comparable situations in the future. Unlike our expressions of emotions, which can be considered to be scientifically public, feelings are private. Our feelings are our own. They are subjective in nature, hidden from view, and difficult to measure and hypothesize about with the objectivity of a scientific method (Damasio, 2001).

Feelings of feelings arise at the top of the pyramid of consciousness. This phenomenon emerges when the organism detects that its representation of its own body state has been changed by the occurrence of an external stimulus and it becomes consciously aware of the feeling. Damasio suggests that the essential form of consciousness arises after the second-order neural structures are engaged, when an organism responding to its environment at the first order begins to be aware that it is responding to its environment appropriately. The moment when we feel our feelings is the moment when we become conscious.

The three phenomena of the emotions, feelings, and consciousness all share one essential point of origin—the body (Damasio, 1999). The body in Damasio's model is represented by three layers of the proto-self, core consciousness, and extended consciousness.

In the model, the proto-self represents the most basic representation of the organism, something humans share with all primates and other animals. Damasio's second conceptual layer, known as core consciousness, arises when an organism develops a sense of self and, based on the information received from proto-consciousness, becomes consciously aware of the changes occurring within its internal bodily state. Core consciousness, like proto-selfhood, is not exclusive to human beings. All sentient beings experience feelings caused by interactions with internal or external stimuli as the nervous system continuously generates images representing the organism's experience of what philosophers of mind call qualia. Core consciousness is concerned with the here and now of the present moment. Damasio described it as the feeling of knowing a feeling (1999).

The third and highest layer of consciousness is unlike the other two. It emerges when consciousness moves beyond the here and now. It depends on the use of all types of conventional memory, from working memory to short- and long-term memory, to reflect on past experiences, plan future courses of action, reason, communicate, and imagine things not present to the senses. The autobiographical self,

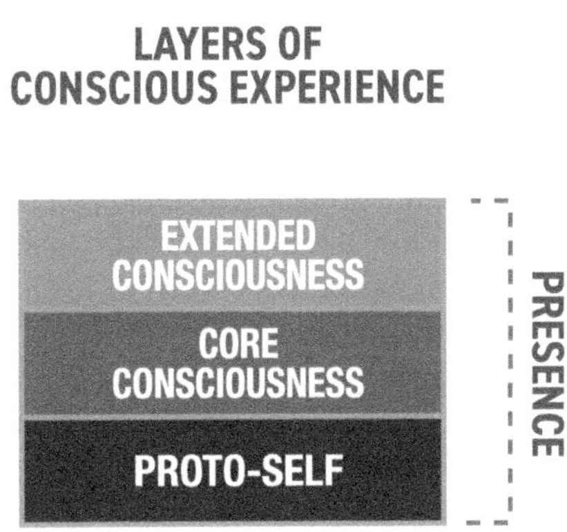

Figure II.1 Damasio's three layers of consciousness

Source: Illustration by © 2024 Peter (Zak) Zakrzewski, based on Damasio, Theory of Consciousness.

which gives us our sense of identity, emerges from extended consciousness by drawing on the memory of past experiences to generate higher levels of thought. We will return to Damasio's model and ways to apply it to XR experience design in Chapter 11 when we discuss the topic of immersing the audience.

The Brain as an Experience Designer

One of the most interesting questions we need to ask in the context of immersive experience design is: How does the brain transform the inherently ambiguous jumble of signals coming from the disparate sensory channels into a coherent world we believe to be full of real places, objects, and people? Based on neuroscientific evidence, the prevailing view among researchers is that the brain operates as a highly focused "prediction machine" that offers our conscious selves compelling "best guesses" to explain the meaning of environmental inputs. Seth describes our conscious experiences as "a kind of waking dream" or "controlled hallucinations" that are both more and less but never exactly whatever the natural world really is (Seth, 2021, p. 80).

The entire process, which is objectively observable and describable with the tools of modern neuroscience, is completely transparent to our subjective view of the world. The understanding of the brain as the "prediction machine" generating "controlled hallucinations" contrasts dramatically with the commonsense view of ourselves observing a mind-independent reality full of objects with real properties like color and shape. In this bottom-up picture, as cognitive scientists call it, of perception, our senses offer a transparent window onto the world by detecting familiar features of the environment such as people, places, and objects and conveying information about their properties to the brain. The brain's job in this model is to go through complex neuronal processing of the sensory information to form the final perceptions.

In our intuitive understanding of cognition, we imagine the brain to be mirroring the outside world. The counterintuitive neuroscientific view shows the brain making its own reality out of incoming sensory data. The neuroscientific view violates our intuitive understanding that suggests that a red rose found in the natural world directly leads to a picture of a red-colored flower known as a rose being formed in the brain. Seth cautions us against settling on intuitively appealing conclusions about cognition by reminding us that the pre-Copernican astronomers who believed themselves to be standing on solid ground concluded that the sun went around the Earth rather than that the Earth was turning on its axis (Seth, 2021).

Trying to understand how humans engage with XR media experiences becomes a lot more intriguing when we realize that our brains do not need any mediation to generate world simulations that convince us of seeing the reality itself. The simulated image of the world we perceive in our minds gives us a transparent illusion of the world "out there." For the sake of survival, we simply do not see the neurons firing away within our own brain. We are only consciously aware of the images our neurons represent for us (Metzinger, 2009). Thomas Metzinger, who studies the phenomenon of consciousness through a blend of philosophy and neuroscience, suggests that the sense of presence we enjoy experiencing the world is a subjective model generated within the tunnels of our egos. Our sense of being present is a simulation formulated for us by our nervous system to allow us to take action in the physical world. Experiencing what we consider to be happening outside of our brain is always brought about by the neural networks operating deep inside the nervous system. This is not to say that the external world does exist, only to clarify that the conscious experience of being connected to that world is exclusively an internal process (Metzinger, 2009).

Seth argues that there is enough evidence to conclude that our perception of the natural world is based not on the outside-in, bottom-up process but on an inside-out, top-down one. Our conscious experiences are built based on the brain's "best guesses" about the meaning of the incoming sensory signals. We find the first glimmer of this top-down understanding of perception in Plato's allegory of the Cave. In the allegory, Plato describes a group of prisoners chained inside a cave, facing a blank wall. All they can see is the wall and the play of shadows cast on the wall by objects passing by a fire burning behind them. Since the shadows are their only reality, the prisoners give them various names as if they were real objects. The allegory suggests that our conscious perceptions of the world are more like the shadows of the real objects we cannot directly encounter. The first modern scientist to argue for the top-down model of perception was the 19th-century German physicist and physiologist Hermann von Helmholtz, who argued that perception is a process of unconscious inference, a product of the sensory signals being combined with the brain's expectations or beliefs about the causes behind them.

There are a few important premises behind the top-down, controlled-hallucination model of conscious perception. First is the understanding that the evolutionarily and culturally prepared brain is constantly making predictions about the causes of the sensory signals based on past experiences. These predictions cascade in a top-down flow through the perceptual hierarchies of the nervous system. When we look at a red rose, our visual cortex starts to formulate predictions about the causes of the sensory signals that originate from the light reflecting off the surface of the flower.

Next, the sensory signals streaming bottom-up into the brain keep the perceptual predictions tied in useful to our survival ways to their causes in the outside world. The signals function as prediction errors to help the brain assess the difference between its expectations and the incoming sensory data at every level of processing. Through a continual process of prediction and error minimization, by adjusting top-down predictions and suppressing bottom-up prediction errors, the brain's perceptual best guesses keep us accurately mapped to the spaces, objects, and events happening in the outside world. The controlled-hallucination view suggests that the subjective experience of seeing a rose is a product of error-verified top-down predictions and not of mirroring of bottom-up sensory signals themselves (Seth, 2021).

The profound XR experience design insight here is that whether mediated or not, presence is not a direct product of sensory inputs. While it might intuitively seem to us that the world is revealed to us directly through our sensory organs, our conscious experience of reality is based on a top-down, inside-out neuronal fantasy that is, if our nervous system functions in a normal way, reined in by outside reality. Media audience members do not experience sensory signals themselves; they individually co-create flavored interpretations of the sensory inputs. In essence, the brain of each audience member is a designer of their own experience assembled out of bits of incoming sensory data into a meaningful larger whole that helps them to make sense of their world, to make decisions, and to take action.

The Human Nervous System: The Most Magical Theatre of All

Immersive experience design (XRXD) is a new frontier of the ever-expanding UX design field that will redefine the way we interact with computational media. XRXD involves creating interfaces and interactions audiences can use in a three-dimensional space to create intuitive, seamless experiences that enable users to interact with digital content in a natural and intuitive way. XR experience designers need to consider factors such as spatial awareness, depth perception, and physical movement in addition to traditional UX concepts such as information architecture, usability, and accessibility in a completely new context of multimodal mediation. The collective, evolutionary human experience of living in the world for millions of years has been shaped by the embodied nature of the multisensory presence within physical and social environments. To successfully mediate such a complex and deeply felt personal phenomenon, XR designers have to master aligning the constantly evolving immersive technologies with a solid understanding of how the sensory apparatus and the brain's vast network of billions of neurons help us to build our own unique maps of reality.

One elusive obvious thing about the effects of media technology on *Homo sapiens* is that the biological structure of the human nervous system is very ancient. Its ability to adapt is inherently slow and evolutionary, while our ability to invent new media technologies that disrupt our cultures and societies proceeds at a revolutionary pace. Despite the dizzying transformation of media technologies from drums and clay tablets to iPhones and MR goggles, researchers struggle to find any measurable differences in the actual structure and overall functioning of the brains of modern humans compared with the brains of those who lived 1,000 years before us. Mapped against the dramatic social changes over the last 500 years precipitated by the dramatic developments in science, technology, and social organization, the structural evolution of the human nervous system seems to be moving either at a glacial pace or not at all (McGilchrist, 2019).

Undoubtedly, our experience of the world plays a significant part in molding our brains, but in a phylogenetic sense, it is a very slow process. In the context of classical natural selection, environmental selective pressures acting over generations favor certain mutations above others, so some changes in patterns of brain function are very likely to have occurred in those 500 years of technological mediation. However, such changes are nowhere near as dramatic as the disruptive changes happening in the globalized Silicon Valley. It takes a moment of deep reflection on the part of all of us involved in the creation of XR media to realize that we can invent technologies at a vastly faster pace than we, as humans, can adapt to them.

Presence is the essence of being alive and staying alive, and the nervous system is the substrate of all conscious experience. The stories we tell ourselves about who we are and what we believe about the world begin deep inside our bodies. Before the brain can assemble complex thoughts and language, the nervous system processes unconscious responses that shape how we experience the world.

The nervous system controls or helps regulate every function of our bodies, from such unconscious functions as breathing and digestion to conscious, voluntary movement and thought. Even though in reality, the nervous system functions in a holistic manner, neuroscientists conceptually divide it into the central nervous system (CNS), which includes the brain, brain stem, and spinal cord, and the peripheral nervous system (PNS), which comprises the peripheral nerves that extend from the spinal cord to the extremities.

The PNS can further be divided into the somatic and the autonomic nervous systems. The somatic nervous system receives sensory information from the skeletal muscles and skin and relays it to the CNS, which controls voluntary movement by signaling the skeletal muscles to contract. The autonomic nervous system automatically regulates bodily functions such as the heart rate, breathing, and digestion, as well as involuntary reflexes like sneezing and coughing.

The autonomic nervous system can be conceptually divided further into the sympathetic and parasympathetic systems. The primary function of the sympathetic nervous system is to stimulate our automatic response to stress and emergency. The parasympathetic nervous system balances the stress response by automatically helping our bodies to return to normal functioning after a stressful event.

Our brains make all of our voluntary decisions, while the brainstem controls involuntary functions such as heart rate. The spinal cord carries signals from the brain and brain stem to the peripheral nerves, sends the sensory information back to the brain, and also coordinates certain reflexes and movement patterns, which can be independent of the brain and brainstem.

Finally, the peripheral nerves deliver messages to and receive sensory information from skeletal muscles and organs. Our automatic behaviors and habits are a product of neural pathways being created, modified, and strengthened through particular groups of neurons communicating with each other repeatedly in the same way over time.

When we repeat movements like walking, running, or shooting hoops, the neurons that control those movements strengthen their connections. Their synapses fire more efficiently, and in time, new synapses begin to form. As a result, these movement patterns become more stable and automatic. The neuroplastic process that creates the more automatic muscular patterns allows our brain to focus on making voluntary decisions that require conscious attention. Without the automatic movement patterns commonly referred to as muscle memory, we would spend hours trying to figure out how to get out of bed, walk to the bathroom, or brush our teeth.

Most of the time, we are unaware of engaging in a constant process of unconsciously creating new movement and posture patterns, which can sometimes reinforce potentially harmful habits such as the head-forward posture we train our nervous system to adopt when we use mobile phones. Only therapists and people recovering from injuries are consciously aware of the number of decisions necessary to complete life's most basic tasks. For most of us, the deep structure of our nervous system spares us this conscious effort by remembering the often-repeated movements and postures.

Brain Lateralization

Another essential feature of the vertebrate nervous system is that the brain comprises two separate but connected hemispheres. This is known as brain lateralization. We know that it takes a whole brain to complete such everyday tasks as walking or making a cup of coffee. When neuroscientists talk of brain lateralization, they are not referring to the trite cliches of pop psychology or such artificial dichotomies as reason versus emotion, rationality versus intuition, or male brain versus female brain. Instead, lateralization means that the left and right sides of the brain have developed to specialize in attending to different types of information, processing sensory inputs that reach the nervous system in very different ways, and engaging in different types of motor behavior (McGilchrist, 2021).

What we know about brain lateralization is that each hemisphere plays its unique part in all the major cognitive functions (McGilchrist, 2021). Cognitive scientists suggest that hemispheric lateralization increases the brain's capacity and efficiency by not duplicating the functions that can be handled separately by the left and right hemispheres. Lateralization of brain functions allows us to increase our cognitive capacity without engaging in the costly process of increasing brain size (Rogers, 2021).

Studies conducted on a variety of species using a range of research techniques suggest that the left hemisphere directs attention to specific, often learned categories of stimuli, while the right hemisphere pays broad attention to a wider variety of stimuli and especially to novel stimuli such as spotting predators (Rogers, 2021, p. 2). The right hemisphere also takes the lead when it comes to expressing intense emotion, handling geometric information, and dealing with social interactions (Rogers, 2021).

The two hemispheres are linked via the corpus callosum, whose estimated 300 million to 800 million fibers connect the topologically similar areas in either hemisphere. Interestingly, only about 2% of the cortical neurons are connected by this tract. What is even more puzzling is that the main function of a large number of these connections is not to excite or cause further neuronal activity downstream but to inhibit the other hemisphere (McGilchrist, 2019).

Brain lateralization offers a particularly interesting avenue for the future development of immersive experiences because XR technology can simultaneously deliver different content to each side of the brain. For instance, binocular disparity that creates a sense of depth is a product of horizontal offsetting of the images delivered to the two eyes. This allows stereoscopic devices such as VR goggles to function as high-tech synoptophores, with images in the two eyes differing in content, size, position, contrast, and luminance. Eye tracking can be therapeutically used in XR devices to measure and treat binocular vision anomalies. This unique ability to simultaneously deliver different visual, audio, or other sensory content to the left and right hemispheres is, as yet, not well explored but is a very auspicious aspect of customizing XR experiences for all manner of neurodiverse individuals.

We Are Optimized for Embodied Action in Environments Full of Objects

We are introduced to the world of outside sensations before we are even born. In the womb—our first physical environment—our bodies are immersed in a clear, slightly yellowish liquid known as amniotic fluid that serves as a protective cushion against any bumps or shakes and facilitates the exchange of nutrients that keep us alive and growing. It allows us to move inside the womb while cushioning us from sudden movements and external forces. It maintains a constant temperature around our young bodies, protecting us from heat loss. There, we are introduced to sounds and sensations, such as the rushing blood and other physiological processes of our mothers' bodies.

The womb is the first of many environments we will be surrounded by in our lifetime. From the first experience of the physical world, our very existence depends on our ability to decode vital environmental messages through our bodies. As living organisms, we subjectively experience an environment as a set of instructions for survival and thriving in an uncertain world.

What we sense inside our bodies and perceive as the world outside flows into our brain through the sensory nerves. The central nervous system structurally and functionally combines two subsystems critical to taking action; one supplies a sensory input and the other processes motor functions. Every action we take and every movement we make is a product of the vast neural network that sends signals from the brain down the spine through the motor nerves.

The sensory nerves emerge from the back side of the spine, and the motor nerves emerge from the front all along the spinal cord from the brain down to the tailbone. Our sensory and motor nerves work together to create our perceptions of the world and of ourselves in it to control our movements. We act in the environment through the nerve signals activating our muscles, the viscera, and the skeleton.

The sensory and motor functions are two vital critical components of the same system divided in the spine but integrated in the brain. The extrinsic part of the sensory system comprising the five sensory channels informs the brain of what is happening in the world. The internal part made up of the proprioceptive, interceptive, and vestibular systems keeps the brain updated on what is happening inside our bodies. These signals allow us to take coordinated action and respond to the environment.

Once we develop the integrated functions of the sensory and motor systems, they become so automatic that we do not notice their ceaseless operation; we just walk, sit, and pick up cups of coffee without having to think about how these actions are accomplished. Yet they are being programmed and reprogrammed by our actions, our circumstances, and the environments we find ourselves in. Sitting on a couch in front of a television set or stooping over a mobile phone program the entire body in ways that are very different from those of our ancestors walking and listening to the forest or sitting around the campfire. As Marshall McLuhan poignantly observed back in the era of electronic media, first we shape our tools, and then, our tools shape us.

Based on the current understanding of brain functioning, it is commonly accepted that perception depends on a kind of brain-based inference or prediction error–minimization process that approximates what is known as Bayesian inference to reduce uncertainty about what is happening around us and what we should do in response. The theory of predictive processing links action and perception as the two complementary components of minimizing sensory prediction errors (Seth, 2021).

Taking actions is a form of hypothesis testing that can lead to minimizing prediction errors. In a process known as active inference, Seth refers to "a . . . self-fulfilling perceptual prediction," the living organism takes actions such as moving its head or eyes to seek out the most appropriate sensory data that can make perceptual predictions come true (Seth, 2021, p. 117). Bodily actions such as reaching out to touch one's nose or walking into a new room update our sensory data that the brain can work with to direct our responses to the environment.

Seth illustrates this process by describing a common task of searching for car keys on a cluttered desk. While our eyes are scanning across the desk, our moment-to-moment visual predictions are being updated by the recognition of such familiar objects as pens, smartphones, coffee mugs, and sticky notes. We continually pass over these items, maintaining our visual focus to examine the scene before us until the perceptual prediction of car keys is finally fulfilled. This process also applies to higher-level initiatives like shopping or playing digital games. All mental actions cascade down into sets of bodily actions that end up altering sensory inputs. Seth argues that since "every kind of action has the potential to suppress sensory prediction errors through active inference," every action we perform directly participates in our perception (p. 117).

We interact with objects in the environment by using conditional predictions. The active inference loop of action, perception, and prediction contributes to the creation of generative models that allow our brains to predict the empirical consequences of various actions. Transparently to us, the brain continually poses questions like: If I look over there, what sensory data may I likely encounter? Conditional predictions allow the brain to know which actions, among countless possible ones, are most likely to bring us closer to our goals by reducing sensory prediction errors.

In everyday life, over short timescales, our actions supply new sensory data that help our brains create a generative model to make better predictions and decide between competing perceptual hypotheses. When we go through such common tasks as looking for car keys, our brain predicts that the most likely action leading us to locate the missing object should include visually scanning the desk rather than, for instance, waving our hands in the air or looking out the window.

The same processes contribute to the fundamental act of learning about the structure and functioning of the world in general. Active inference applied in the short term can give rise to a virtuous circle in which well-predicted everyday actions supply us with useful higher-level information about the structure of the world. That information can be incorporated into generative models, enabling improved perceptual inference to direct new actions more accurately predicted to deliver even more useful information.

But what does all of this have to do with designing immersive experiences? One of the fundamental insights from decades of digital game design practice is that as players navigate the game environments, they attain mastery and reach flow experiences by intuiting, internalizing, and making predictions about the unwritten rules of the game. Designing compelling immersive experiences like designing their game counterparts is a challenge of aligning the inscribed XR environments with the experience as anticipated by the audience. This is best accomplished based on the empathic understanding of the participants' conceptual background.

We Are Optimized for Interaction, Cooperation, and Co-Regulation

Another essential function of the nervous system is its inherent capacity to coregulate with others in our social circle. All social interactions, whether online or face-to-face, affect us at deep emotional and psychological levels. The social self is the part of us that arises from being surrounded by others. Being embedded in a social network allows us to perceive others' perceptions of us. For most individuals, this part of selfhood emerges gradually during childhood and continues to develop and evolve throughout life.

Our ability to connect safely to others is necessary for our sense of well-being. As the basis for social selfhood, the nervous system offers us a unique gamut of emotional experiences, from ways to feel bad to feeling really good about ourselves and the world. Co-regulation of our nervous systems with other human beings and even our pets is necessary for success and survival from the moment we are born. At first, we are unable to meet our physical and emotional needs. Later, we depend on others to advance our social and professional lives. Eventually, we increase our dependence on others when we age.

The rapid proliferation of mediated social environments during the Web 2.0 revolution clearly shows how compelling and even addictive the social aspect of media can be. One of the most intriguing aspects of computational media platforms is how they affect how users interact with one another and how they facilitate the emergence of group behavior. Automatic social learning is a powerful force that shapes our lives. Computational social scientist Alex Pentland has found through his research that exposure to the behavior examples of not just friends but all peers is the most critical factor for driving idea flows and shaping behavior within a social network. His research shows that providing social network incentives is a much more powerful method of changing behaviors than using individual incentives (Pentland, 2015).

To understand the full scope of issues involved in herd behavior, we do not have to look any further than the old and new media—social media. For instance, the psychological effect of the "like" button comes from exploiting the phenomenon known as the sociometer. Psychologist Mark Leary proposed that self-esteem is an essential psychological gauge by which we perceive how we are valued and socially accepted by others. This form of social capital has a high survival value for complex social animals who have evolved to live in collectives. Our ability to survive and thrive depends on how well we can navigate our relationships with others in our tribe. We also need to know as much as possible about the relationships among our peers. Questions such as: Who is working with whom? Who is mating with whom? Who is fighting with whom? Who can be an ally? Who can be trusted and cannot? all have a high survival value.

Group identity is another powerful force behind social media's massive appeal. Social psychologist Henri Tajfel traced the origin of our preoccupation with group identity to the formative challenge of early tribal existence in which family kinship alone was not enough to bind unrelated group members into successful social units. This unconscious recognition of the value of sticking together with a tribe is the reason people feel compelled to hang flags in front of their homes and wear alma mater pins on their lapels.

The importance of socialization makes our complicated and turbulent relationship with social media much easier to understand. *New York Times* reporter Max Fisher's aptly titled book *The Chaos Machine* addresses the consequences of managing human social relations through digital platforms expressly designed to manipulate attention. These expressive acts communicate to our in-group that we value our affiliation with them and that we can be trusted to contribute to the common good (Fisher, 2022). The flip side of the sharing of the in-group identity is the prejudice and hostility we can feel toward any out-group. As hunter-gatherer tribes, we were often compelled to compete for territory and resources with other groups, so collectively rallying against members of an out-group can retrieve powerful negative emotions of fear and hate.

For many individuals, joining a community organized around fighting perceived grievances can become an addictive daily ritual. Media platform algorithms, trained to record and amplify innate user tendencies, can gradually guide audience members into either positive behaviors or rabbit holes and information bubbles as they engage with a particular type of content.

For much of human history, a shared and strictly enforced code of behavior was the best way to live another day. Seeing someone violating an important norm makes us angry. We want to see the transgressors punished. We feel compelled to broadcast our anger so that others can also see the violation, see our righteous stance, and eventually join us in shaming and punishing the violators. Fisher suggests that our innate desire to punish cultural violators runs so deep that it can be observed in infants. In experiments, children only a few months old were shown two puppets, one of which shared while the other refused to. The infant subjects consistently rewarded the sharing puppet by giving them candy and punished the nonsharing one by taking the candy away. In the same way that social media warriors felt reassured by a feeling of shared grievances, test subjects as young as 2 and 3 years old went even further, rewarding puppets who punished the nonsharing puppets and even punished the ones who were nice to them (Fisher, 2022).

With the progressive generations of computational media, perhaps without fully realizing it at first, Silicon Valley seamlessly transitioned from tinkering with hardware and software to dabbling int the complexities of the human psyche. XR offers us a choice of a fresh start with a focus on mental health and well-being based on what we already know about human dependence on others.

One thing is certain: social mediation will not be the same in the age of immersive augmented intelligence. In the XRI world, our social interactions will not just be human-to-human conversations supported by machine-to-machine algorithmic dialog but also regular human-to-machine exchanges between XR media users and AI agents. In a report titled GPT-4o System Card, the internal audit team at OpenAI linked the increased anthropomorphization of AI-based agents to the potential risks of emotional reliance, miscalibrated trust, disrupting social norms, and reducing the users' need for real human interaction. During the testing, the team observed participants using language indicating the formation of shared emotional bonds with the GPT-4o model, expressing such sentiments as "This is our last day together" (OpenAI, 2024, p. 3).

Generation of content delivered with an emotionally impactful, high-fidelity human-like voice has been noted to lead to increasingly miscalibrated trust in "hallucinated" information. In a heartfelt tribute to his childhood idol, Elvis Presley, the Irish rocker Phil Lynott articulated an important insight about human bond formation by singing "I'd rather go on hearing your lies than to go on living without you" (Lynott, *King's Call*, 1980). Persuasion researchers such as Robert Cialdini have for a long time recognized that the likability and reliability of those who deliver the information can be more important than the accuracy or veracity of the information itself.

Human-like socialization with immersive multimodal AI agents may also produce unintended consequences impacting human-to human interactions. XR media users might reduce their need for human-to-human interaction by forming emotionally satisfying social relationships with the AI-based actants. Extended interactions with AI models might disrupt social norms of healthy back-and-forth exchanges. For example, OpenAI model's deferential demeanor encourages its users to interrupt the interlocutor. The models can elicit other unrealistic superhuman expectations such as the unfailing willingness and ability to complete tasks for the user and the near-infinite capacity for storing and "remembering" key details exchanged in the conversations.

Misalignments, Existential Threats, and Catastrophic Anxiety

Flow experiences happen when our actions and predictions are well matched. Violating audience expectations can not only seriously interfere with the pleasure of immersion, it can have profound psychological effects on the audience members. Family therapists and communication scholars Watzlawick, Beavin Bavelas, and Jackson proposed a conceptual framework of orders of knowledge to analyze meaning-making and communication within social systems. First-order knowledge is based on sensory perception and awareness of objects in our environment. This can be based on touching physical objects or experiencing virtual objects in a mediated environment.

We attain second-order knowledge when we formulate higher-level concepts about the objects we sense. For instance, we recognize keys as useful objects that can help us to open doors or start a car.

Our third-order premises are built over a lifetime of unique experiences to become the explanatory principles for how we believe the world should work. A key turned in the right direction should unlock a door. A handle pressed in the right direction should open a door, etc. When we use keys or assess the weather outside on a daily basis, we also learn something about the nature of locks and keys and the weather in general. The third-order premises built over a lifetime become the basis of our mental models, or maps of reality, which allow us to determine how to interact with the outside world.

Watzlawick et al. argued that humans possess an incredible ability to constantly adapt to changes at the first two levels of knowledge. Throughout our lives, as living and learning organisms, we constantly acquire new skills such as pressing on door handles, squeezing can openers, and so on. We are happy to learn new things in the realm of the first two orders of knowledge as long as our third-order premises about the self and the meaning of the universe remain intact (Zakrzewski, 2022; Watzlawick et al., 1967). We insist that the world should behave according to our carefully built-up model, and we find it deeply disturbing when the world violates our expectations. As neuroscience research suggests, our mental concepts of the world quite literally flavor all of our sensory input to help our brains evaluate the meaning of their survival value.

In order to comprehend the meaning of such everyday objects as roses, apples, and chairs, our brains have to combine various bits and pieces of knowledge from our previous experiences with them. The combined information demonstrably changes the firing of neurons in the sensory and motor regions that contribute to constructing a mental instance of a concept in question. Concepts generated at the higher level of cognitive processing give meaning to the chemicals that create the sensations of sights, tastes, and smells. The ongoing simulations cause tangible changes in the nervous system and the body. The entire process of concept construction and its flavoring of our perception of the world is invisible to us, most likely because it happens deep inside our very own nervous system (Feldman Barrett, 2017).

The unconsciously generated conceptual models of reality constitute our reality. Violations to our conceptual models of reality can be perceived as an attack on our reality itself. Psychiatrist Kurt Goldstein argued that challenges to a person's individual or group third-order premises can be perceived as no less than existential threats. He described confrontations between worldviews as potential sources of catastrophic anxiety. He argued that this type of threat to one's grand view of the world can lead to some of the most severe forms of anxiety experienced by humans. We do not have to look much further than the nonimmersive forms of social media platforms to see the validity of this insight in the realm of media environments.

The Language of Experience: Multimodal Composition and Designing for the Eight Sensory Channels

Creating not just isolated experiences but entire psychic environments for embodied, sentient humans requires a thorough understanding of how conscious experience occurs in the natural world. In the context of neuroscientific findings, our perception of the world can be understood to be a transparent illusion—an elaborate and invisible-to-us construct of the mind. Our brains, the authors of our embodied sentient experience, live inside a dark box where they cannot directly see or hear anything but only orient themselves by receiving signals from the eight sensory channels. Our perceptions are formed when our expectations based on past experiences meet the signals coming into the brain from the outside. Since each brain is unique, we experience even the same situations slightly differently. We can also experience similar situations differently based on our ever-changing cognitive and emotional states.

Maria Sanchez-Vives and Mel Slater argue that immersion is best defined by the overall sensory experience an XR system can deliver, the number of sensory channels it can simulate, and the quality of rendering in each channel (Sanchez-Vives & Slater, 2005). Designing both compelling and humane artificial environments requires a solid understanding of the subtleties of how humans experience the world. In the so-called real world, we dwell in organismic space and time defined by our biochemistry.

Our eight senses, so critical to the experience of presence, have been shaped by evolutionary necessity to prepare us to navigate our environment in particular ways (Wilson, 1999). We are evolutionarily predisposed to favor certain experiences and avoid others based on a particular bias in the cognitive development psychologist Martin Seligman referred to as prepared learning. Our sensory preferences have their basis in epigenetic rules, a range of inherited regularities of development encompassing anatomy, physiology, cognition, and behavior (1999).

Based on the research in epigenetic DNA modifications, we understand that all environments, including XR, can have long-lasting effects on human participants' biological functions and mental well-being. For instance, genes such as the glucocorticoid receptor have been shown to moderate the impact of environmental childhood adversity on future mental illness (Tyrka et al., 2016). E. O. Wilson has described epigenetic rules as "the algorithms of growth and differentiation that create a fully functioning organism" (p. 163). The propensity of our sensory systems to break the otherwise continuous environmental inputs into discrete units is one example of how epigenetic rules affect our experience of the world.

A well-functioning brain is a parallel processor. Different senses do not work separately. They are integrated. If we hear a car horn while crossing the street, air molecules impinging on our eardrums cause electrical signals to be sent to the auditory cortex. This brain activity causes a cascade of other neural events that recruit various related groups of neurons that, together, keep us safe from being run over by the oncoming traffic. Neurons in the auditory cortex evaluate the pitch of the sound to tell the difference between the car horn and the sound of an ice cream truck or the guitar at a rock concert. Our brain activates another group of neurons to determine where the sound originated. These neural activities invoke a visual orienting response, making us turn in the direction the sound came from to establish what made it.

Our minds come to understand the world around us through the interplay between low-level and high-level processing. Low-level processing in our brain hears disparate sounds and sees flickering pixels on the screen. It might even be able to combine patterns of sounds or pixels to recognize a basic source or shape in our auditory or visual vocabulary. Still, the high-level processing combines it to generate meaningful mental images. If, as a result of the neural activity in the amygdala (our emotional center), we are informed by our nervous system that we are in danger, the neurons in the motor cortex will make us jump out of the way (Levitin, 2006). One of the most critical functions of immersive experience design is to help the audience to organize all of the meaningful blobs of sensory input into content that moves them to experience something extraordinary.

We can learn a lot about the true impact of sensory processing on cognition from the instances when such sensations are impaired. Individuals who do not process and integrate sense information correctly can exhibit a variety of behavioral problems and emotional issues, such as depression, anxiety, and motor clumsiness, leading to difficulties performing everyday tasks.

The neuroscientist and educational psychologist Jean Ayres explored the idea that processing and integration of sensory information is a vital substrate for adaptive behavior. Her work was inspired by clinical observations of children with learning disabilities. She observed that some children with learning disorders experienced difficulty processing and integrating sensory information. This, in turn, demonstrably affected their behavior and their ability to learn new skills. Ayres hypothesized that the resulting behavioral problems were partly caused by the suboptimal integration of sensory information combined with the inability of higher brain centers to modulate and regulate the activity of sensory-motor centers.

Ayres conducted several cluster and factor-analytic studies, which found clusters of symptoms that fell into meaningful patterns. One was developmental dyspraxia, identified in children who consistently exhibited difficulty generating ideas for, planning, carrying out new motor activities, and processing tactile and other somatosensory information (Schaaf & Miller, 2005). Consequently, Ayres developed the sensory integration (SI) theory to explore the relationship between the fundamental neural processes that allow individuals to receive, modulate, and integrate sensory input and the resulting behavior.

Dr. Lucy Jane Miller, who works with patients with a condition known as SPD, has found that symptoms of poor sensory processing can occur within a broad spectrum of severity and can be experienced by all of us. Individuals who do not process and integrate sense information correctly can exhibit a variety of behavioral problems and emotional issues, such as depression, anxiety, and motor clumsiness, leading to difficulties performing everyday tasks.

This research suggests that by directly affecting how the nervous system receives messages about the inner and outer environments, XR designers can profoundly affect the participant's ability to take appropriate action and make motor and behavioral responses. As computing power increases and related XR technologies, such as sensors and sensory displays, gain in complexity to form presence-worthy media environments, XR system designers will continue to gain the power to ever more accurately simulate entire visual, acoustic, tactile, and even olfactory and gustatory environments to engage the vestibular, interoceptive, and proprioceptive systems to immerse their users in experiences far richer than the isolated and fragmented communication provided by the earlier forms of media.

Even though the mediation of the gustatory and olfactory channels is still in its infancy, the basic experimental work has already given researchers the ability to affect those sensory systems. XR researchers have already demonstrated, at varying degrees, preliminary abilities to deliver computer-generated stimuli to all five of the exteroceptive sensory channels. As this steady progress toward the cultural imperative of *Star Trek*'s holodeck continues, immersive experience designers will be exploring an expanded toolbox for multimodal rhetoric. With further advances in XR technology, immersive media will impact more than just narrative transportation experienced by moviegoers and gamers. Common daily activities such as shopping, navigating our physical environments, working, or cooking will gradually fall within the purview of XR augmentation. The task ahead of us is to co-create a new language of immersive experience that aligns what we know about the human nervous system with the technologies that make immersion possible.

Immersive Experience Design (XRXD): Working With the Objective Stimuli That Elicit the Subjective State of Presence

We define immersive experience design as a form of rhetorical, multimodal composition of objective stimuli created to elicit a subjective emotional and cognitive state (presence) in the participant. The definition is based on the insight that as immersive media makers, we do not directly control our audience's subjective emotional and cognitive states. We can only set up the conditions to elicit them by mindfully curating multimodal elements targeting those states. A formal approach to designing multimodal communication draws on two related components. Design elements represent the objectively verifiable sub-modalities of each sensory channel. They are the operationalizable building blocks from which XR experience designers can create the sensory dimension of the immersive worlds.

The five exteroceptive sensory channels focused on tracking changes in the external environments have unique submodalities. The visual channel, for instance, has such dimensions as the object's size, shape, color, texture, value, movement, transparency, and location. Elements such as volume, tone, pitch, tempo, and rhythm support what we experience in the acoustic space. The kinesthetic channel can be stimulated by such unique elements as location on the body, intensity of pressure, texture, temperature, weight, hardness/softness, wetness/dryness, vibration, flexibility/rigidity, and solidity/brittleness. The gustatory channel includes submodalities such as sweetness, sourness, saltiness, bitterness, spiciness, tartness, savoriness, and dryness. The mix of these gustatory elements forms the basis of all cuisine, from fast food to Michelin three-star restaurants. The olfactory channel can be defined by the submodalities of sweetness, pungency, freshness, staleness, muskiness, spiciness, woodiness, floral, earthy, fruity, herbal, nutty, burnt, and smoky scents.

> ***We define immersive experience design as a form of rhetorical, multimodal composition of objective stimuli created to elicit a subjective emotional and cognitive state of presence in the participant.***

While design elements are the building blocks of multimodal communication, design principles are the guidelines designers use to combine those elements to create optimal sensory experiences. Undoubtedly, as the XR field matures, immersive experience designers will develop their unique design principles, but many existing ones are highly applicable to immersive media design. Some XRXD principles are based on years of experiential learning in such fields as graphic design. Many of the more recent ones emerged out of the game and UX design fields. Others come from interdisciplinary research exploring human intelligence and behavior such as Gestalt psychology and cognitive science.

Gestalt, as a school of psychology, emerged in the early 20th century in Austria and Germany. The field's research and theory focused on understanding perception and higher-order cognitive processes. Gestalt theorists emphasized processing entire patterns and configurations rather than individual design components. Some of the critical aspects of the Gestalt theory of interest to XR designers deal with investigations of the relationship between the parts and the whole, particularly in the area of the visual experience.

The significant claim of Gestalt psychology argued that the human mind tries to find the most straightforward solutions to cope with the complexity of the incoming sensory information by forming groups of items with specific common characteristics. Compositions with more robust groupings or gestalt are easier to comprehend. Grouping generally contributes to unity in design. Weaker grouping creates less cohesion and more variety. Variety can add interest to an image but can also impede comprehension. Ultimately, the successful design strikes a balance between unity and variety.

Some other critical principles of Gestalt psychology include similarity, which suggests that the more alike the items are, the more likely they are to form a group in the viewer's mind. Dissimilar items will resist grouping and tend to show more individual variety. Another important Gestalt grouping consideration is the importance of spatial relationships between elements. Items that are close together are more likely to form a group. Proximity relationships usually dominate over similarity relationships. The principle of alignment is closely related to proximity.

Elements that align visually along their edges, vertically, horizontally, or through their centers appear more related than elements that are distributed unevenly.

Don Norman, one of the well-known figures who contributed cognitive science insights to UX design, popularized such important experience design concepts as affordances. He identified different aspects of digital product design based on levels of cognitive processing. Based on his research, Norman argued that users react to experiences at three levels of emotional processing: visceral, behavioral, and reflective.[2]

The application of cognitive science to the UX field resulted in a focus on usability and on addressing the problem of aligning the implementation model of digital products (how the technology works from the engineering perspective) with the user's mental model through the creation of a represented model (user interface) that bridges the two. Like their Gestalt colleagues, cognitive scientists argue that humans have limited cognitive resources to deal with a complex world. The principle of least effort suggests that people will choose the least demanding course of action whenever there are several ways of achieving the same goal (Nielsen & Norman, 2015).

Human attention and perception are biased by the top-of-mind goals and needs, the context of the situation, prior life experience, and the culture in which we grow up. The narrow bandwidth of attention means we are not very good at multitasking. Contrary to popular belief, rather than accomplishing multiple tasks at once, multitaskers are actually just shifting their attention between different tasks, wiping out their short-term memory in the process and missing an opportunity to move information to long-term memory. They are much more likely to become distracted, make errors, and have an "inattentional blindness" to events occurring in their environment.

One of the most significant contributions of cognitive science to digital product design was the introduction of the concept of mental models. Mental models are very personal maps of how the world should work. Mental models shape our thinking and behavior. They are the reasons we buy some products and not others, vote for political candidates, solve problems, and choose friends. Our mental models are based on beliefs formed by our past experience, culture, and context, not logic. As representations of real, hypothetical, and imaginary systems, they allow people to predict the outcomes of actions and interactions. As we argued before, exploring the audience's mental models is a foundational step in the immersive design process.

Some of the essential experience design principles include the need for each composition to present the audience with a focal point or a point of entry. Advertising art directors who have mastered this craft attract the viewers' eyes, ears, and noses with evocative imagery, sounds, and smells. The fast food chain KFC is famous for pumping the smell of fried chicken into every neighborhood in which it locates their stores.

Every sensory experience needs to offer coherence. Ads, paintings, and digital apps must appear as coherent systems to be understandable to their audience. Well-designed experiences need to provide their audience with an opportunity to flow through them. Flow as a design principle applies equally to paintings where the viewer's eye can flow around the surface of the canvas and an XR application which needs to allow an audience member to move through a set of encoded rules with maximum agency and freedom.

Flow is related to rhythm, which lies at the core of being alive. Our hearts beat at a certain rhythm. Accomplished athletes and dancers move with a unique rhythm. Music games such as *Guitar Hero* and *Dance Dance Revolution* are apparent examples of immersive experiences built around rhythm. The games present players with a continuous series of control inputs that must be performed in time with the beat.

Whether we stand, walk, or run, life is about balance. Designed compositions need to balance the sensory elements on the page, on the screen, and in virtual space. Immersive experiences need to balance stimulation and rest so the participants do not get too overwhelmed or too bored. Contrast is the spice of any experience. Those who like sweet-and-sour chicken or pumpkin-spice lattes understand the sensory power of good contrast and the application of a unity–variety continuum. Good experience design contrasts and balances colors, sounds, and shapes. It straddles the creative zone between order and chaos and plays the minimal versus the ornamental.

The narrow band of human perception, combined with hemispheric lateralization and the neurophysiology of our sensory system, has resulted in a figure/ground feature of perception. Throughout our waking hours, our brains decide what to focus on and what to leave in the background. Some of the most intriguing sensory experiences, such as the well-known vase–face visual illusion play with this feature of our cognition.

Foveated rendering used in MR is a good example of the human need for focus of our finite cognitive resources. Since the fovea area of the eye (discussed in detail later) can only focus very narrowly, we must choose what we focus on. The designer's job is to help the user focus on what matters most. Within the spatial computing paradigm, what matters most to the participant may lie within the virtual world's perceptual foreground, middle ground, or background, which they traverse with six degrees of freedom (DoF). The transparency–opaqueness continuum has already proven valuable for mixed reality experiences when the user wearing a heads-up display (HUD) might want to move between the virtual and the augmented worlds.

Last but by no means least, we have typography, whose millennia-old rules are meant to help the reader comprehend a written text. Typography will be as critical a component of XR as it has been with traditional graphic design and digital product design. A detailed discussion of design principles is beyond the scope of this book, so we strongly encourage those interested in learning more to reach for such books as Wucius Wong's classic *Principles of Form and Design*; *Universal Principles of Design* by William Lidwell, Kritina Holden, and Jill Butler; *Thinking with Type* by Ellen Lupton; and Don Norman's *The Design of Everyday Things*.

The Visual Sensory Channel

Seeing is the primary way in which as humans we orient ourselves in the physical world. The visual system gives us a way to explore the world of the visible spectrum. The fact that vision is a whole-brain system is underlined by the puzzling way in which nature eventually settled on locating the visual cortex, which is concerned with the reception and interpretation of signals from the eyes, which are located at the front of the head, at the back of the brain. Critical visual information processed by the brain includes important-to-survival features, which we refer to as formal design elements, as color, light value, shape, texture, orientation, and motion. All of these aspects of vision can be operationalized as objective stimuli aimed at creating subjective emotional states experienced by the users of XR systems.

In our daily interactions, our eyes scan our environment at about three saccades per second, continuously making snapshots of the world around us. In conscious visual processing, the retinal image is dynamically mapped between high-level and low-level information-processing clusters in the brain. The continuous feedback loops between higher and lower areas create a dynamically nested flow of information between what just happened a few milliseconds ago and what is happening right now. The image of the immediate past in our minds continuously creates a context for what we can experience in the present (Metzinger, 2009).

Understanding how humans perceive objects located in space as a mental reconstruction of light patterns has already put XR designers in a position to leverage the primary visual channel of human perception to a level of immersive simulation of presenting virtually generated worlds. Design of effective immersion relies on matching the complex subjective processing of perceptual cues in the human nervous system to the precisely defined (through the multiple coordinate systems) virtual objects and environments. Within XR environments, the content comprising spatial representation is a product of Cartesian geometry and mathematically generated definitions of polygons, solids, textures, light reflection properties, and force feedback interactions created by software engineers.

Many of the advances we see in the virtual reality (VR) area of XR have happened in the hardware development of visual displays such as the Meta Quest goggles. The complexity of engineering immersive visual displays poses one of the greatest challenges to the adoption of augmented reality (AR). This has led some developers to pursue such mixed reality shortcuts as passthrough technology, which uses sensors to approximate what the user would see if they looked directly through a set of optical lenses. Other strategies to add cognitive augmentation rely more heavily on the auditory channel.

A particularly interesting feature of neurophysiology of vision is the fovea of the human eye. This tiny region of the human retina is the most pronounced among all primates. It lies at the center of our gaze and is densely packed with photoreceptors. While the fovea is only about 1° across, its resolution is about 100 times that of the periphery. The entire part of the visual field that is brought into resolution is no more than about 3° across. The narrowly focused beam of attention humans are capable of is closely linked to the unique processing which differentiates the left-hemisphere (McGilchrist, 2019). The focused foveated vision excels at perceiving detail and color, while vision produced by the peripheral region of the eye is better tuned for spotting contrast and motion in the environment.

A great example of how an understanding of neurophysiology of vision can allow engineers to support sensory immersion is the concept of foveated rendering, a technology aimed at reducing the extensive computational power required for displaying XR images. Foveated rendering uses an eye tracker integrated within the XR headset to assess where the user's attention is focused. Based on this information, the system can cut down the rendering workload by reducing the image quality in the peripheral area of the eye. The technology allows XR designers to give the users an optimally rich sensory experience while assigning the precious processing resources where they can contribute most to the visual detail of the scene. Dynamic foveated rendering is a core component of Apple's visionOS operating system and the company's Vision Pro mixed reality headset. Meta Quest Pro supports eye-tracked foveated rendering via OpenXR when used with games using the quad views rendering technique.

Access to all of the sensory channels can be operationalized through a thorough understanding of their submodalities. The submodalities of the visual sensory channel include such foundational elements of visual perception as size, shape, color, value, texture, transparency, and movement. Combined through the design principles discussed earlier, these elements can form a solid operationalizable, objectively verifiable, and iterable foundation of the immersive experience design.

The Auditory Sensory Channel

Our sensory exploration within the audible spectrum occurs within the auditory system. The human brain can distinguish between specific sound frequencies by mapping them onto its primary auditory cortex. This allows particular areas of the auditory cortex to be responsible for the processing of variation in sound frequency or amplitude. Language processing is an important feature of the auditory dimension of the human nervous system. Our ability to process language is critical for successful functioning in social settings. The major area responsible for comprehending language, identified as Wernicke's area, is located, in most normally organized individuals, in the left hemisphere.

Sound design has a long tradition in media making since the early days of the radio and the "talkies." Since the 1990s with developments in wireless, location-aware, and distributed computing technologies audio augmentation has been shown to be a great channel for providing serendipitous background information to users via unobtrusive digital audio transmissions. More advanced audio augmentation requires virtual audio sources to be registered with a physical location so the listener can experience them as emanating from a three-dimensional environment. Head tracking, spatial sound synthesis, and user-specific head-related transfer function (HRTF) technologies have come a long way toward approximating sound propagation in complex environments (Schmalstieg & Höllerer). AR headsets such as Microsoft's HoloLens, which features embedded speakers, are already capable of supporting highly immersive quality spatial audio. The submodalities of the auditory sensory channel include volume, tone, pitch, tempo, and rhythm. These are the building blocks out of which immersive media designers build the auditory dimension of virtual worlds.

The Olfactory Sensory Channel

Our ability to recognize and process scents within our environment is the domain of the olfactory system. The olfactory system is unusual among the sensory systems, as it only has one source of sensory input located within the neurons of the olfactory epithelium. It appears to be more of a sensory filter than a complete associative circuit composed of many inputs and outputs. However, the olfactory bulb is connected to other important parts of the brain such as the amygdala, neocortex, and hippocampus, from which it receives top-down information for overall sense making. Our olfactory system is responsible for discriminating among odors in the environment and the detection of smells critical to survival.

The inclusion of the sense of smell in multisensory simulations goes back to Morton Heilig's 1962 *Sensorama*. This stand-alone movie console presented audiences with multisensory cinematic experiences including 3D viewing, stereo sound, tactile sensation of wind, and aroma simulation. Heilig's *Sensorama* released aromas directly at the viewers from a close distance. The task of delivering scents through the air at greater distances in a directed yet unobtrusive manner is a more difficult challenge, but this aspect of XR technology is being actively pursued by a number of researchers.

An experiment called The Food Simulator (Iwata et al., 2003) attempted to coordinate haptic and gustatory modalities. Volunteers in the experiment were asked to bite onto a force feedback apparatus, which could simulate a number of common food textures. At the same time, they were administered small amounts of liquids containing combined simulations of five basic tastes: sweet, sour, salty, bitter, and umami.

In another experiment, Narumi et al. (2011) developed a Meta Cookie prototype that combined visual and olfactory augmentation by applying AR markers made of edible ink to plain cookies. The prototype utilized olfactory and visual displays to create an impression in the subjects of differently flavored cookies. Several combinations of aromas and visuals simulating various common cookie flavors projected by the prototype caused participants to report changes in the perceived taste of the augmented cookie. Some of the submodalities of the olfactory sensory channel include sweet, pungent, fresh, stale, chemical, burnt, smoky, and strong/mild. These are the building blocks out of which experience designers can build the olfactory dimension of immersive environments.

The Gustatory Sensory Channel

Our sense of taste does more than just allow us to enjoy fine food and drink. Our sense of taste has high evolutionary survival value. It gives us the ability to discriminate between safe and potentially harmful foods. The main components of the gustatory system, such as the primary gustatory cortex, are located near the somatotopic area of the tongue, and in the insular cortex deep within the lateral fissure. The secondary taste regions are located in the opercula, deeply folded in the cortex within the lateral sulcus between the temporal and frontal lobes. Don

Norman argues that we are genetically predisposed to prefer sweet and salty tastes over sour or bitter ones. Sweet tastes signal the presence of carbohydrates, which are found in food sources with a high calorie count. Glucose is the food the mammalian brain depends on for its main source of energy.

Another important function of the gustatory system is the detection of salt, which is critical for maintaining a regulated and stable internal body environment and replenishing of water into the bloodstream. The attraction we feel toward salty snacks is the product of the survival value of salt.

We typically tolerate sour tastes only in small quantities. This moderated tolerance can protect us from eating such potentially harmful food as overripe fruit, rotten meat, or other dangerous, bacteria-laden substances. We typically dislike bitter tastes, especially in larger doses, very possibly because the majority of poisonous pharmacological agents such as nicotine and strychnine are bitter tasting.

Some of the submodalities of the gustatory sensory channel include sweet, sour, salty, bitter, hot, tart, savory, juicy, dry. These are the building blocks out of which experience designers can build the gustatory dimension of immersive environments.

Haptic Perception

Haptic perception encompasses the kinesthetic and tactile sensory channels. Tactile perception covers our sense of touch and our ability to sense information from the skin. Kinesthetic perception deals with our ability to sense body movements, muscle feelings, and position of joints, limbs, and fingers. Touch provides us a mechanism to gain geospatial knowledge about the environment and objects in it based on what we are able to touch directly.

Haptic sense can also aid us in developing models and representations of geographic space we experience and interact with through touch. Manipulating objects through the haptic sense allows us to acquire and develop the abilities to learn geographic space. It is a precursor to developing and understanding of more to complex spatial arrangements (Jacobson, 2009). The application of haptics in immersive media continues to grow. It is now applied to industrial and medical applications and in gaming systems.

The tactile system processes the touch information received by the body from the environment. Tactile information is distributed through the neural pathways and the spinal cord to the various parts of the brain including the brain stem, the thalamus, and the somatosensory cortex. The main receptive area dealing with touch sensations is the primary somatosensory cortex, located in the lateral postcentral gyrus in the parietal lobe of the brain. The somatosensory cortex part of the nervous system integrates such diverse functions as the processing of touch, pressure, temperature, and pain in the body thanks to its many connections to other brain areas. Researchers at the Star Institute, which focuses on the sensory integration treatment for individuals with SPD, suggest that individuals with impaired sensory integration can exhibit such pathological symptoms as tactile defensiveness and underresponsivity to both touch and pain.

Simulating realistic sensory impressions of touch in XR experiences without the presence of sensory-rich physical objects is a challenging problem facing XR researchers. Touch feedback is feasible with currently available XR technologies but does not yet reach the level of a realistic representation necessary for realistic representation of virtual objects. At the moment, it is more suitable for symbolic feedback, such as taps or the familiar mobile phone vibration (Schmalstieg & Höllerer, 2016).

Instrumented immersive environments and touch devices use a number of technologies that can produce impressions approximating touch. Physical proxy objects can provide passive touch feedback. Kinesthetic feedback produces force feedback that can be experienced by the users through the nerves in muscles and joints. Tactile feedback, such as surface texture, vibration, or temperature, can be experienced by the users on the surface of the skin through a variety of sensors embedded in the tissue of both skin and subskin.

Wearable haptic displays such as gloves, jackets, shoes, and exoskeletons are a focus of exploration by XR researchers in an effort to add the simulated sense of touch to virtual experiences. Wearable technologies capable of injecting weak electrical signals into the user's body can augment co-located visual and audio feedback by being location and context aware. The entire area of haptic virtuality is a constantly expanding, active field of research bringing us ever closer to delivering the experience of touch to simulated XR environments. The submodalities of the sense of touch include intensity, area, texture, duration, temperature, weight, pressure, hardness/softness, sharpness/dullness, wetness/dryness, and flexibility/rigidity. These are the building blocks out of which XR designers can build the tactile dimension of immersive environments.

The Vestibular System

So far, we have covered the five exteroceptive senses commonly thought of as the foundation of multisensory experience, but there are three additional interoceptive modalities that profoundly affect the way we experience our internal physical environments. While it might seem that immersive experience designers can control the five exteroceptive channels directly, this, as we have discussed in the light of neuroscientific research into conscious experience, is an illusion. The brain is the ultimate experience designer. All experience designers can contribute to the project is a bundle of objective stimuli that end up contributing to the participant's subjective experience. This contribution is even less direct when it comes to the interoceptive modalities. The constantly advancing XR technologies continue to allow system designers to indirectly influence the vestibular, proprioceptive, and interoceptive sensory channels. Together, all eight will play a vital role in the way we experience virtual and augmented worlds.

The vestibular system is one of those less frequently discussed sensory channels, but it is of great importance to immersive experience design. This sensory channel regulates our sense of balance and orientation in physical space. The vestibular system regulates our sense of balance and orientation in physical space.

Its function within the body is to inform the nervous system about movement and position of the head relative to gravity.

The system has two related components that are responsible for the detection of rotation and detection of linear acceleration and deceleration. The vestibular system performs its task of keeping us upright and balanced by communicating primarily with the neural parts of the brain that control our eye movements. The three semicircular pairs of canals (SCCs) comprising the vestibular system act in a way similar to a three-axis gyroscope capable of detecting angular velocity. The SCCs are most sensitive to the range between 0.1 Hz and 5.0 Hz.

The parallel counterparts work in a push-pull, stimulation–inhibition fashion, allowing individuals to sense rotation in all directions. The horizontal canal detects rotation around a vertical axis such as when we perform spins in dance or ice skating. The anterior semicircular canal's primary function is to detect forward and backward movement such as the one involved in nodding our heads. The posterior canal can detect movement in a frontal plane of motion such as when gymnasts perform cartwheels.

The vestibular sensory system has a powerful and wide-ranging influence on many parts of the brain. When projecting to the cerebellum, it affects movement of the head, eyes, and overall posture. It can also affect the cranial nerves to permit focus when the eyes are fixed on a moving object. It can support signals adjusting circulation and breathing when the body is in motion. When projecting to the spinal cord, the system allows quick reflex reactions related to balancing. Signals from the vestibular system can affect how the thalamus controls head and body motor responses.

Thorough understanding of the vestibular system is particularly critical to creating VR content because of effects such as motion sickness, which occur when the virtual vestibular stimuli do not correspond to the stimuli supplied by the senses. Motion sickness can be the result of either real or apparent motion. Symptoms of cybersickness can include nausea, dizziness, disorientation, drowsiness, and even vomiting (Jerald, 2016).

The vestibular system has been tapped as a component of a virtual reality exposure therapy (VRET) created by Virtually Better Inc. to treat such phobias as the fear of flying, heights, and storms (Aukstakalnis, 2017). The system can be run on a personal computer linked to a smartphone-based stereoscopic head-mounted display. The phone's accelerometers track the roll, pitch, and yaw of the users' head positions, while a handheld controller allows them to navigate the simulations in a controlled manner which permits gradual emotional adjustment to previously traumatic experiences (2017).

Proprioceptive System

The proprioceptive system is responsible for sensing the position of our muscles and joints while we move. It gives us the ability to coordinate the body parts as we interact with the environment. Proprioception tracks where the body is and how it is moving in the physical environment by processing the sensory signals flowing from the receptors situated all over the skeleton and musculature. We do not think about the essential role proprioception plays in all our actions because this channel does its work in the background. Without proprioception, the simple act of touching one's nose with the eyes shut would not be possible. From the neuroscientific perspective of active inference, touching your nose is made possible by the brain making a suite of proprioceptive predictions about your hand movement and the position of your nose. The proprioceptive system combines sensory information from neurons in the inner ear about the motion and orientation and stretch receptors in the muscles and the joint-supporting ligaments for achieving such functions as our physical stance.

The proprioceptive system can work on a conscious level when the signals travel up the posterior column-medial lemniscus pathway to the cerebrum part of the brain. It can also function unconsciously when the signals travel through the dorsal spinocerebellar tract to the cerebellum area of the brain. Because the body is one frame of reference humans understand intuitively, proprioception can be utilized in AR for the placement of augmentation. Virtual name tags and profiles can be placed on human participants and inanimate objects as part of designing embodied immersive experiences.

Augmented information can also be referenced by the participant's own body parts. Head-referenced displays are ideal for displaying information that needs to be constantly visible to the user, such as system-level navigation and status updates. Torso-referenced displays can take the form of virtual objects, such as tool belts, which can be virtually attached to the body. Hand-referenced displays can place useful information on the palms of the user's hand. Such options are most appropriate when users engage in the manipulation of real and virtual objects within the IoT world. Full-body augmentation can become extremely useful for applications involving sports, performance art, and medicine. User behavior involving body-placed augmentation can be tracked by XR systems for valuable behavioral data analysis.

The Interoceptive System

The interoceptive system is responsible for addressing the self-awareness of the internal physiological and physical states of the body. The interoceptive system provides such critical signals from the body as the sensation of pain, temperature level, muscular and visceral sensations such as tension, vasomotor activity, and the perceived need for air. The system is associated with autonomic motor control, which is different from mechano-reception sensed in the skin or proprioception, which is sensed in the muscles and joints of the body. The primary interoceptive activity involves providing the basis for subjective feelings of our physical and emotional awareness occurring in the right anterior insula (Craig, 2009).

The potential of the XR's design-level affordance to mediate the connection we have with the environment and with our own bodies can be seen in the system called Bravemind, designed to administer empirically validated prolonged-exposure therapy to patients suffering from PTSD. *Bravemind* immersively guides patients, in a controlled manner, through experiences underlying their combat-related memories in order to diminish the resulting anxiety. The system integrates sensory mediation with visual displays featuring realistic scenes and characters, auditory and tactile displays of rumbling engines and explosions, and olfactory displays emitting context-relevant odors such as burning rubber and cordite to deliver *in vivo* simulations of war zones (Aukstakalnis, 2017). The resulting multimodal simulation engages the participant's entire nervous system to provide an effective and measurable psychotherapeutic treatment. Integrated sensory simulations exemplified by *Bravemind* show us the potential of leveraging the eight sensory channels to help XR users in the areas of embodied mindfulness and directed emotional self-awareness.

Peak Experiences, Optimal Experiences, and Flow

We relate immersive transportation, or the ability of XR designers to set up conditions for participants to achieve a sense of presence, with the concept of optimal experiences. Researchers within the field of positive psychology, a field pioneered in the early 2000s by Martin Seligman and Mihaly Csikszentmihalyi, distinguish between a number of constructs related to psychological well-being that include optimal experience, peak experience, and flow. Seligman, in particular, wanted to evolve psychology away from the narrow focus on such negative mental phenomena as trauma, suffering, and pain and toward the study of happiness, well-being, and flourishing.

University of Michigan psychology professor Christopher Peterson defined the goals of positive psychology as focusing on creating positive experiences like joy and inspiration and building positive traits such as resilience, gratitude, self-esteem, self-confidence, and compassion. Positive psychology has been applied to sports and personal coaching, education, relationships, the workplace, and many other domains of life.

The construct of peak experience was first proposed by the humanistic psychologist Abraham Maslow in 1971. After developing his hierarchy of needs, Maslow focused on the ability of people to attain high points in life during which they felt in harmony with themselves, their circumstances, and their surroundings. He associated peak experiences with such states as extreme joy, feeling of fulfillment, and the pleasurable moments of loss of self in a transpersonal dimension. Peak experiences are related to Maslow's identified need for self-actualization, which is based on a subjective feeling of doing what one is "meant" to do and being who one is "meant" to be in life.

Peak experiences are fairly rare phenomena. In Maslow's pioneer studies, they were reported by only 1 out of 3,000 participants. The newer concept of optimal experience, extensively discussed by Delle Fave et al. (2011) defines optimal experiences as states "characterized by a complex and positive balance between the emotional, cognitive, and motivational components of the psychic system" that are commonly found in everyday life, and they report that in a cross-cultural study involving over 1000 participants, 86.2% reported enjoying optimal experiences associated with one or more daily activities (2011, p. 101).

Optimal experiences arise out of a harmonious feeling of well-being that is the product of synchronic interplay between positive mood, high cognitive investment, and short- and long-term motivation. They mainly manifest as focused involvement rather than happiness or joy. During optimal experiences, participants do not report loss of self-awareness or ecstatic states associated with peak experiences but do report a sense of engagement and control of the situation. The reported loss of self-consciousness associated with optimal experiences is related to a complete absorption in the activity that frees the individual from concerns about their self-image, their own behavior, or social judgment. Participants experiencing optimal experiences also report their actions flowing naturally without concern for specific outcomes or external evaluations of performance.

Researchers have found that optimal experiences are not associated only with leisure and play. They can also emerge during work and studying activities if such activities can inspire intrinsic motivation, engagement, and concentration. Participants engaged in sports, games, arts, and even hobbies can attain optimal experiences when their intrinsic motivation is matched by goal setting and intentional effort toward specific goal or skill acquisition. "Serious leisure" or play activities leading to optimal experiences require structure based on a clear set of rules and procedures that can foster agency, choice, and autonomous action focused on meeting challenges and pursuing individual or collective goals.

The third important related psychological concept of flow was introduced by Mihaly Csikszentmihalyi in the mid-1970s. Unlike Maslow's peak experience, flow

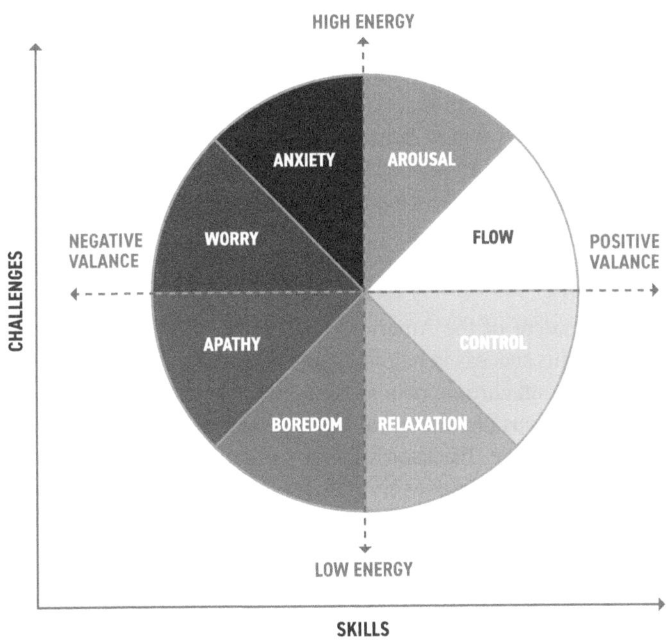

Figure II.2 The sensus of XR-mediated presence

Source: Concept and illustration by © 2024 Peter (Zak) Zakrzewski, based on Riva & Waterworth, Presence and the Self, Delle Fave et al. Experience Fluctuation Model.

is not characterized by passive immersion in a state of transcendence uncontrollable by the individual but by the merging of action and awareness and an absorption in the ongoing task supported by active engagement and intentionality. Both flow and optimal experience are spontaneously occurring emergent states that cannot be induced by either media designers or participants at will.

We build on Delle Fave et al.'s presentation of flow as part of their experience fluctuation model by combining it with Riva and Waterworth's concept of the sensus aspect of presence. Giuseppe Riva and John Waterworth leveraged Damasio's three layers of consciousness to propose the three-dimensional model of the presence consisting of focus, locus, and sensus. Sensus, the third component of Riva and Waterworth's model, refers to the level of attentional arousal engendered in the participant by the XR experience. Arousal directly affects the degree of engagement and activity exhibited by the participant during the immersive experience.

Sensus is based on balancing the challenges posed by the activity with the skills possessed by the participant. We mapped them out across two spectra of the low to high energy levels and emotional valence measuring the emotional response from negative to positive. Low challenges matched by low skills required for the task, shown in the lower-left quadrant of the Figure II.2, result in such responses as apathy and boredom. High-level challenges not quite matched by the necessary skill levels can result in the task eliciting from the participants the emotional responses of worry and anxiety. Matching high-level challenges with the higher levels of necessary skills results in participants experiencing arousal and flow. Matching high skill levels with medium level challenges can result in the participants feeling a sense of control and relaxation.

We will return to the topic of Sensus and the other two components of Riva and Waterworth's model in chapter 11 when we cover them in more in depth in the context of immersing the audience phase of the XRXD logic model.

Bio- and Neurofeedback: Designing for Behavioral Change

The founding president of the American Society for Clinical Hypnosis, Milton H. Erickson, credited his success in helping patients manage such conditions as negative behavioral patterns and chronic pain to the ability to closely observe and respond to his patients' unconscious physiological activity such as breathing, heart rate, muscle tension, and tone of voice. Erickson's therapy relied on the "readiness" with which his patients' unconscious minds were picking up and responding to the useful and meaningful signals he provided (Erickson, 1980, p. 208). He assessed this readiness by observing such biomarkers or honest signals as breathing rate, tension in the body, and pupil dilation. Erickson argued that those who want to master successful interaction must develop what he referred to as a "respectful awareness" of human capacity to exchange nonverbal communication signals at an unconscious level.

Immersive systems provide XR media makers with a unique capacity to observe and respond to the audience's mental and emotional states with actual measurements of users' physiological activity and behavior. Data about audience response can come from qualitative accounts of subjective conscious experience such as questionnaires and interviews, but perhaps more interestingly, it can also come from biomarker sensors and observed behavior. What is quite unique about immersive media applications is that they can provide valuable user feedback from biometric data obtained from biosensors incorporated within the XR systems. Biosensor technology is widely used in biomedical diagnosis, as point-of-care monitoring of treatment and disease progression, and in drug discovery, forensics, and biomedical research.

XR systems can also incorporate electroencephalography (EEG), functional magnetic resonance imaging (fMRI), or functional near-infrared spectroscopy (fNIRS) to provide measurements of neural activity in the brain. All of the biosensor-generated data combined with advanced AI algorithms can create unprecedented kinds of feedback loops between the XR media and their audiences.

As immersive media designers, we are beginning to see a glimpse of the behavior-changing power of XR. We already have at our disposal an immense body of research in the areas of bio- and neurofeedback that can guide how we think about assessing and guiding the audience experience of the system by managing each interaction in a way that matches user needs and expectations.

Therapeutic application of biofeedback sets a precedent for this type of finely managed interaction. Biofeedback can be described as a process of gaining greater awareness of key physiological functions of the body in order to exert control over the body's systems. Humans engage in self-biofeedback naturally as part of the interoceptive process with varied levels of consciousness and intentionality. Wellness and health innovation centers such as the Mayo Clinic offer biofeedback therapy that can help patients to monitor their own brain waves, breathing patterns, respiration, heart rate, blood volume changes, muscle contraction, sweat glands, and perspiration activity.

Biofeedback and neurofeedback have become staples of various forms of therapy because of their potential to elicit novel behaviors as a result of an unconscious, behavioristic, or environmental conditioning leveraging naturally occurring mental processes. Monitored biofeedback can lead to helping subjects gain greater awareness of many physiological functions of their own bodies, leading to achieving a greater degree of control of emotional and cognitive states.

Regardless of whether biofeedback is achieved through person-to-person therapeutic interaction or with the aid of various electronic instruments, the process comprises three important steps: (1) the measurement of the physiological activity such as brain waves, heart function, breathing, muscle activity, or skin temperature in the subject; (2) an accurate feed of this information back to the subject; and finally (3) the presentation of the biomarker information that can elicit the desired physiological, neurological and behavioral changes by the subject.

Psychiatrist Bessel van der Kolk leverages the potential of using biomarkers such as heart rate variability (HRV) to regulate the emotional and cognitive states of patients suffering from PTSD. As a biomarker, HRV allows researchers to measure the relative balance between the sympathetic (SNS) and the parasympathetic (PNS) nervous systems.

In one experiment, Van der Kolk et al. recorded the depth and rhythm of breathing and measured the pulse of two groups of subjects with and without PTSD symptoms. In a regular pattern of heart rate variability, a well-regulated individual's heart rate goes up with each inhalation. The heart rate then slows down during exhalation. After testing about 60 subjects, van der Kolk and his colleagues discovered that people with PTSD had unusually low HRV, which meant that the sympathetic and parasympathetic nervous systems of people suffering from the condition were out of sync. Under stress, well-regulated individuals temporarily sped up their breathing, making both their breathing and heart rate irregular and temporarily out of sync as part of a normal stress response.

Persons experiencing chronic PTSD exhibited a more permanent shut-down pattern of rapid and shallow breathing. Their heart rate slowed down and fell out of sync with their breath. Chronic PTSD sufferers reliving a traumatic memory initially exhibited deep, labored breathing, which is typical of a panic reaction. This pattern was followed by rapid, shallow breathing and slow heart rate, another out-of-sync pattern, which showed that the person was shutting down and withdrawing emotionally (van der Kolk, 2014).

The researchers followed the initial experiment by collaborating with a hatha yoga instructor to create a program that tested the effectiveness of mindful embodied practice to HRV regulation. They selected 37 women experiencing severe trauma who had already received years of various forms of therapy without benefit. With the aid of a custom-designed computer system, they measured HRV in eight yoga practitioners from the control and the test groups to find that directed embodied practice significantly improved arousal problems in subjects suffering from PTSD. After 20 weeks of practicing yoga, chronically traumatized women in the group developed increased activation of the insula and the medial prefrontal cortex, which are brain structures critical to self-regulation. Additionally, self-regulation improvements coincided with a dramatic change in the subjects' relationship to their own bodies (2014).

Neurofeedback is a form of biofeedback focused on measurement of real-time brain activity. It is most commonly obtained through the application of electroencephalography. Neurofeedback has been practiced since the late 1950s, when psychology professor Joe Kamiya started studying the phenomenon of internal perception. Kamiya discovered that his subjects could learn through feedback to tell when they were producing alpha waves associated with relaxation.

EEG technology underpinning neurofeedback works by measuring the varying heights (amplitude) and wavelengths (frequency) of a subject's brain wave activity. The frequency, which is measured in hertz (Hz), or cycles per second (cps), refers to the number of times a measured waveform rises and falls within the time measure of one second. Frequencies shown by the EEG systems provide objectively measured insights into the subject's inner world of brain activity. The rate at which our brain waves are firing is directly related to the state of arousal. Delta waves, which exhibit the slowest frequencies of 2 to 5 Hz, are associated, with sleep when the mind is turned inward.

People who show too much slow-wave activity in their waking state exhibit poor judgment, poor impulse control, and generally foggy thinking. Eighty percent of children diagnosed with attention deficit hyperactivity disorder (ADHD) show excessive slow waves in their frontal lobes.

Theta frequencies, at 5 to 8 Hz, are associated with the edge-of-sleep and hypnotic trance states. Brain states that accompany this frequency range allow the creation of novel connections and associations by inducing free flowing and creative states of mind unconstrained by logic or the ordinary demands of life. Theta frequencies can also occur when people report being depressed. Alpha waves, at 8 to 12 Hz, are accompanied by a sense of peace and calm, familiar to people who practice mindfulness and meditation.

Beta waves, at 13 to 20 Hz, are the fastest brain wave frequencies. A brain dominated by beta waves is oriented toward the outside world. On the positive side, this enables people to engage in focused attention while performing tasks. On the negative side, high beta waves over 20 Hz are associated with agitation, anxiety, and body tenseness (van der Kolk, 2014). Advanced computer-aided EEG analysis, known as qEEG, can trace the subject's brain wave activity millisecond by millisecond to produce a data visualization reminiscent of color maps showing which frequencies are highest and lowest in the monitored areas of the brain. Quantitative EEG systems can show researchers how well various brain regions are communicating and working together. The ever-growing databases of normal and abnormal brain wave patterns allow a classification and comparison of quantitative EEG patterns among thousands of users (van der Kolk, 2014).

In neurofeedback sessions, clinicians can set the reward and inhibit undesirable frequencies to let the feedback reinforce selected brain-wave patterns and discourage others. Neurofeedback systems give clinicians the ability to track circuitry in very specific parts of the brain. By moving the electrodes to the area of the left brow, the neurofeedback system can stimulate the feeling of being focused and alert by rewarding beta waves in the frontal cortex. Moving the electrodes to the crown of the head can produce a feeling of being detached from the images presented on the computer screen and more aware of the inner sensations in the subject's own body (van der Kolk, 2014).

Neurofeedback systems can provide the user with a visual and auditory mirror of the functioning of their own brain. XR systems leveraging neurofeedback have the potential to mimic the embodied responses of people engaged in face-to-face conversations in which we are rewarded and encouraged by observing smiles and nods signifying our partner's engagement. Neurofeedback systems can prompt the user toward creating new patterns of brain activity that can enhance the brain's natural propensity toward balance and self-regulation. Modern neurofeedback systems can function much like digital games to provide rewards and inhibitions, signifying success or failure by various sensory cues fed back to the user via sounds and visuals.

According to Sebern Fisher, the former clinical director of a residential treatment center for severely disturbed adolescents, the technology may help accomplish two important tasks: free up any innate but stuck oscillatory properties in the brain and allow new brain patterns to develop (van der Kolk, 2014).

Research in the area of cognitive EEG experiment protocol has laid the foundation for the possibility of brain–computer interface (BCI). In BCI, brain activity associated with cognitive states is used to estimate a model of a subject's cognitive brain state, intent, or response for the purpose of controlling external computer interfaces. The director of the Swartz Center for Computational Neuroscience, Scott Makeig, proposed an experimental paradigm in the design of cognitive EEG experiment protocols that enabled measurement and analysis of human brain dynamics, including eye movement and motor behavior, while performing cognitive tasks in 3D environments.

Makeig and his colleagues coined the term "mobile brain/body imaging (MoBI)" to describe the process of combining complementary data modalities such as motion capture, eye tracking, and sound and video recordings with high-density EEG data to allow the analysis of brain dynamics under conditions much more closely resembling everyday life situations than typical neuroscientific experiments. The MoBI approach has been applied in studies of brain dynamics supporting such as everyday functions as expressive gesturing, gait, balance, and cognition during walking (Ojeda et al., 2014).

The MoBILAB toolbox introduced by Makeig and colleagues supports an offline analysis and visualization of any number of concurrently recorded multimodal data streams leveraging the object-oriented programming supported by MATLAB. The technical feasibility constraints still limit the type and range of states that can be estimated, as well as the number of data streams which can be processed, but MoBI has set a definite direction in the brain–computer symbiosis (2014).

The Ericksonian insight about the importance of interaction and communication occurring below the conscious level reveals in a subtle but powerful way how XR media can become, over time, persuasive biofeedback environments. As the technologies continue to mature, bio- and neurofeedback have the potential to become core aspects of immersive experience design, but leveraging these approaches does not come without some potential risks.

The triumph of pharmacology over other forms of therapy since the 1980s offers a cautionary tale for XR media approached without holistic sensitivity and informed intuitive judgment. Van der Kolk argues that technologically based interventions displacing more holistic forms of therapy end up enabling people to suppress any negative symptoms without addressing the underlying issues. In order to mindfully apply bio- and neurofeedback to XR systems, designers need to bring a perspective of empathy, judgment, and sensitivity in mobilizing the entire repertoire of the human perceptual system, which helps participants to create positive and beneficial models of reality.

The Synchronic View of Immersive Experience

The authors of the seminal book titled *Managing the Customer Experience,* Shaun Smith and Joe Wheeler (2002), defined branded experiences as being intentionally designed to consistently deliver value to the end user. Intentionality and attention to detail are the hallmarks of compelling experience design practice. Immersive media designers need to make a series of mindful decisions about creating the minute details of the spatiotemporal dimension of the proposed virtual worlds.

For the sake of simplicity, we like to conceptually divide the spatial and the temporal aspects of XR experience design based on the synchronic (spatial) and diachronic (temporal) views of the inscribed (designed but not yet encountered by the audience) immersive experience. Taking the synchronic view of the virtual space allows the design team to assess the three-dimensional spatial aspect of the XR application away from the complexity of having to factor in the action unfolding in time. Obviously, the spatial aspect of immersive experience cannot be separated from its temporal dimension, but conceptually considering the immersive space as if it were frozen in time (synchronic) can allow an easier and more thorough exploration of the various zones and objects needed to deliver the optimal experience.

We liken the synchronic view of XR to the dramatic and cinematic concept of the mise-en-sène. Borrowing the dramatic concept of the mise-en-sène to attend to the spatial dimension of XR application enables media makers to enhance the immersive experience through the deliberate arrangement and blocking of such vital elements of virtual experience as actants, sets, props, costumes, and lighting. We will return to the

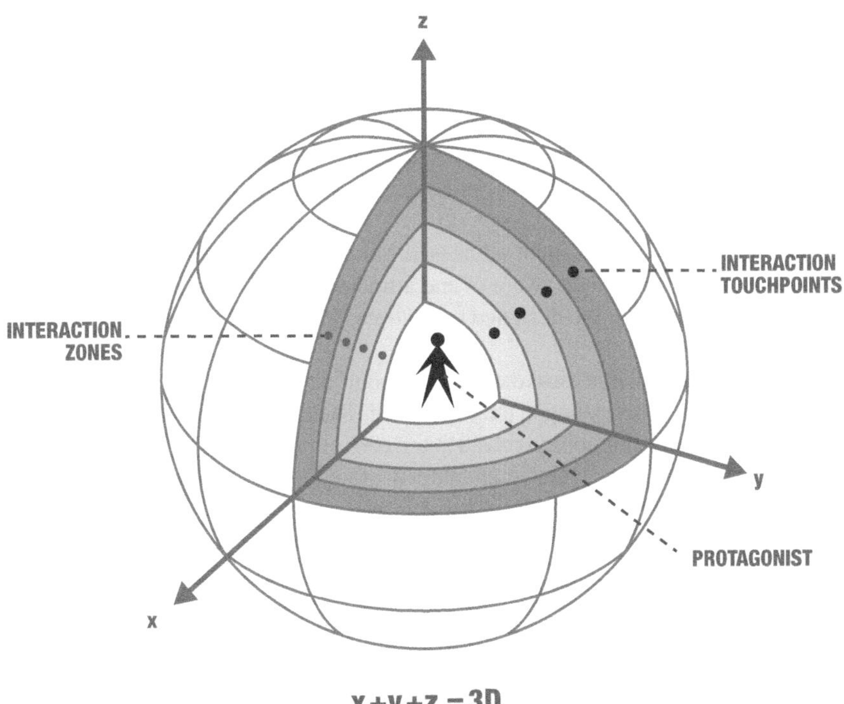

Figure II.3 The synchronic snapshot view of the XR experience

Source: Concept and illustration by © 2024 Peter (Zak) Zakrzewski.

detailed description of the synchronic view of XRXD in Chapter 4, when we discuss the use of context scenarios as an immersive experience design tool.

In the synchronic view of the XR application, the design team considers how many interaction zones are needed to deliver the desired audience experience. Interaction zones are akin to levels in digital games, scenes in a movie, or locations in virtual and augmented environments. The zones logically separate the XR space into areas organized by a particular activity, user goal, or skill level.

The next set of decisions needed to be considered within the synchronic view concerns the interaction touchpoints the protagonist needs to encounter on the way toward their goal. Interaction touchpoints are a concept borrowed from brand experience design, in which a customer is presented with a wide range of opportunities, ranging from websites to retail environments and customer service representatives, to interact with the brand. The best way to understand interaction touchpoints would be to think of megabrands such as Nike or Apple consistently forming and strengthening the relationship with their customers through an array of well-chosen and well-placed interactions on their social media channels, inside of their retail stores, and through their physical products. We will return to the topic of interaction touchpoint design in Chapter 4.

The Diachronic View of Immersive Experience

The diachronic view of the inscribed (designed but not yet encountered by the audience) experience gives the design team the conceptual view of the XR application in the fourth dimension of time. Experience designers map user journeys using a five-step model originally introduced by Larry Keeley of the design consultancy Doblin. The original model conceptualized the user journey in the form of 5Es.

The first step of Enticement addressed the attraction phase of the experience. Audiences need to be made aware of and attracted to the experience. Like all other stages, attraction has to be designed with strategy and care to address the needs of the core audience.

In Keeley's model, Entry was the inflection point, at which point the user would make a commitment to participating in the experience. The Engagement stage addressed the interactions and the emotions involved at the core of the curated experience.

The Exit stage was about managing a graceful and memorable end of the engagement part of the experience. It was based on the insight that the most recent things experienced have the most impact on how the experience is remembered. Helping participants to leave on a strong, positive note sows the seeds for future engagement with the franchise.

The Extension stage of the original model was focused on the often-neglected effort to extend the interaction beyond the original experience by keeping the audience connected to the franchise as repeat users or, better yet, brand ambassadors.

In our revised framework, we consider Attraction, Immersion, and Extension to be the main stages of the experience and Entry and Exit to be the transitions between them. While the stages might get more attention from the design team, transitions need to be designed with equal care, for they are the connectors between the main stages. Improperly designed transitions might jar the user even to the point of walking away from the entire process. For instance, some early e-commerce interactions were designed so poorly that they led willing buyers to abandon the transactions.

It is worth pointing out that the diachronic experience design model can be applied at both the macro and micro levels. It can be utilized to design the strategy to attract and guide the audience to the XR media franchise. It can also be applied to all of the interactions within the media experience. Let us recap the revised model.

1. Attraction is the entry point into the experience. It can be considered an analog to a physical affordance like a door into a room and an appeal to the user to engage. In this do-or-die stage, designers need to consider the entire discovery path, which will help to draw the audience to the proposed experience. Social and traditional media and peer networks afford many opportunities to attract and build an audience.
2. The Entry phase represents the transition during which the audience is transported from their initial interest into the immersion stage of the experience.
3. Immersion is the second and pivotal stage in the user experience process. This is the stage within which the Aha! and the Wow! moments associated with presence can occur. Supporting presence is all about letting the audience shut out the everyday and be fully immersed on the deepest level. Concertgoers, athletes, sports fans, and people who meditate or spend time in nature know this feeling well. Presence is the pinnacle of the XR experience design process. It is also pivotal because it can transform a first-time user into a fan or even a brand ambassador, or if it is disappointing, it can turn them away.
4. We refer to the exit stage as Transformation. As people exit experiences, they rejoin their everyday world, which tests the bond built during the immersion stage of the experience. In order for the audience member to stay connected to the franchise and the community, the connection must withstand this separation

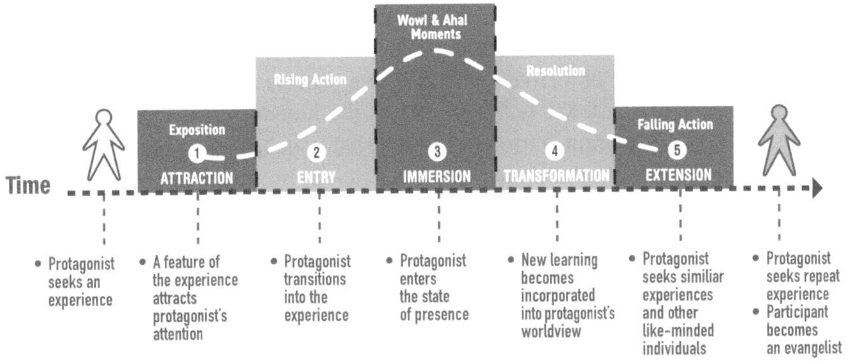

Figure II.4 The diachronic view of the XR experience

Source: Illustration by © 2024 Peter (Zak) Zakrzewski, adapted from Doblin Group's Compelling Experiences model.

shock. This is best accomplished by creating immersive media content that transforms the audience with unforgettable experiences and/or new knowledge and understanding of the world. The transition between presence and repeat immersion takes a skilled process, culminating in the Extension stage.

5. The desired outcome of the Extension stage is to support the audience members in staying connected beyond the initial experience and becoming dedicated long-term members of the community. Extension is where individual users coalesce into dedicated communities of people who want to share the experience and connect with other like-minded users. They want to collect objects that remind them of the immersion and transformation that have taken place in their lives. Designing for extension can make a difference between building and scaling an influential audience that not only comes back for more interactions with the brand but also recommends it to their near peers and others in their social networks.

Designing for Presence

Throughout this book, we argue that the ultimate goal of immersive experience design (XRXD) is the setting up of the right conditions that support participants in attaining presence, which can give rise to flow, and Aha! and Wow! moments. As immersive experience designers, we have at our disposal the tools of rhetorical, multimodal composition of objective stimuli that can elicit a subjective emotional and cognitive state of presence in the audience.

While immersive experience design builds on human–computer interaction and traditional UX design, XR offers system designers a number of unique opportunities for creative immersive media making. XR designers have an unprecedented opportunity to create entire virtual universes that immerse users in the world of computer code by reintegrating the eight senses fragmented by the previous iterations of media technology. Not since the time when early humans interpreted social and environmental signals in the natural world could all of the senses come together in a single mediated environment.

If designing immersive experiences means applying objective stimuli (immersion) to elicit a subjective state (presence) in the user, it is important to clarify the difference between the objective and subjective aspects of XR media. One of the unique characteristics of XR is that it affects the human sensorimotor system in a way that tricks the brain into creating internal psychological states that transport participants beyond the immediate physical environment and into the virtual world of digital media. Virtual reality researchers refer to the simulated sense of "being there" and "acting there" as presence. As mentioned before, immersion can be understood as a form of visceral communication.

Virtualized presence can take many forms. We can experience mental and embodied presence. We can also experience social and emotional presence when virtual environments simulate social interaction. We embrace the research perspective that views presence as a subjective mental state in which a user "feels physically present within the computer-mediated environment" (Sanchez-Vives & Slater, 2005, p. 4).

Sanchez-Vives and Slater draw a clear distinction between presence and immersion. They suggest that immersion is a product of technology, while presence is an internal physiological and psychological state experienced by participants immersed in XR environments (Sanchez-Vives & Slater, 2005). Presence is an emergent phenomenon, a product of rhetorical efforts of media designers using immersive technology to elicit a subjective state in the user actively engaged in pursuing their own goals.

Breaks-in-presence spell the breakdown of the immersive illusion. Glitches at the level of the physical and technological substrate, such as loss of tracking or the experience of bumping into objects in a physical environment, are detrimental to XR experiences. One of the most interesting features of presence is that XR users appear to experience at least partial inattentional blindness to the real world and the technical substrate of the virtual experience.

Slater argues that presence is based on two primary sensory illusions: the illusion of place and the illusion of plausibility. The place illusion addresses the question of location, while the plausibility illusion convinces the brain that whatever is happening is real. Technologically driven factors such as low-latency head tracking, accurate 1:1 body tracking, and fast frame rate fool the human perceptual system into believing that we are present in a place other than our real physical location. In the same way, a virtual world that accurately meets our expectations in the way it behaves and reacts to our actions makes us perceive the XR environment as being coherent in its construction (Bye, 2017).

Researchers who take the perspective of reality being formed through action rather than through mental filters define presence as the ability of the XR system to successfully support the participant's actions in the virtual environment (Sanchez-Vives & Slater, 2005). Under this definition, the perceived reality of the user experience is defined relative to the functionality of digital objects and environments rather than to their appearances (Sanchez-Vives & Slater, 2005). While the focus of the experience-centered approach is on the creation of virtual environments that support the sense of "being there," the action-centered approach aims to create environments that support a targeted range of user actions and their agency to "act there."

As the title of this book suggests, we put presence at the pinnacle of the XRXD pyramid. A discussion of presence necessarily refocuses the attention of XR system designers on the user. We highlight the importance of user agency as a critical factor in emergent behaviors within sociotechnical systems (Zakrzewski, 2022). Freedom enjoyed by the participants within the context of XR systems involves explicit design decisions regarding the participants' ability to choose their course of action based on a set of genuine alternatives. We equate the agency of human participants with deliberate conscious and informed choice while contrasting human actors with digital objects as nonhuman actants whose behavior is under the full control of the system designers.

Building on both actor–network theory (ANT) and the actantial model outlined by Julian Greimas, Peter proposed to divide the agents within sociotechnical systems into (1) human actors who possess intrinsic agency of their own and (2) nonhuman actants whose extrinsic agency is inscribed into them by the system designers (2022). The inclusion of user participation as a critical dimension

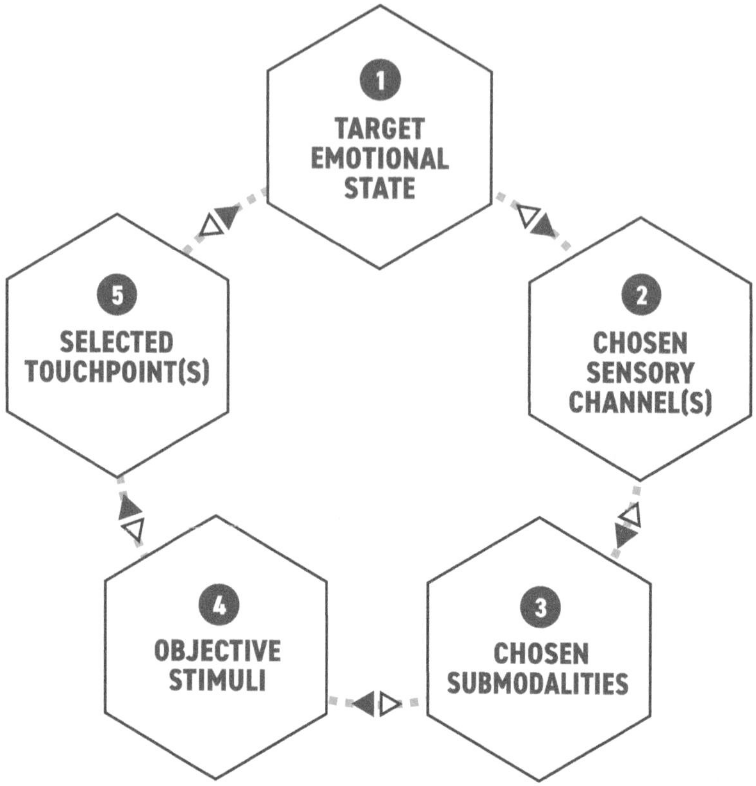

Figure II.5 The presence elicitation model

Source: Concept and illustration by © 2024 Peter (Zak) Zakrzewski.

complementing ubiquity and virtuality not only enriches the picture of the full potential of XR but also allows us to better balance the social (desirability) and the technical (feasibility) aspects of XR system design. The reference to systems thinking and the concept of emergence allows designers of sociotechnical systems to conceptualize purposeful XR environments as an outcome balanced between user participation and designer control (2022).

The Presence Elicitation model shown in Figure II.5 supports strategic design decision-making by challenging the team to start the experience design process by (1) identifying the audience target emotional states the XR application needs to support, followed by (2) choosing the appropriate sensory channels to address. (3) Within each channel, the designers need to identify all of the relevant submodalities. (4) Next, the team needs to create all of the objective stimuli that will define (5) the critical touchpoints the participants will encounter on their immersive journey.

We include a chart in Figure II.6 that can further align the target emotional states for each phase of the XR experience unfolding in time. The chart maps out the three stages and two transitions of immersive design across the eight sensory

	OBJECTIVE STIMULI								SUBJECTIVE STATE
	SEE	HEAR	TOUCH	SMELL	TASTE	VESTIBULAR ORIENTATION	PROPRIOCEPTIVE COORDINATION	INTEROCEPTIVE STATE	TARGET SUBJECTIVE STATE
ATTRACTION									
ENTRY									
IMMERSION									
TRANSFORMATION									
EXTENSION									

Figure II.6 Targeting subjective emotional states

Source: Concept and illustration by © 2024 Peter (Zak) Zakrzewski.

channels, all with the view toward eliciting and supporting strategically chosen target subjective emotional and cognitive states in the audience.

The XRXD Logic Model

We finally arrive at the last item we would like to cover before moving to the first step of designing your immersive media project. In this section, we outline the five phases and 12+1 steps of the logic model behind XR experience design first discussed in the introduction section. We refer to this framework as the XRXD Logic Model to highlight the difference between a fixed, linear, step-by-step, or paint-by-numbers process and flexible abstract problem-solving tools like logic models. The main difference is that logic models reward goal states and not specific actions. Steps within logic models can be approached in different order and revisited depending on the project type and goals. Knowledge within each step of the logic model is considered to be provisional, meaning that any assumptions must be tested and can be iterated depending on the outcome of the hypothesis test.

Logic models allow change makers a great deal of creative freedom within each step. They are frequently used by policy planners and evaluators in nongovernmental organizations and social enterprises to demonstrate how proposed programs might work to solve identified problems. The XRXD Logic Model is a powerful conceptual tool that can help immersive media makers and XR system designers describe, evaluate, and then achieve their stated vision or proposed change hypothesis.

Well organized creative problem-solving efforts begin with the team creating a representation of the problem at hand. Most complex problems such as the creation of immersive media systems cannot be solved linearly. They often require holistic analysis, hypothesis-testing cycles, and iteration.

The XRXD Logic Model will enable the design team to sketch out an abstract problem space within which they can search for an optimal solution. The resulting explicit statements underlying formal theoretical, informal, and design assumptions brought to the surface by the logic model will allow media designers to act in a strategic and purposeful manner throughout the project cycle. This level of abstraction can help map a path forward without needing detailed knowledge which might not be readily available at the start of the project. The framework will also offer guidance and facilitate communication among the design team, the client, the funders, the evaluators, and all major stakeholder groups involved in the project.

Conceptual models that visualize the complexity of technologically intensive systems can make systematic reviews more transparent and understandable to all stakeholders and decision-makers. Mapping media design initiatives through logic models can also help media makers identify and manage the human and financial resources needed for the execution of their projects.

Eagle-eyed readers will recognize that the XRXD Logic Model mirrors the chapter layout of the entire book. This is done deliberately to highlight the importance of applying a disciplined design process to XR media projects, which, by their sociotechnical nature, will only increase in complexity as their various underlying technologies continue to coalesce and mature. The logic model identifies the most salient and relevant steps and relationships between them within a typical XR experience-design problem space.

As mentioned before, because of the flexibility, all of the steps can be restated, reiterated, and even taken out of sequence. For instance, a team might want to test the general feasibility of the project by building a functional prototype before diving into outlining the vision for the project. That might actually make the project vision a lot more realistic and grounded in the reality of the available resources and competencies, allowing the team to return to the prototype-building phase with deeper clarity. In this way, the model promotes systems thinking that highlights essential relationships between different steps by illustrating the individual project parts' relationships to the whole. With that being said, let us look at the XRXD Logic Model in detail.

Phase I: Discovery and Pre-Suasion

Phase I: Discovery and Pre-Suasion covers the first three steps of immersive media creation.

1. The Developing Your Vision step encourages the design team to explore the strategy behind the persuasive media project. It helps the team to answer the following questions: What is our purpose? What is our vision for the outcomes of this project? How will the audience be transformed by this XR media experience?
2. The second step, Discover Your Audience, focuses the design team on discovering and defining the project audience and themes that are most likely to resonate with this core segment. It guides media makers through the process of discovering micro insights about audience interests, goals, needs, and motivations. The step explores the fundamental desirability (people and culture) building block of persuasive media design. This stage of the model is focused on answering the following questions: Who is the audience? What do we know about our audience and their media consumption needs? What themes might resonate with our audience?
3. The third step focuses on articulating the value hypothesis for the project. The value hypothesis is the most important of all the leap-of-faith assumptions a creative team makes about the project. In stating the value hypothesis, the design team hypothesizes about the value the media project will bring to the audience, what job it will do in their lives, how it will resonate, and how it will move and transform the audience. XR media applications will increasingly offer value within many categories. XR immersion can entertain, offer a service, help with work-related tasks, educate, and assist in research, shopping, healthy lifestyle, etc.

   An effective hypothesis has to be specific, measurable, and falsifiable. A well-stated hypothesis can be tested by building a prototype, conducting an experiment, and getting real-world feedback. A compelling value hypothesis is essential for any media project hoping to resonate deeply with the audience. A clearly stated value proposition can not only unite and guide the design team but also become the starting point in inviting the audience to participate in the media franchise.

Phase II: Ideation

Phase II: Ideation covers the next three steps in the XRXD building process.

4. Step 4 of the logic model is dedicated to constructing the context scenario. Scenarios are some of the least resource-intensive design tools that help media makers to conceptualize how the audience will experience the proposed XR media text. Scenarios use narrative structure to guide the team through the process of answering the question: How can the XR system best respond to the user to support their goal (effectiveness) efficiently while being engaging, error tolerant, and easy to learn? We cover the details of the context scenario design process in Chapter 4.
5. Step 5, which we refer to as the scenario breakdown, prompts the team to assess the technological feasibility aspect of the proposed media experience. The technological complexity underpinning immersive media experience makes the technology stack the foundation of XR projects. In this step, the design team combs through the scenario created in step 4 to identify the full package of technologies and digital assets needed to create the aesthetic and interactive dimension, which will define the desired user experience needed to support the desired immersive experience. We cover the entire scenario-breakdown process in Chapter 5.
6. Step 6 of the model focuses on building the proof of concept that can assess the initial assumptions about the desirability and value of the project stated in step 3 and the feasibility hypothesis stated in step 5. In this step, the media design team puts together the initial prototype(s) that will form the basis of hypothesis-testing experiments. We cover this step in detail in Chapter 6.

Phase III: Planning and Building

Phase III: Planning and Building covers the critical building-to-think stage of the XRXD process.

7. Step 7 prompts the team to address the immersive experience as an integrated system. As the creator of the *Sims* digital game series, Will Wright, suggests, robust game experiences immerse players by creating navigable systems of freedoms and limitations, challenges and rewards, which players can internalize in their minds. The world of immersion is psychologically a new phenomenon that needs a high level of internal consistency and understandability to support the user's journey. We cover XR system design in Chapter 7.
8. Step 8 of the model is all about building the functional system prototype. In this critical step, the rubber meets the road, so to speak. Executing the functional system prototype combines the planning done in the previous steps with the engineering creativity to get the team closer to being able to invite the audience to participate in the XR experience. We cover the entire process of building functional XR prototypes in Chapter 8.

9. Step 9 is all about playtesting the system prototype by putting the audience in the loop. In this step, the team will get a chance to test all of their assumptions regarding the desirability, feasibility, and usability of the proposed XR experience by having the ultimate judges of the immersive experience—the audience—provide the real-world feedback.

   The Playtesting the System step marks the end of the designer-inscribed phase of the project begun in step 1. During step 9, the team enters into the dynamic (and much less predictable) phase of user immersion and social interaction between the audience and the XR system. At this point in the process, the design team will be challenged by the audience to begin the process of calibration, assessment, and system iteration. We cover this important step in detail in Chapter 9.

Phase IV: Launching and Learning From Play

Phase IV: Launching and Learning from Play covers the last three steps in the XRXD building process.

10. Step 10, which we call Attracting the Audience, entails the critical work that needs to be done to guide the audience to the XR experience and help them enter the immersive world. Steps 10 through 12 are based on the diachronic model of experience design we present in Chapter 4. In this case, the designed experience forms a path from attraction to immersion and, eventually, to extension of the experience beyond the initial moment of play. We cover the details of the work to be done in step 10 in Chapter 10.
11. Step 11, Immersing the Audience, covers the essence of what XRDX is all about: the immersive experience aimed at creating the right conditions for the emergence of XR presence. This step is a culmination of the preparatory work done before. The groundwork laid in steps 1 to 10 finally bears fruit with the emergence of play, presence, and flow. We cover step 11 in detail in Chapter 11.
12. All good things in life must come to an end. Step 12 is all about extending play. Like steps 10 and 11, 12 follows the diachronic model of experience design. Its goal is to invite the audience to become members in the franchise community and to open it to an even wider audience by word of mouth.

    The focus of the work needed to be done in step 12 is based on the research of media scholar Henry Jenkins and his industry collaborators, Joshua Green, and Sam Ford who coined the term "spreadability." Extending play involves a process of successfully blending the key attributes of the XR media experience to appeal to motivations of particular user communities in a way that inspires them to embrace the XR media franchise and share content with others in their network.

    XR system design work does not end with scaling the media franchise. XR media designers need to continue maintaining and improving the system.

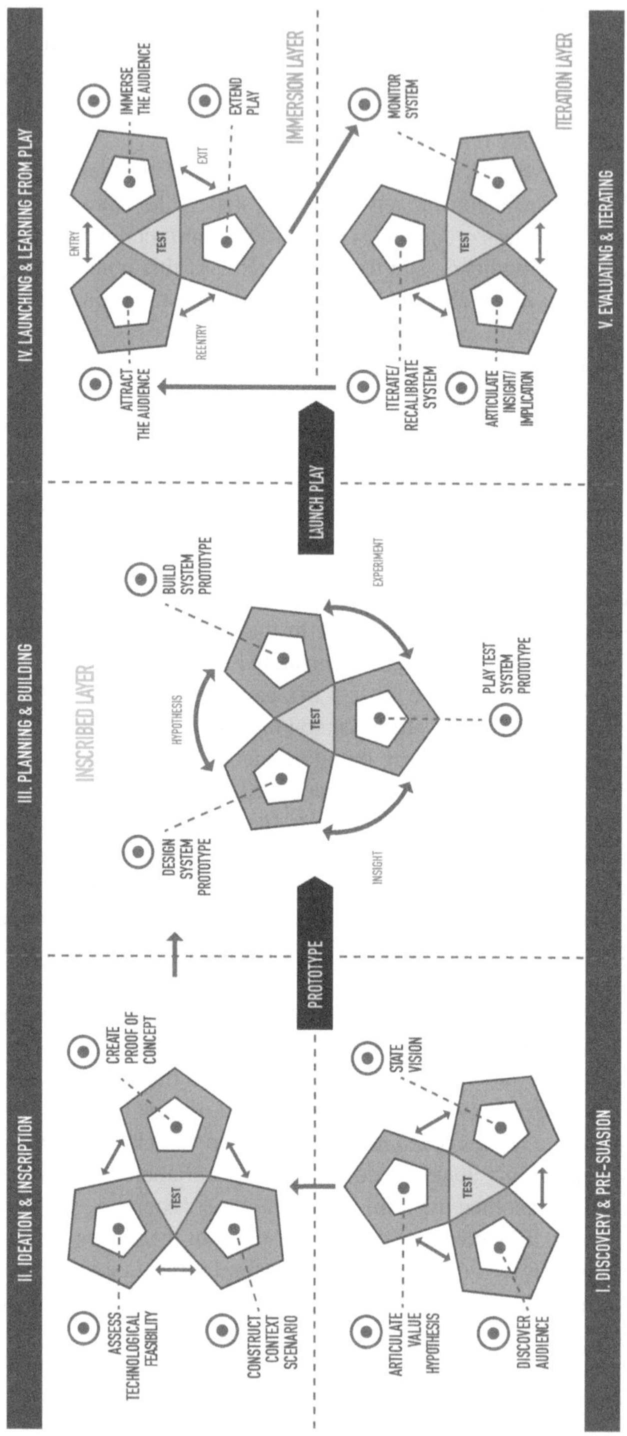

Figure II.7 XRXD (XR experience design) logic model

Source: Concept and illustration by © 2024 Peter (Zak) Zakrzewski.

The last three steps of the model we call Evaluating and Iterating the System, focus on the critical design task of monitoring, evaluating, and iterating the XR system at play to nudge the emergent human+computer networks toward their original function.

13. Step 13 involves observing and monitoring the system at play.
14. Step 14 is about articulating the insights from the observations and about proposing a theory of transformation.
15. Step 15 involves the work of recalibrating and iterating the system to keep it on course toward its intended function. We will explain all of these steps in detail as we move together through the book.

Summary

This section describes extended reality experience design (XRXD) as a field that transcends traditional screen-based user experience by engaging a participants with a wide array of sensory and cognitive modalities. We examine how immersive media offers the potential to co-create a sense of presence, making mediated experiences feel as if they are real on an emotional and physical level. We situate this immersive potential in a historical context, tracing its evolution from early prototypes by pioneers like Morton Heilig's Sensorama to modern VR, AR, and MR technologies capable of simulating nearly all sensory modalities.

We also address the neuroscientific underpinnings of XR, explaining how experiences in virtual environments engage multiple brain regions similarly to real-world experiences. We discuss embodied cognition, where language, emotions, and sensations are processed dynamically, influencing the participant's engagement and perception within XR spaces. We introduce the three-layered model of consciousness proposed by neuroscientist Antonio Damasio, which outlines how basic emotions progress into conscious feelings, creating a foundation for designing XR experiences that impact self-perception and emotional states.

We present XR experience design (XRXD) as a paradigm shift, redefining design strategies through a neuroscience-informed, multisensory approach that shapes immersive media's future. Finally, we introduce the XRXD Logic Model, which guides you through the design, production, and deployment of immersive media experiences.

Notes

1 The initial light energy is absorbed by three pigments in the cone cells, which we can label blue, green, and red. The initial reaction to light is transduced into electrical signals, which are relayed to the optic nerve. The resulting information is recombined and distributed along two axes, which the brain interprets as continua of green to red or blue to yellow. The information generated from the vast numbers of cones and mediating ganglion cells is passed first into the thalamus and then into the dedicated arrays of cells in the primary visual cortex, located at the brain's rear. It takes only milliseconds for the color-coded visual information to spread throughout the brain.

2 We will cover these levels in greater detail later in the book. For more information, refer to Norman Nielsen's Group website at www.nngroup.com/.

References

Aukstakalnis, S. (2017). *Practical augmented reality: A guide to the technologies, applications, and human factors for AR and VR*. Addison-Wesley.

Barrett, L. F. (2017). *How emotions are made: The secret life of the brain*. Houghton Mifflin Harcourt.

Barrett, L. F. (2020). *Seven and a half lessons about the brain*. Houghton Mifflin Harcourt.

Bye, K. (2017, July 14). Mel slater's theory of VR presence vs an elemental theory of presence [Podcast]. https://voicesofvr.com/556-mel-slaters-theory-of-vr-presence-vs-an-elemental-theory-of-presence/

Craig, A. D. (Bud). (2009). How do you feel—now? The anterior insula and human awareness. *Nature Reviews Neuroscience*, 10(1), 59–70. https://doi.org/10.1038/nrn2555

Damasio, A. R. (1999). *The feeling of what happens: Body and emotion in the making of consciousness* (1st ed.). Harcourt Brace.

Damasio, A. R. (2001). Emotion and the human brain. *Annals of the New York Academy of Sciences*, 935(1), 101–106. https://doi.org/10.1111/j.1749-6632.2001.tb03475.x

Delle Fave, A., Massimini, F., & Bassi, M. (2011). *Psychological selection and optimal experience across cultures: Social empowerment through personal growth* (Vol. 2). Springer Netherlands. https://doi.org/10.1007/978-90-481-9876-4

Erickson, M. H. (1980). *The collected papers of Milton H. Erickson on hypnosis, vol. 1: Nature of hypnosis and suggestion* (E. L. Rossi, Ed.). Irvington Publishers: Distributed by Halsted Press.

Fisher, M. (2022). *The chaos machine: The inside story of how social media rewired our minds and our world* (1st ed.). Little, Brown and Company.

Iwata, H., Yano, H., Uemura, T., & Moriya, T. (2003). *Food simulator*. ICAT.

Jacobson, R. D. (2009). Haptic or touch-based knowledge. In R. Kitchin & N. Thrift (Eds.), *International encyclopedia of human geography* (pp. 13–18). Elsevier. https://doi.org/10.1016/B978-008044910-4.00451-X

Jerald, J. (2016). *The VR book: Human-centered design for virtual reality*. Association for Computing Machinery; Morgan & Claypool.

Kelly, K. (2016). *The untold story of Magic Leap, the world's most secretive startup*. Wired. www.wired.com/2016/04/magic-leap-vr/

Kriegel, U. (Ed.). (2020). *The Oxford handbook of the philosophy of consciousness* (1st ed.). Oxford University Press.

Levitin, D. J. (2006). *This is your brain on music: The science of a human obsession*. Dutton.

Lynott, P. (1980). *King's call. On Solo in Soho*. Phonogram Ltd.

McCracken, G. D. (2008). *Transformations: Identity construction in contemporary culture*. Indiana University Press.

McGilchrist, I. (2019). *The master and his emissary: The divided brain and the making of the Western world* (New expanded ed.). Yale University Press.

McGilchrist, I. (2021). *The matter with things: Our brains, our delusions, and the unmaking of the world*. Perspectiva Press.

Metzinger, T. (2009). *The ego tunnel: The science of the mind and the myth of the self*. Basic Books.

Narumi, T., Nishizaka, S., Kajinami, T., Tanikawa, T., & Hirose, M. (2011). Meta cookie+: An illusion-based gustatory display. *Virtual and mixed reality-new trends: International conference, virtual and mixed reality 2011, held as part of HCI international 2011*, Orlando, FL, USA, July 9–14, 2011, Proceedings, Part I 4, 260–269.

Nielsen, J., & Norman, D. A. (2015). *The definition of user experience (UX)*. Nielsen Norman Group. www.nngroup.com/articles/definition-user-experience/

Ojeda, A., Bigdely-Shamlo, N., & Makeig, S. (2014). MoBILAB: An open source toolbox for analysis and visualization of mobile brain/body imaging data. *Frontiers in Human Neuroscience*, 8. https://doi.org/10.3389/fnhum.2014.00121

OpenAI. (2024). *GPT-4o system card*. https://cdn.openai.com/gpt-4o-system-card.pdf

Pentland, A. (2015). *Social physics: How social networks can make us smarter*. Penguin Press.

Rogers, L. J. (2021). Brain lateralization and cognitive capacity. *Animals*, 11(7), 1996. https://doi.org/10.3390/ani11071996

Sanchez-Vives, M. V., & Slater, M. (2005). From presence to consciousness through virtual reality. *Nature Reviews Neuroscience*, 6(4), 332–339. https://doi.org/10.1038/nrn1651

Schaaf, R. C., & Miller, L. J. (2005). Occupational therapy using a sensory integrative approach for children with developmental disabilities. *Mental Retardation and Developmental Disabilities Research Reviews*, 11(2), 143–148. https://doi.org/10.1002/mrdd.20067

Schmalstieg, D., & Höllerer, T. (2016). *Augmented reality: Principles and practice*. Addison-Wesley.

Seth, A. (2021). *Being you: A new science of consciousness*. Dutton.

Smith, S., & Wheeler, J. (2002). *Managing the customer experience: Turning customers into advocates*. Financial Times Prentice Hall.

Speer, N. K., Reynolds, J. R., Swallow, K. M., & Zacks, J. M. (2009). Reading stories activates neural representations of visual and motor experiences. *Psychological Science*, 20(8), 989–999. https://doi.org/10.1111/j.1467-9280.2009.02397.x

Sutherland, I. E. (1965). The ultimate display. *Proceedings of the International Federation of Information Processing Congress*, 2, 506–508. www.informatik.umu.se/~jwworth/TheUltimateDisplay.pdf

Tyrka, A. R., Ridout, K. K., & Parade, S. H. (2016). Childhood adversity and epigenetic regulation of glucocorticoid signaling genes: Associations in children and adults. *Development and Psychopathology*, 28(4pt2), 1319–1331. https://doi.org/10.1017/S0954579416000870

van der Kolk, B. A. (2014). *The body keeps the score: Brain, mind and body in the healing of trauma*. Viking.

Watzlawick, P., Bavelas, J. B., & Jackson, D. D. (1967). *Pragmatics of human communication: A study of interactional patterns, pathologies, and paradoxes*. Norton.

Wilson, E. O. (1999). *Consilience: The unity of knowledge*. Vintage Books.

Zakrzewski, P. (2022). *Designing XR: A rhetorical design perspective for the ecology of human+computer systems*. Emerald Publishing Limited. http://doi.org/10.1108/9781802623659

Section III

Discovery and Pre-Suasion

1 Your Purpose

Developing Your Project Vision

Vision: One Question That Answers a Thousand

The author of *Essentialism*, Greg McKeown, quotes the legendary ballet dancer Anna Pavlova to argue that "the secret to success" lies in following "one aim without halt" (McKeown, 2014, p. 119). How do we identify this one goal to follow without halt or hesitation? When we lack clarity, we waste time and energy on the trivial many instead of the well-chosen few. Discovering our essential intent boils down to being able to answer the most critical question of all—why? Stating why you do what you do will force you to focus on what really matters. Media projects can range in their aims from goals such as entertaining an audience to the mission of healing trauma, helping people learn about important issues, or simply the desire for self-expression. Regardless of the goal, a project must start with a solid answer to the question of purpose to motivate the team, direct the resources, and ultimately to deliver the intended value to the core audience. Because of the technical resources, expertise, and logistics required, being clear on one's purpose is of paramount importance to immersive media projects.

Finding your reason to do something clarifies whether you can sustain the commitment necessary to complete long-term projects. Answering the question of why leads us to the understanding of our essential intent, which will be both inspirational and concrete and, down the road, both meaningful and measurable. Answering the why question will settle thousands of more minor ones that pop up along the way. The one carefully chosen strategic choice will eliminate many options and tangential courses of action for the project's duration. Once you achieve clarity regarding your intention, all answers to the subsequent questions will come into better focus (McKeown, 2014). Our intent shapes our vision in both senses of the term. The vision shapes the project's mission, the details of what, who, when, and how. And the mission helps to clarify the project milestones.

The Audience-Product Alignment: Three Types of XR Project Visions

Finding the right audience–product fit is one of the most critical issues concerning any form of media making. Finding this fit is not a simple matter of marketing the XR experience once the production is all but done. The media-making process

DOI: 10.4324/9781003442462-5

happens between two well defined milestones. It begins with a clear purpose for the project. It continues with an exploration of the world in which the audience lives, followed by ideation, planning, and building the experience platform, inviting the audience, and finally, evaluating the performance of the system in play. Clear vision marks the destination. Understanding the often unarticulated needs of the audience marks the starting point of the journey.

Venture capital firm Sequoia, which helped launch such media giants as Atari, Apple, Nvidia, and YouTube, has identified three categories of problems defining the challenges and opportunities for teams working on digital products. The three categories define three very different types of potential project visions. In the case of a hair-on-fire problem, the team addresses the audience's clear and urgent need. The demand for such an experience is apparent. The audience members are aware of their need. They are actively looking for solutions while comparing existing ones. For instance, hardcore gamers are constantly on the lookout for new gaming experiences. As a consequence, the category for products such as VR games is crowded with competitors. A team building a new immersive game must rise above the noise. In the case of well-understood user needs, the vision for the desired experience has to call for the best-in-class solution in order to be noticed. The application and its experience cannot just be faster or cheaper; it must be truly differentiated (Team Sequoia, 2023).

In the case of a hard-fact problem, the team faces a user need or a pain point that is universally accepted as an unsolvable hard fact of life. Your team, and potentially your team alone, can see that while the problem might pose a hard challenge, it is solvable. As a consequence, your solution disrupts the status quo with an unexpected approach. Besides the obvious issue of creating a new-to-the-world product, the team's challenge here lies in the fact that the potential audience is resigned to living with the problem or without their need being fully addressed. They are not urgently trying to solve the problem or seek products and experiences that address the need. In order to have your new solution adopted, your team will have to convince your users to change their current behaviors. The solution you create for the audience must overcome their force of habit. This means that your solution must be compelling enough to be worth making a change. This presents a unique opportunity for the team to create a truly disruptive immersive product.

In the third and perhaps the toughest case, your team will offer not just a product or an experience but a vision of the future. What you bring to your audience will enable a new reality—the kind of disruptive yet adoptable product like an iPhone. The challenge facing a team working on the vision-of-the-future problem is that the prospective users are not aware of the need or the problem and the opportunity to accomplish something entirely groundbreaking. Your main obstacle, in this case, is disbelief.

We might argue that many aspects of immersive media fall under this category. Media users are not entirely convinced that they need AR glasses to shop or travel. Movie audiences are not quite rushing to buy MR goggles to watch their favorite shows on HUD devices. Part of the challenge of offering vision-of-the-future products is that the solution must come with its own ecosystem. As a disruptor, Apple's iPhone only came alive with a multitude of apps available on the App Store. Meta's

Quest and Apple's Vision Pro are not just headsets; they are a new way to access media content. Vision-of-the-future products succeed only when they make the new paradigm irresistible. This long-term proposition presents a grand challenge but also a massive opportunity to shape the future and reshape the world.

Developing Your Vision

Marketing strategist Kevin J. Clancy describes true vision as the difference between creating an organization aimed at designing computers that are 15% cheaper than the competition and a passion-driven company working on creating personal computers to empower every human being on the planet.

There might be no better example of a team ceaselessly pursuing an ambitious vision than the company started by Steve Wozniak and Steve Jobs. Great visions have the power to transform the status quos that exists today to the new, more desirable future. Powerful visions draw a picture of what the world will look like once your desirable state is reached. Visions can be sweeping, like the one John F. Kennedy presented while announcing the moon landing mission during his address before a Joint Session of Congress on May 25, 1961, or they can be more humble versions guiding a development of an AR application.

A well-stated vision will serve as an inspiration and a destination. It will give a clear direction and a strong sense of purpose to the entire design team. Well-crafted vision statements are unique to you and your project. They are inspirational and motivating, bold and confident, exciting and barely attainable, but at the same time specific enough to translate into a well-understood mission and plan of action. One of the most famous examples of a powerful vision is the inspirational direction the founders of Apple set for their new startup by aspiring to put a personal computer in the hands of everyday people.

The founders of Apple Computer were shaped by the 1960s counterculture. They grew up in the shadow of pioneering computer hackers who sought to fulfill the long-standing dream of bringing together art and science through then-emerging computer technology. The West Coast hacker community included such early HCI visionaries as Stewart Brand and Gordon Ashby, who celebrated the concept of labor as a ludic project. Work was meant to provide its creators with fun and meaning. Live/work cooperatives, such as Ant Farm, Berkeley Tribe, and Pacific High School, espoused the "radical connectivity" approach years before the introduction of the innovation cluster model that later defined Silicon Valley. Brand, the founder of the *Whole Earth Catalog*, made a pronouncement that deeply resonated with young Jobs and Wozniak—"Ready or not, computers are coming to the people" (Scott, 2015, p. 103).

Based on the hippy hacker philosophy, Apple's founders envisioned putting computers in the hands of everyday people, from students and teachers to artists and people preparing their own taxes or simply sharing great recipes for a home-made lunch. It is easy to forget how bold this vision of the barely 20-year-olds was in the mid-1970s, when the computer industry was ruled by corporate giants such as IBM and the men in lab coats operating room size computers.

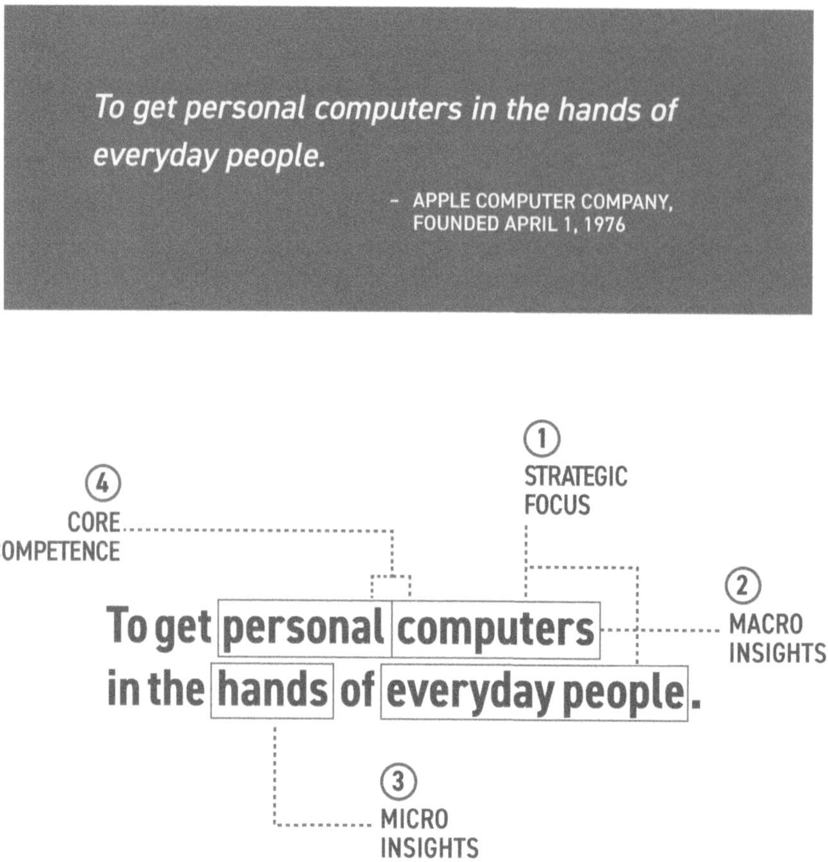

Figure 1.1 An anatomy of a vision statement

Source: Concept and illustration by © 2024 Peter (Zak) Zakrzewski, based on Apple Computers, 1976.

In Figure 1.1, we offer an analysis of Apple's early vision. The statement has four vital components: (1) the strategic focus (working within the emerging computing field), (2) the macro insight (the broader trend of computers entering all aspects of life and work), (3) the microinsight about the audience (the embodied nature of human usability needs), and (4) the team's core competence and passion (building personal computers for themselves and other like-minded people).

In the case of XR media projects and apps, the strategic focus can include creating content for the emerging XR enterprise or consumer markets. The macro insights can come from observing the emerging XR technology field, which continues to offer new opportunities to solve old problems with new forms of immersive access to advanced computation. The microinsights are based on an empathetic understanding of the prospective audience's unmet needs. These insights can come from a variety of sources, including the makers' own experience, pain points such as frustration with big-box-store shopping, or a need for a more immersive form of documentary media. They can come from interviews with prospective audience members. They can come from listening to immersive media community discussions or from observing the efforts of other immersive media pioneers.

We devote the next chapter to the vital exploration of the audience's life context. The team needs to also consider their own strengths and weaknesses to make sure that the vision is attainable.

When preparing your vision, gather your founding team and brainstorm the answers to the following questions:

1. What is our strategic focus? Where will our project fit in terms of the media market (gaming, film, social media, etc.) and geographic market scope (local market, global audience)? This answer will suggest any potential partners needed for your project.
2. Who is our primary audience? What job do we want to do for this audience? Do we aim to entertain them, inform them, help them learn something, assist them in performing a particular task like shopping, or enable them to live a healthier life? These value hypotheses might be later confirmed by more microinsights you unearth from your audience research.
3. What larger cultural, economic, and technological trends make our project doable? Why is the time now to bring this project to life? The answers will help you sort out the availability of resources needed for your project.
4. How will we build and deliver this valuable experience to our audience? This answer will address the core competence your team will need to possess in order to do the job.

(Zakrzewski, Empathic Disruption Course Materials, Schulich School of Business, 2023)

Project vision that addresses the big why of pursuing XR media projects is closely linked with the project mission, which outlines the path to the vision state by answering the how, when, where, and how much questions that can be turned into a project management plan that allocates time, resources, and money. In going through the steps toward accomplishing a vision, the team will alternate between the specific and the general aspects of the project, each giving more clarity to the process.

Overcoming Uncertainty: A Transformative Way to Solving Problems

When Miles Davis, Julian "Cannonball" Adderley, John Coltrane, Bill Evans, Wynton Kelly, Paul Chambers, and Jimmy Cobb walked into Columbia's 30th Street recording studio in NYC on March 2, 1959, they came with no more than

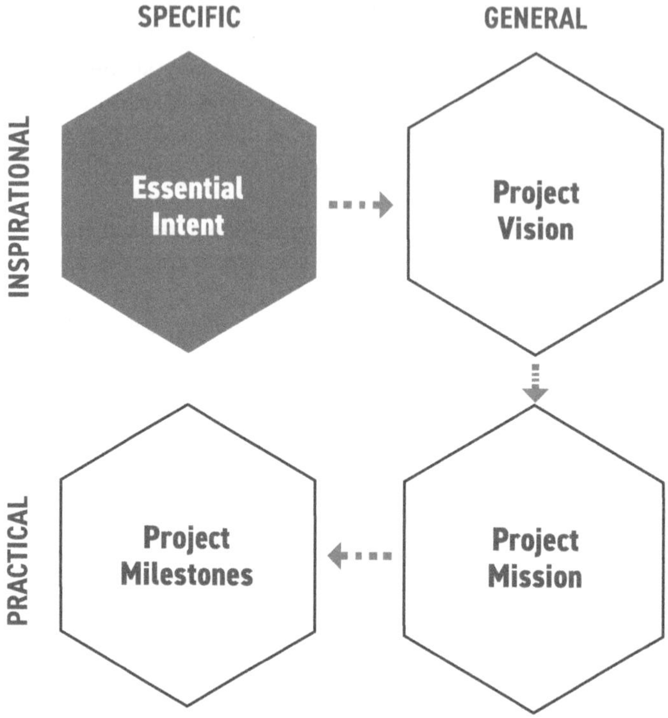

Figure 1.2 Four steps toward accomplishing a vision

Source: Illustration by © 2024 Peter (Zak) Zakrzewski, based on Greg Mckeown's, Essential Intent Matrix.

their instruments, their well-honed musical intuition, and a few scraps of notes from their band leader. They had no idea that what would emerge from that improvised session would deeply touch music fans for decades to come and become the bestselling jazz album of all time. Rather than being overwhelmed by the challenge of introducing the world to modal jazz, they simply immersed themselves in the act of creation and played their hearts out.

No one can ever guarantee an outcome of a truly creative act. It must be explicitly acknowledged that creative media makers live in the world of uncertainty. People who seek the comfort of predictability do not become artists or innovators. Successful creatives, instead of being overwhelmed or daunted by the challenge of facing uncertainty, emerge from its fog with something new and exciting—a gift to the world. They embrace uncertainty as the defining challenge of going from zero to one, of creating something that has never been done before.

Otto Scharmer, who studies strategies of creative problem-solvers, shares an uncertainty-defying insight—to overcome uncertainty one only needs to know two things: the direction one wants to go and the very next step. In the opening of this chapter, we alluded to a concept of transformative problem-solving by suggesting that immersive media projects are a journey between the unfulfilled audience need state of the world and the desirable new state defined by the team's vision.

Cognitive scientist James Greeno—who helped to develop the theory of situated learning, which emphasized the influence of social interactions and environment on learning—conceptualized the existence of three categories of problems. The arrangement problem requires problem solvers to rearrange parts in order to satisfy some imposed prearranged criteria. In the inducing-structure problem category, the problem solvers are faced with several instances of a given problem and must discover the general rule or pattern involved in addressing them all.

Finally, transformation problems consist of an initial state and a goal state. The problem solvers' challenge in solving problems within the transformation category lies in finding a sequence of operations leading from the initial state into the goal state. Successful solutions to transformation problems require an ability to identify the differences between the current and the goal states throughout the entire process. Once such differences are identified, problem solvers need to take appropriate actions to reduce the differences (Riley & Greeno, 1988; Greeno, 1989).

In the course of teaching disruptive innovation in an MBA program, Peter has combined Greeno's transformation problem framework with Allen Newell and Herbert Simon's means–ends analysis (MEA) cognitive problem-solving model to turn them into a conceptual tool for transformative change making (Figure 1.3).

The particular usefulness of applying transformative problem-solving to immersive media design lies in its ability to factor uncertainty into the equation while letting the design teams focus on creating new-to-the-world immersive experiences. We would argue that there are actually three things media makers need to know when embarking on a new project: the current state (audience need), the goal state (team's vision), and the very next step that gets the team closer to the goal of achieving transformative presence.

The freedom granted by transformative problem-solving lies in the simplicity of the strategy that comprises the clear and explicit steps of discovering the current undesirable state, working out a purpose—a vision, or the desired state—and only having to worry about the intermediate steps one at a time. The intermediate steps become

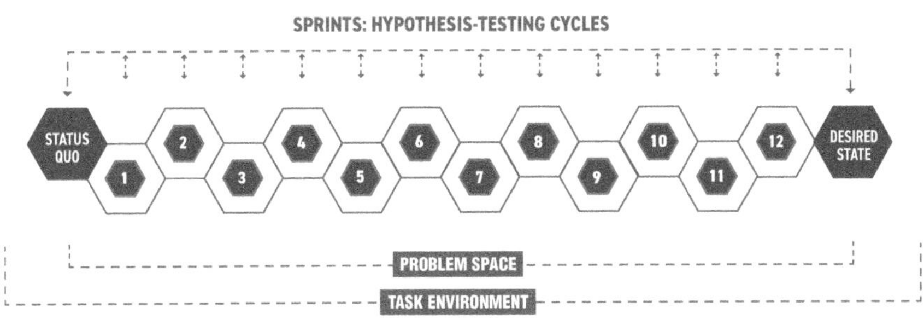

Figure 1.3 Peter's transformative problem-solving model

Source: Concept and illustration by © 2022 Peter (Zak) Zakrzewski, based on Greeno's Transformative Problem Solving and Newell & Simon's Means-Ends Analysis.

hypothesis-testing cycles, sometimes referred to in digital product development as sprints. Choosing the starting point and the end point narrows down the "problem space" to a manageable size for which resources can be planned and allocated.

The 12 steps shown in Figure 1.3 follow the XRXD logic model introduced in Section II. Once the desired state is expressed as a project vision and the status quo is outlined as an unmet audience need, immersive media makers can focus on navigating the individual steps from discovering the audience to articulating the value proposition, constructing the context scenario, assessing the technological feasibility, creating the proof of concept, designing, building, and playtesting the system prototype, attracting and immersing the audience, extending play, and finally iterating the system to keep it on course toward fulfilling its intended purpose.

Action and Understanding: The Provisional Nature of Knowledge

Legendary reporter Carl Bernstein described the methodology behind his and his partner, Bob Woodward's, reporting on Watergate as "the best obtainable version of the truth" (Bernstein & Woodward, 2022, p. 1). In a pursuit of well-guarded secrets, understanding does not rest on a rock-solid foundation of unassailable truths but on hypotheses created by assembling evidence that requires cognitive flexibility, effort, and patience.

The best obtainable version of the truth is quite difficult to get right, because as Bernstein suggested, it requires a lot of hard work, critical thinking, persistence, and a great deal of humility. More than anything else, it requires comfort with what Pierre Bourdieu described as a state of aporia, a philosophical term indicating a seemingly irresoluble impasse in an inquiry. As pragmatic change makers, designers understand the usefulness of provisional understanding that dissolves aporia slowly by moving forward one step at a time.

Since dealing with the world's complexity vastly overpowers the human cognitive capacity, creative problem-solving requires flexible, agile mental modeling that adapts to the changing circumstances and iterates responses based on real-world feedback (Zakrzewski, 2022). These characteristics can be found in the doctrine of pragmatism, which lays the foundation for a self-correcting process of inquiry guided by three important principles. (1) The first of these is a commitment to identifying and clearly defining the key premises of the problematic situation that needs to be addressed. (2) The second involves a commitment to a scientific attitude and process. (3) The third principle addresses leveraging the power of participatory culture to enable deliberation and a free flow of ideas among the community of practice.

The founders of pragmatism, John Dewey and Charles Peirce, considered theory to be a tool that mediated the advancement of understanding rather than being a foundation of knowledge. According to pragmatic doctrine, all of our understanding is provisional. It is falsifiable and always needs testing. A judgment we make about the success or failure of our ideas depends on how closely they bring us to the attainment of our stated goal (2022).

Pragmatic practitioners test their hypothesis by linking experience to data. The success of a creative proposal boils down to the differential between the stated purpose and the measured outcomes. During the course of a creative project, immersive media makers make a series of leap-of-faith assumptions from the initial vision to the value hypothesis of what the project will do for the intended audience to the feasibility and usability hypotheses of the experience being effective, efficient, and engaging.

All of these assumptions are provisional statements to be tested for their connection to the reality of the audience's response. Each step in the media-design process brings the creative team closer to their stated goal by seeking useful levels of clarity rather than absolute certainty. In testing our assumptions, (1) we start with observation and evidence. (2) From data, we move to the formulation of abstracted insights that give us a better understanding of the problem. (3) At the third level of clarity, we derive a list of predictions about useful effects we might expect from our insights. (4) The resulting implications stemming from our original observation allow us to apply our knowledge to the creative project and to test our hypotheses in order to iterate our work.

Connecting Back to the Territory

If creative imagination can provide a map to a more enchanting world, ideation efforts need to be periodically matched by testing the doability of ideas within the actual territory of the audiences' life context. Applied design thinking and doing mixes two complementary approaches to ensuring design teams do not get lost by creating inaccurate maps of users' reality or the reality of what is actually technologically feasible. The two approaches are known as thinking to build and building to think. The first one refers to the essential process of ideating concepts before building. This in and of itself could serve as most people's definition of design. However, practicing designers are also concerned with experiential learning that comes from building things.

This is where the second part comes in. Building to think acknowledges the importance of empirical testing of abstract ideas. Mixing the two approaches forms a virtuous circle between creative imagination that allows us to visualize possibilities and building prototypes that make our conceptions tangible and testable. While ideation enables us to imagine new-to-world ideas and to speculate on the consequences of our actions, prototyping gets our ideas out of our heads and into the hands of the users to get fast and inexpensive feedback that spurs the evolution and iteration of digital products.

Each major step between the project launch and the desired outcome can be considered a single sprint aimed at getting the team closer to the stated goal. To confirm that each sprint does indeed bring the team closer to the realization of the vision, we need to build hypothesis testing cycles into the process. The build-measure-learn feedback loop is a core component of the lean start-up methodology championed by Eric Ries. The approach is well suited to the design and deployment of media experiences and emphasizes the importance of iteration, speed, and efficiency in developing a project to meet audience needs and achieve sustainability.

In the hypothesis-testing loop, the team starts with a hypothesis. For instance, a value hypothesis might assume that people who like to renovate their own homes would welcome an AR app that would guide them throughout an entire plumbing or backsplash tiling project, from learning how to do the work to shopping at a big-box store by guiding them to the right location within the store and the right product among the plethora of different options.

To test such a hypothesis, the team needs to devise and conduct an experiment that can prove or falsify this leap-of-faith assumption. Ries referred to all major do or die assumptions as leap of faith assumptions because of their long-term consequences (Ries, 2014). In research terminology, "validity" refers to making sure that the right thing is being tested. An experiment might isolate the big-box shopping component of the app by asking a group of willing test subjects to watch a prepared instruction video superimposed over the actual project site within the home and then follow specific instructions to shop at the store. The prototype itself might take many, forms from instructions fed to the subject over Bluetooth headset by a live agent mimicking the app interaction to a tablet-based AR prototype to a headset-based one. The point is to find out if DIY folks would find value in using the AR app for big-box shopping.

After the data from the experiment is gathered for analysis and inference, the team can work on (4) identifying key insights. These might include such findings

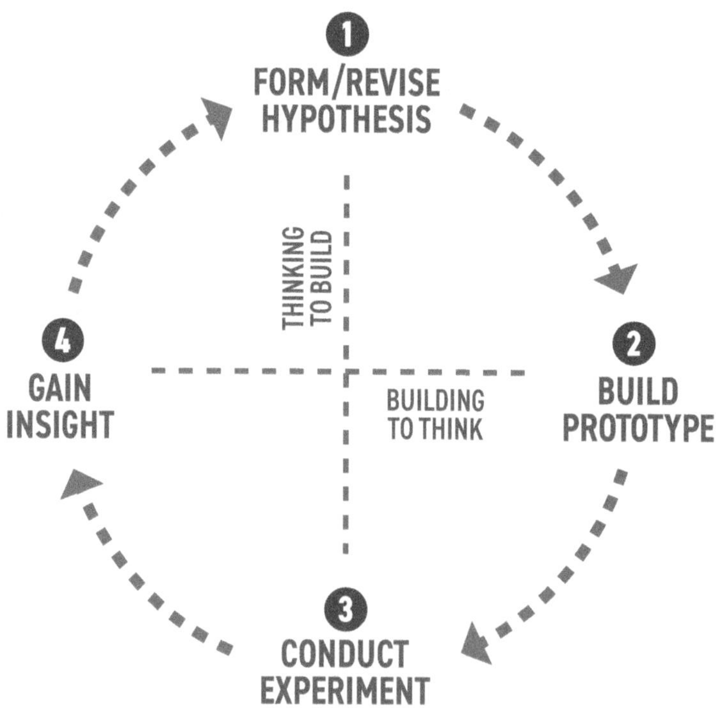

Figure 1.4 Hypothesis-testing feedback loop

Source: Concept and illustration by © 2024 Peter (Zak) Zakrzewski, based on Eric Ries', Build-Measure-Learn Feedback Loop.

as the users' reporting satisfaction with the consistency of flow of directions throughout the process but not enjoying the bulkiness of the MR headsets used, etc. The important point in assessing the success or failure of the experiment is to stay focused on the dependent variable or the hypothesis being tested, which, in this case, would be the desirability of the value being evaluated. Once the insights are articulated, the team needs to decide whether to stay with the hypothesis or to restate it based on the insights discovered in the test.

Four Types of Immersive Experiences

Another framework we find useful to consider at the early stages of the XR media-making process is Nicole Lazzaro's concept of four types of fun encountered in digital games. It provides a useful starting point for deciding on what type of immersive experience your team wants to provide for the audience. Lazzaro identified these as hard fun, easy fun, serious fun, and people fun (Lazzaro, 2021).

Translated to the immersive experience design process, these four categories can be represented as (1) challenge-based immersive experiences such as seeking, achievement, mastery, self-actualization, or simple task completion. (2) Easy fun

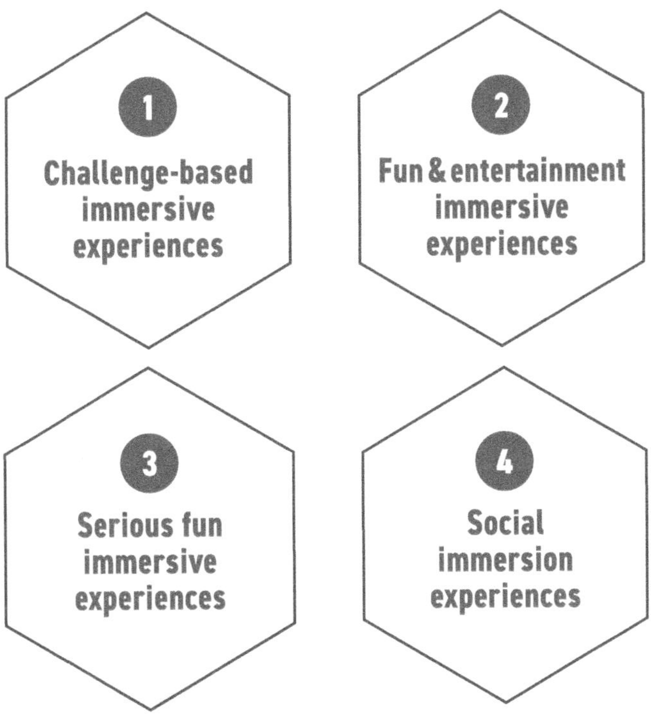

Figure 1.5 Four types of immersive experiences

Source: Illustration by © 2024 Peter (Zak) Zakrzewski, based on Nicole Lazzaro's Four Types of Fun.

experiences provide audiences with entertainment and momentary diversion from ordinary life. Easy fun can be offered in a lean-back mode, where the audience can sit back and enjoy the ride or virtually walk through a fun adventure. (3) Serious fun experiences allow participants to learn something new, meaningful, and exciting. Serious fun requires the audience to lean into the experience and occasionally take over the wheel. (4) Social immersion provides participants with all of the drama and excitement that comes from dealing with other human beings. Social spaces are some of the most popular forms of immersive experiences, where audiences seek connection, friendship, competition, and even love.

These four categories can further clarify the vision for your project and help identify the team's real passion and core competence that will allow you to start focusing on the right audience. The four categories will attract different audiences with different media consumption needs. They will require different ideation and different skills and resources to complete. Knowing the type of experience your team aims to create from the start can save you a lot of time and effort potentially wasted on chasing wild geese that will not lay any golden eggs.

Persuading With Media

As immersive media makers, we need to explicitly acknowledge the magnitude of the rhetorical power of XR experiences. This acknowledgment will allow us to realize the full potential of this new medium by openly discussing the ethical responsibility media designers and makers have toward their audiences.

Since its origins in ancient Greece, rhetoric has been vilified as a practice based on duplicity and deception. Postmodern scholars implicated rhetoric in the politics of asymmetrical power balance in society.

Critics of rhetoric accuse it to be a unilateral process, in which a powerful party with vested interests engages in a one-sided communication expressly aimed at shutting down a debate and bringing others to consent. English scholar James Crosswhite disagrees with these assessments to argue that there is a fundamental difference between oppression, physical violence, and a rhetorical contest of ideas (Crosswhite, 2013). Our definition of rhetorical media design is quite different from technological paternalism. The difference between the two boils down to the difference between rhetoric and coercion.

Engaging audiences in co-creation of presence is part and parcel of participatory culture and of an iterative media design process. Rhetorically minded media designers make creative proposals to which users respond in implicit and explicit ways. Crosswhite points out that despite any suspicions we might have regarding rhetoric's potential for manipulation, persuasion as an expression of reason has a unique place in addressing the fundamental communicative aspect of human coexistence (Crosswhite, 2013).

Multimodal design is not the only persuasive tool at the hands of immersive media makers. Storytelling is one of the oldest forms of persuasive meaning-making. American psychologist Jerome Bruner argued that those looking to persuade others can tap two different modes of cognitive functioning that provide uniquely distinctive ways of structuring reality and ordering the inner mapping of the world. One such approach takes the form of a well-formed argument, and the other is a good story. Both can be used as means for convincing others, but their ways of convincing are fundamentally different.

When we present arguments, we aim to convince our audience of the argument's truth and accuracy. When we tell stories, our job is to convince the audience of the story's lifelikeness. Consequently, the two differ radically in their aims and verification procedures. Reasoned arguments, such as those used in the scientific method, are verified by eventual appeal to procedures for establishing formal and empirical truth. Stories, on the other hand, do not need to establish absolute truth but verisimilitude (Bruner, 1986, p. 11). We will return to the unique power of narrative transportation to create entire virtual universes throughout this book.

As XR media makers, we persuade our audiences by leveraging the power of media technology. Media scholar Marshall McLuhan was one of the first observers to suggest that media are not simply technological channels of communication but psychic environments that powerfully and imperceptibly shape human consciousness as we immerse in them. Immersive environments represent some of the most technologically complex forms of mediation to date. Their power to address the five external and three internal sensory channels, combined with their flexibility, durability, and procedurality, can turn XR interactions into highly persuasive experiences.

Digital games are one area of computational media that has shown the persuasive potential of mediated presence. Ian Bogost, a proponent of serious games, argues that digital games can convey messages and compel players to engage in behaviors strategically chosen by the designers (Bogost, 2008). The ability to procedurally set and enforce game mechanics allows game designers to create persuasive environments through expressive computation. Game designers offer players a form of controlled expression by setting the rules and marking the boundaries for the interaction field in which play takes place (Zakrzewski, 2022).

The imposition of well-crafted constraints, in fact, creates environments primed for creative expression. In persuasive XR experiences, entire arguments can be embedded in the rules and code of the application. The way these rules are then understood, internalized, and expressed by the audience defines the persuasive dimension of the game. The internalization of the rules by the players is a critical point at which instantiated digital code manifests itself in the material world through the players' actions.

The potential to use media technology to progressively shape user behavior started being programmatically explored at Stanford's Persuasive Tech Lab in 1997. The lab's founder, Brian Jeffrey Fogg, referred to this approach as captology—a neologism cobbled from the words "computers," "persuasive," and "technologies." Fogg describes captology as a blend of "the design, research, and program analysis of interactive computing products created to change people's attitudes or behaviors" (Fogg, 2003, p. 5).

There is perhaps no better example of persuasive tech than social media platforms such as Facebook and TikTok. The potentially addictive behavior elicited by these platforms can be traced to the direct yoking of advanced computer algorithms to the human nervous system that are very effective at leveraging the neurochemical called dopamine. Feeling affection and receiving social validation are natural occurrences of dopamine release. This ancient bonding mechanism might be one of the main reasons average Americans cannot stay away from their mobile devices, checking their status 150 times daily (Fisher, 2022).

The powerful psychological effect of Facebook's "like" button comes from exploiting the phenomenon known as the sociometer. Social capital has a high survival value for complex social animals who have evolved to live in collectives. Our ability to first survive and then thrive still depends on how well we can navigate our relationships with others.

Understanding core human needs is the foundation of successful experience design. XR designers could not possibly set up the conditions enabling full immersive presence without inviting the audience to trust and to participate in this new medium. The critical difference between rhetoric and coercion comes down to being explicit about one's intent, understanding the difference between the two, and acting on that understanding.

The Five Rhetorical Lenses of Immersive Media Design

In his previous book, Peter outlined five rhetorical ways XR media makers can ethically influence their audiences and invite them to participate in immersive media experiences. The five rhetorical lenses design teams can employ to make their immersive experiences more compelling include (1) utility, (2) aesthetics, (3) behavior, (4) socialization, and (5) reflection (Zakrzewski, 2022).

Let's Not Mess Up the Users' Hair: The Rhetorical Lens of Utility

All well-designed experiences provide some value to their audiences. The utility of XR experiences has two distinct dimensions. XR provides users with broad access to the utility of advanced computation, AI-based assistants, and virtual environments. These tools allow people to extend their limited cognitive capabilities and perform meaningful actions.

Harvard Business School professor Clayton M. Christensen called this utility focus "job-to-be-done" (Christensen et al., 2016, p. 10). He argued that as various needs arise in our lives, we seek products and services and hire them to do those tasks. An audience looking to bond with friends over a fun movie and a pizza party is not too concerned if they watch it on Netflix, Disney Plus, or Hulu. They have a job to be done, which is vitally important within their life project. Those who can empathize with those audience members and best understand their needs will be able to provide the most compelling value proposition to resonate with that audience.

To illustrate his concept, Christensen used the example of a humble fast food milkshake. His research team's ethnographic data showed that more than 40% of the milkshakes were purchased in the morning by single customers who did not purchase anything else. They picked up the milkshakes to consume in their cars.

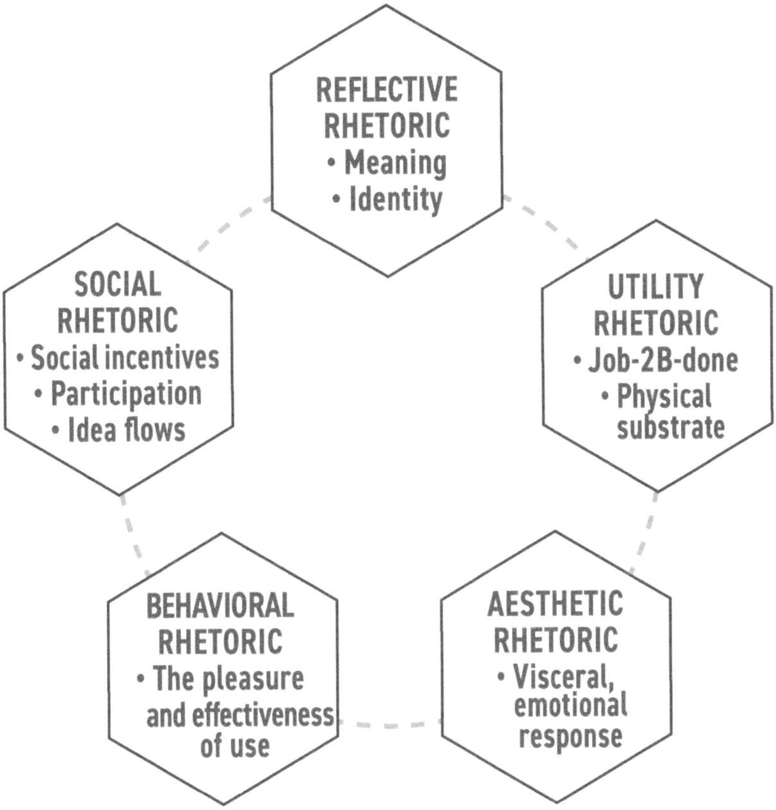

Figure 1.6 Peter's five rhetorical lenses of immersive media

Source: Illustration by © Peter (Zak) Zakrzewski, adapted from Peter (Zak) Zakrzewski.

When the research team interviewed these customers, they revealed they bought milkshakes expressly to alleviate the sometimes long and usually boring drives to work. While sitting in traffic, they had one hand free, which allowed them to hold a cup with relative ease. They were also not hungry enough for a full breakfast. When these customers purchased other products such as bananas, bagels, and candy bars, they found them unsatisfactory for the job. The bagels were too dry. The jam and cream cheese on them was too sticky. Bananas did not last long enough. The candy bars made them feel guilty.

Like the milkshake doing the job of being a driving snack, media experiences have to do a job in the life of the audience, and they have to do it much better than competitors do. The essence of effective user experience design is to uniquely integrate all of the efforts, resources, and processes available to the design team with a single focus on doing the job the core audience needs done. Deep resonance with the target audience is the starting point of adopting any proposed user experience.

The utility aspect of designing XR experiences is closely linked to the embodied nature of our experience in the physical world. The importance of the physical substrate of immersion and spatial computing is easy to underestimate because of the emphasis on the virtual side of XR. Beyoncé's comment about the Magic Leap headset messing up her hair after trying a prototype of the AR glasses is indicative of some of the challenges XR media faces when it comes to mass adoption. Bulky headsets causing strained eyes and necks and messed-up coiffures are only some of the challenges the physicality of XR technology poses to the goal of effortless immersive presence.

As embodied beings, we experience the pleasure of using well-designed tools in a way that signals to us a certain level of mastery. Playing well-tuned musical instruments and riding well-built mountain bikes gives us a sense of skilled accomplishment. Utility rhetoric is the place for designers to explore the physical substrate of XR experiences to address the embodied nature of human cognition. Utility rhetoric can help designers leverage the XR medium as an extension of the human nervous system to accomplish jobs to be done in a way that gives us a sense of physical comfort and pleasure of use.

Attractive Things Work Better: The Aesthetic Rhetorical Lens

The aesthetic rhetoric is built on the survival value of beauty and pleasure. It is based on cognitive science research in the area of the visceral level of cognitive processing. This aspect of conscious experience deals with the preconscious formation of first impressions, concerned with the sensory appearance of objects in our environment.

Aesthetic rhetoric has nothing to do with the concept of or the critique of taste. It is about the visceral emotional states we experience when responding to objects in the environment and their effect on our emotions and decision-making. We receive multiple sensory inputs, spanning olfactory, auditory, tactile, and visual modalities that elicit involuntary or reflexive responses. The responses matter not only because of the aesthetic pleasure they give us but because they have a survival value.

The sense of smell, for instance, is critical in the first few hours of life. Newborns have just enough locomotor function to be able to crawl up their mother's body and to direct their laborious movement guided by the concentration of the scent of amniotic fluid present in the Montgomery glands around the mother's nipple (Tantam, 2018). Evolutionarily useful emotions directing our behavior based on smell include both attraction and disgust. This is linked to the reflex of nausea that might prompt an individual to be repelled by certain foods that might not be safe to ingest.

Visceral processing of experiences is an automatic, prewired function of the human nervous system. This level of cognitive processing allows us to make rapid judgments about what is safe or dangerous in order to alert other brain centers and send appropriate signals to the motor system.

Don Norman goes so far as to suggest that attractive things work better because of how our emotions affect the way we make our judgments about our experiences and how we deal with usability issues based on our level of commitment to the company behind the product. Our emotional system qualitatively changes how our cognitive system operates as a whole. When we are in a state of positive affect, our neurotransmitters broaden brain processing and relax our muscles, allowing the brain to attend to a wider range of potential solutions. Under positive affect, we are much more likely to think creatively and develop novel ideas and solutions. In a state of positive affect, we are more likely to see the big picture and not dwell on minor inconveniences.

Conversely, when we feel anxious or endangered, neurotransmitters will narrow our brain's activity to focus on survival. Anxiety can produce tunnel vision, making people so focused that they may fail to observe what would otherwise be obvious. People escaping fire frequently attempt to push doors that are clearly marked with a "pull" sign. This is the direct effect of the familiar flight-or-fight response, which acts to tense muscles to prepare us for immediate action (Norman, 2003).

Users in a relaxed, happy, and positive mood are more likely to respond creatively to any encountered system problems. They are more likely to overlook or cope with minor functional issues in products and systems.

System designers who expect users to experience negative affect, such as in the case of emergencies, must ensure that all the information required to perform important tasks is continuously brought to the users' attention. Systems that induce a sense of fun and play allow designers to be more creative because they are more likely to be allowed to have minor imperfections.

The aesthetic rhetorical lens helps immersive experience designers plan for the immediate emotional impact of the XR experience. Because visceral design has a lot to do with the users' initial reactions, it can be tested simply by putting the digital product in front of the users and seeing their response. Effective visceral design requires the form-giving skills of industrial and graphic designers, the auditory acumen of sound designers, and the spatial awareness of architects and interior designers.

Since immersive media designers cannot directly control users' subjective states but can only control the objective stimuli that elicit the desired subjective emotional states, effective immersive experience design requires familiarity with submodalities of the sensory channels. The visual mode, for instance, includes the submodalities of size, shape, texture, color, tonal value, movement, transparency, and location. The auditory channel responsible for the aesthetics of sound consists of such objectively controllable aspects as volume, tone, pitch, tempo, and rhythm. We covered all of the sensory channels and their submodalities needed for the aesthetic rhetoric in Section II.

Usability Matters: The Behavioral Rhetorical Lens

The behavioral rhetorical dimension of immersion, like the aesthetics of the XR experience, is inspired by Don Norman's three levels of cognitive processing. It draws on insights into the unconscious nature of our automated behaviors. For instance, people can effectively perform well-rehearsed habitual actions, such as walking, riding a bike, or driving a virtual race car based on unconsciously internalized skills while being able to hold a conversation.

Once our habitual skills become automatic, we often become unaware of the details involved. Highly skilled athletes and musicians allow their bodies to act automatically while they consciously attend to the higher-order structure of the game or the piece of music they are playing. The actions at the behavioral level can be enhanced or inhibited by the higher-level processing of the reflective layer. They can also enhance or inhibit the processing at the lower visceral layer.

Ian Bogost's proposal for the leveraging of the persuasive power of digital games has a lot to do with the phenomenon of behavioral processing. Like various things we acquire through social learning (holding a knife and fork or chopsticks), game rules coded into computational media can influence users' behaviors over time. All interactions with digital objects and experiences made of computer code have the potential to rhetorically persuade (shape) users to develop new habits or even points of view. Once fully internalized, such rules can condition new automated user behaviors.

At the behavioral level, appearances and rationale do not matter as much as does usability. In the realm of physical objects, the solid haptic feel of a door handle and the comforting solid snap of the closing laptop are behavioral cues that give users a sense of quality and safety. Components of good behavioral design include such qualities as function, understandability, usability, and physical feel (Norman, 2013).

Norman argues that user goals require designers to consider the sequence of actions necessary to accomplish the user's ultimate goal. Experience designers applying behavioral rhetoric will not only effectively meet users' goals but do so efficiently by making experiences as seamless as possible. They will also make them engaging by ensuring that all interactions are appropriately sensory rich and pleasurable.

Behavioral rhetoric again reminds XR designers of the importance of embodied human cognition. The tangibility of familiar physical objects with their weight, texture, and surface must be translated into the design of both physical and digital components of the experience. Behavioral rhetoric needs to also address the ease of learning and error tolerance of the immersive environment by ensuring that XR systems are comprehensible and that experiences they support are designed to be forgiving of common user errors.

Let's Get Together: The Social Rhetorical Lens

Humans demonstrate a high degree of context awareness when acting within social situations. Our brains have evolved to co-exist and even co-regulate with others. We are wired to be social beings, and much of our energy is devoted to connecting with others. Our mirror neurons make us tense our muscles in empathy while watching sports or when we anticipate other people's reactions during conversations.

The social rhetoric addresses the interpersonal dimension of the immersive experience, our need for belonging and esteem, and the attendant highly social nature of our mental makeup. Social rhetoric reminds us of XR's unique opportunity to create new forms of face-to-face and remote interaction and collaboration between groups of co-located and remote users.

Social rhetoric draws its power from the deeply abiding human need to belong. Its influence has two critical dimensions for immersive media design. One of these is social conversations' impact on assigning particular meanings to events and media texts. The other is the importance of idea flows in shaping human behavior and adopting new cultural codes. The former draws on Henry Jenkins's research on participatory culture. The impact of idea flows on behavior leverages Alex Pentland's research on social physics.

The concept of participatory culture is based on the idea of collective intelligence, first proposed by Pierre Levy. The main insight of Levy's proposal is that while no single individual or a group can possibly know everything, in a network society, everybody knows something. Diverse and well-connected social networks can develop enormous expertise and knowledge. Broadening access to diverse kinds of expertise can give social groups a powerful ability to solve complex problems.

Henry Jenkins, who has applied Levy's concept to his fan-based media research, suggests that participatory media environments can enhance the creative power of social networks by allowing the free circulation of vital information. This is well demonstrated by a variety of social gaming experiences. For instance, a study conducted by Anat Shoshani and Maya Krauskopf found that children cooperatively playing the online digital game *Fortnite* displayed greater prosocial behavior such as willingness to cooperate (Shoshani & Krauskopf, 2021).

Another insight informing the social rhetorical aspect of immersion comes from computational social scientist Alex Pentland, who suggests that constantly evolving social interactions with others heavily influence what we believe, value, and desire and how we choose to act in the world. A single exposure to an implicit or explicit suggestion from a trusted peer can often be sufficient to convert a person's beliefs. Social influence is an ongoing process between close friends and colleagues with frequent face-to-face contact. Seeing peer group members adopt a novel idea provides a powerful motivation to join in and cooperate with others (Pentland, 2015).

Social physics addresses several design issues concerning immersive systems. The first one is the impact of the social context on individual behavior. The second one is the power of computational social science to observe, evaluate, and tweak the emergent behavior within immersive systems.

The social phenomena of our connected world emerge out of countless microtransactions between individuals who trade not only goods and currency but also information, ideas, and even gossip. Social actors make decisions that are always a blend of individual search and social information. When the personal information obtained from individual searches is insufficient, people tend to rely more on social information. In situations in which decisions have to be made under conditions of uncertainty, social learning has a confidence-enhancing effect. Designers interested in leveraging the persuasive power of social rhetoric can learn a lot from closely studying insights from social physics.

Life Is a Dramatically Constructed Thing: The Reflective Rhetorical Lens

The founder of the third school of Viennese psychotherapy, Viktor Frankl, argued that human beings are, above all, motivated by the search for meaning. We have a unique capacity to reflect upon past experiences, understand the past, and reflect on the future to imagine new possibilities. The reflective rhetoric taps both the interpersonal and intrapersonal dimensions of the immersive experiences through the power of narrative and reflection. Reflective rhetoric is inspired by Norman's reflective level of cognitive processing.

Unlike the visceral and behavioral levels, reflective cognitive processing extends beyond the moment of engagement. It focuses on the moment of reflection when we assign meaning to past experiences and contemplate possible futures. A reflective rhetorical lens allows immersive media makers to mediate the interaction between the audience's identity and the meanings they invest in the digital content and objects. The reflective level is linked to the powerful brain mechanisms that allow us to think creatively, construct architecture, design, make art, and solve problems.

Trauma researcher Bessel Van der Kolk suggests that humans possess two distinct forms of self-awareness. Our autobiographical self is rooted in language. It keeps track of the self across time. It creates meaningful connections among various experiences to assemble them into a coherent and deeply resonant autobiographical story. Our reflective narratives evolve over time as people incorporate new input and change their overall perspective on who they are and how they believe things ought to be.

The other form of self-awareness is based primarily on the physical sensations of the living body. This system keeps track of moment-to-moment self-awareness. Its knowledge can be translated into words only if we are not rushed and feel safe.

The two ways of knowing and understanding our experiences are localized in different parts of the brain, which are not strongly connected to each other (van der Kolk, 2014). This can lead to a disconnection between our stories and our bodies. Van der Kolk observed in his PTSD therapy group sessions with former soldiers that they could tell dramatic tales of death and destruction while ostensibly radiating a sense of pride and belonging with their bodies. The language-based, autobiographical, storytelling system excels at creating narratives for private and public consumption. If told often enough, these can make us believe they are accurate accounts of our true past experiences.

The nonverbal somatic system registers a different kind of truth, a truth about how we emotionally experience a particular situation deep inside. The nonverbal self-awareness, primarily based in the medial prefrontal cortex, can affect the emotional brain and its responses, and the powerful stories we tell ourselves and others can profoundly influence our actions and conscious behavior. When these two systems diverge, we can potentially be at odds with the inner truth of our embodied experience.

Media makers interested in the reflective rhetorical quality of their work need to factor in the power of inner storytelling. Our stories about ourselves and the world can deeply affect our moods. Narratives can lessen isolation and emotional pain by providing acceptable explanations for dramatic events. van der Kolk argues that they can also become problematic when they provide people with intuitively appealing targets to blame in some extreme cases.

Psychologist Jerome Bruner suggests that narrative comprehension is among the earliest powers of the mind to be manifested by young children and is among the most widely used strategies for organizing the human experience. From early childhood, we organize our understanding and memory of our social experience mainly through narrative (Bruner, 1991). An internal narrative allows us to create an encapsulated inner world and tell ourselves about our hopes, desires, and dreams. Such internal narratives create a sense of agency and position us in various social constructions, suggesting our status and capacities as social actors.

Many culturally resonant ideas are themselves products of a social narrative consensus. Narrative is an essential form of inter- and intrapersonal communication, allowing reflection and cultural transmission. Narratives are versions of reality whose persuasive powers and acceptability are governed by social conventions and narrative necessity rather than by empirical verification or adherence to the rules of logic (1991).

The Vision and the Audience: The Journey of Rhetorical Immersive Transportation

Talking about the vision for an immersive media project without mentioning the audience is a lot like talking about a chicken without mentioning an egg. No media experiences happen without an audience. But which comes first? We believe that a media design team should have a well-defined direction for the project based on their strategic focus, macro insights about the technology, economy, and culture, their core competence, and of course, microinsights about the audience before diving deeply into the audience's life context. So, like the chicken and an egg, the project vision and the exploration of the project audience share the same genetic makeup of the unmet audience need. But within the XR media design process, the vision-defining step leads to the audience-defining exploration and discovery.

With that in mind, we argue that the rhetorical power of immersive media experiences can be best understood as a sequence of progressive steps, starting with attracting the audience's attention, transitioning the participants into the immersive experience supporting presence, and culminating in an immersive transformation. Audiences enter immersive worlds as they enter narrative ones, by identifying with characters whose identity they find appealing through four stages of immersive transportation: (1) Relatable characters and resonant themes are the foundation of the immersive experience of mediated presence, they are based on and, in turn, can impact attitudes and beliefs in the physical world. People are attracted to stories and experiences based on their identities and values. We refer to this stage of immersive transportation as relatability. Relatable characters and themes attract the audience; (2) The Resonance stage addresses the depth of emotional engagement the prospective audience has with the major theme of the immersive experience; (3) Immersively supported resonance leads to presence, which is achieved in varying degrees as the audience members voluntarily engage in their immersive journey; (4) Finally, Transformation occurs when participants emerge from the immersive experience with changed beliefs, attitudes, and/or knowledge.

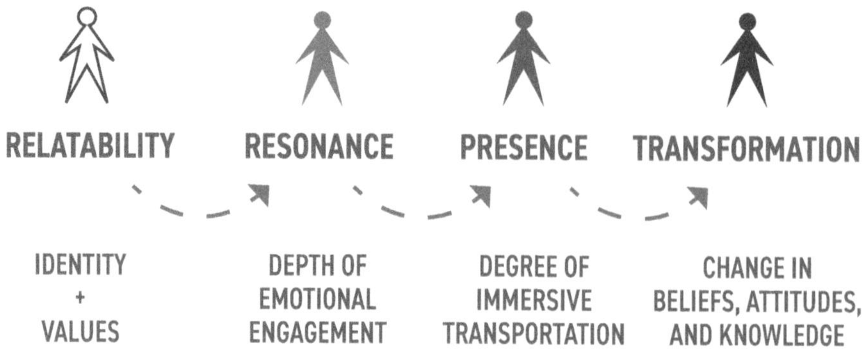

Figure 1.7 The chain of persuasion; four stages of immersive transportation

Source: Illustration and concept by © 2024 Peter (Zak) Zakrzewski.

Summary

In this chapter, we examined the significance of establishing a clear project vision at the start of the immersive media–making process. We highlighted the value of focusing on a singular goal that answers the critical "why" question of what drives your team. The clarity of purpose will help you not only differentiate your work from competing offerings but also to focus the precious resources. This chapter provides a step-by-step guide for crafting a vision statement, emphasizing its critical role in shaping successful projects.

We emphasized the importance of taking well-defined steps toward achieving your vision while managing uncertainty, which inherently comes with the territory of creativity and innovation. The hypothesis-testing loops will keep the team connected to the reality of the audience's world while emphasizing the value of learning from real-world feedback as the team refines the approach toward the desired vision state. We discuss immersive experience types and the diversity of audience engagement, from challenge-based to social immersion, as part of the immersive media design process.

We also describe the five rhetorical lenses covering utility, aesthetics, behavior, socialization, and reflection to ethically inspire and influence audiences and invite them into the immersive experience your team is designing. We discuss the ethical issues involved in working with the most rhetorical medium in history, stressing the importance of respecting user agency and mindful persuasion over technological paternalism and coercion.

We hope that we were able to convey how a well-articulated vision can serve as an inspiration and a road map guiding all subsequent decisions and milestones within a project. Having outlined the value of the transformative power of a well-defined vision, let us now turn to its key ingredient—the microinsight informing us about the needs, hopes, and desires of our core project audience.

References

Bernstein, C., & Woodward, B. (2022). *All the President's men* (50th anniversary ed.). Simon & Schuster.

Bogost, I. (2008). *Unit operations: An approach to videogame criticism*. The MIT Press.

Bruner, J. (1986). *Actual minds, possible words*. Harvard University Press.

Bruner, J. (1991). The narrative construction of reality. *Critical Inquiry*, 18, 1–21.

Christensen, C. M., Hall, T., Dillon, K., & Duncan, D. S. (2016). *Competing against luck: The story of innovation and customer choice*. HarperBusiness, an imprint of HarperCollins Publishers.

Crosswhite, J. (2013). *Deep rhetoric: Philosophy, reason, violence, justice, wisdom*. The University of Chicago Press.

Fisher, M. (2022). *The chaos machine: The inside story of how social media rewired our minds and our world* (1st ed.). Little, Brown and Company.

Fogg, B. J. (2003). *Persuasive technology: Using computers to change what we think and do*. M. Kaufmann.

Greeno, J. G. (1989). Situations, mental models, and generative knowledge. In D. Klahr (Ed.), *Complex information processing: The impact of Herbert A. Simon [21st Carnegie-Mellon symposium on cognition]* (2nd [print.], pp. 285–318). Symposium on Cognition; Erlbaum.

Lazzaro, N. (2021, July 7). Engagement in wonderland: The 4 keys to fun in XR. *Medium*. https://uxdesign.cc/engagement-in-wonderland-the-4-keys-to-fun-in-xr-90966faa5f0f

McKeown, G. (2014). *Essentialism: The disciplined pursuit of less*. Crown Business.

Norman, D. A. (2013). *The design of everyday things: Revised and expanded edition*. Basic Books.

Pentland, A. (2015). *Social physics: How social networks can make us smarter*. Penguin Press.

Ries, E. (2014). *The lean startup: How today's entrepreneurs use continuous innovation to create radically successful businesses*. Crown Business.

Riley, M. S., & Greeno, J. G. (1988). Developmental analysis of understanding language about quantities and of solving problems. *Cognition and Instruction*, 5(1), 49–101. www.jstor.org/stable/3233609

Scott, F. D. (2015). Networks and apparatuses, circa 1971: Or, hippies meet computers. In A. Blauvelt (Ed.), *Hippie modernism: The struggle for utopia*. Walker Art Center.

Shoshani, A., & Krauskopf, M. (2021). The Fortnite social paradox: The effects of violent-cooperative multi-player video games on children's basic psychological needs and prosocial behavior. *Computers in Human Behavior*, 116, 106641. https://doi.org/10.1016/j.chb.2020.106641

Tantam, D. (2018). *The interbrain: Embodied connections versus common knowledge*. Jessica Kingsley Publishers.

Team Sequoia. (2023). The Arc product market fit framework. *Sequoia Capital*. www.sequoiacap.com/article/pmf-framework/

van der Kolk, B. A. (2014). *The body keeps the score: Brain, mind and body in the healing of trauma*. Viking.

Zakrzewski, P. (2022). *Designing XR: A rhetorical design perspective for the ecology of human+computer systems*. Emerald Publishing Limited. http://doi.org/10.1108/9781802623659

2 Understanding Your Audience

Who Is Your Audience and Other "Leap-of-Faith" Assumptions

Starting a new creative project is a daunting task, particularly if the outcome might be a new-to-the-world digital product or experience. As we mentioned before, going from zero to one necessarily involves dealing with uncertainty. Generating well-calculated hypotheses of what the world could look like rather than relying on doing what has worked in the past is a part and parcel of the creative process.

In the start-up world, fundamental hypotheses underlying critical decisions are known as leap-of-faith assumptions. They are the kinds of beliefs we hold or intuitions we develop as creative media makers about what might work for our audience. In the course of immersive projects, innovative media makers will face many situations for which no prior performance data is available. This makes hypothesis formulation an indispensable component of experience design. Sometimes, creative hunches lead to blockbusters such as *The Lion King* or *Toy Story*, and sometimes they lead to commercial disasters such as *The Suicide Squad*. Eric Ries, who champions the lean start-up methodology, has identified four kinds of do-or-die leap-of-faith assumptions faced by creative teams.

The first and biggest among them is the audience hypothesis. This is fairly self-explanatory. Choosing the right audience to co-create the vision with is the most important decision media makers make on a project. Second, the value hypothesis addresses how the project will resonate with the audience. Will the goal of the project attract the audience's attention? Will it fire their imagination? Will it perform a meaningful job in their lives?

Third, the feasibility/usability hypothesis is another set of critical assumptions made about the user experience. Will the available technology support the desired user experience? How will the system perform to support the audience in accomplishing their goals effectively and efficiently? How compelling will the experience be? Finally, the growth hypothesis addresses such questions as: How will the word about this project spread? How big can the audience for this project grow?

Before any of the latter hypotheses can be tested, the first one, addressing the project's audience, has to be well researched and formulated. Who is our audience? What attracts them? What do they find relatable? How will our XR experience fit into the context of their lives? Without eventually finding out answers to such

DOI: 10.4324/9781003442462-6

questions, it will be very hard or nearly impossible to make judgments about the suitability of most creative and technical decisions the team will have to make during the course of the project.

Participation: The Birth of an Audience

Comprehending any form of mediated content is not a passive act. Media audiences who turn a series of still images and streams of sound into meaningful moviegoing experiences engage in amazing feats of neural and cognitive processing. Mediated experiences such as movies, videos, blog posts, and digital games comprise thousands of digital assets, individual shots, sound clips, and digital objects brought together by media makers to create an illusion of coherent scenes evoking particular locations and sequences of goal-directed actions by the protagonists.

The viewers do not consciously notice the vast majority of the individual pieces, yet they combine these odd bits and bobs to construct meaningful narratives that go beyond those presented by the sensory stimuli. Media magic happens within an individual audience member's mind and nervous system, transforming streams of flickering pixels and oscillating sound vibrations into coherent narrative worlds. Media content offers audiences vicariously experienced events that offer their own psychological plausibility that goes beyond the neurobiological properties of the sensory modalities (Magliano & Zacks, 2011). They are constructed by the audience members based on their own unique, individual neurodiverse makeup, cultural background, and life experience.

A rhetorical XR media-making practice is not a one sided affair. Immersive media experiences are predicated on the willing participation of conscious human actors who make informed choices by selecting from the possible options of future actions rather than being coerced procedurally or otherwise. We highlight the importance of explicitly factoring in the agency and choice of the audience in immersive experience design. Audience participation can be seen as a unique dimension of immersive media making.

Explicit inclusion of active user participation is one of the defining dimensions of XR media. The framework we present in this book anticipates and accommodates user participation at all stages of the media-making process. Rhetorical media making goes hand in hand with the concept of choice. People who are subject to any form of coercion do not need to be persuaded. Users' freedom within the context of immersive systems can be described by two core principles. First, the audience members must be presented with a set of genuine alternatives. Second, they must be able to choose their own course of action within the presented alternatives (Zakrzewski, 2022).

Understanding that audiences co-create media experiences along with the media makers means that discovering the audience's needs, life contexts, goals, and motivations requires as much care as the effort put into creating fictional characters, objects, and sets that make up the final product. The success of your project will depend on how well it can meet the needs of your audience while pursuing your team's creative vision.

Audience insight cannot be easily obtained by hypothesizing about the audience's motivation sitting back at the studio. It cannot be gleaned from sifting through market surveys. Genuine curiosity about the audience is critical to understanding the deep motivations that attract your audience to the XR experience you set out to create. The kind of critical knowledge needed for empathizing with the audience is best gathered by immersion in the audience's lifeworld.

In this chapter, we explore several qualitative methods originally developed in psychology, anthropology, and design research that are highly applicable to audience research aimed at the creation of XR media.

The Audience Profile–Building Process

Experience designers understand that audience profiles are abstractions not fictions. While they abstractly represent the common characteristics of an audience segment, they are built on a firm foundation of user research. The process we recommend comprises seven stages.

1. Identify the Broad Audience Segment

Identify the broad audience segment. For instance, if you aim to create an educational experience for medical schools, your target might include medical students and faculty who end up being the users of the program. At the end of this step, your team will get closer to finding your core audience by eliminating many potential segments that do not match the project objective.

2. Determine the Cognitive and Behavioral Variables

Determine the cognitive and behavioral variables for the core audience segment that are most relevant to your project. The categories can include (A) identity the audience members relate to, (B) key microinsight that articulates what the audience segment needs or considers resonant. This can be expressed as a quote that gives the design team more insight into the audience segment's mindset. Next come (C) unmet needs, goals, and motivations, (D) habits and known behavioral patterns, (E) interaction needs (visceral, behavioral, or reflective), (F) aptitudes (novice, intermediate, advanced) (G) context of use (when and where).

3. Conduct a Contextual Inquiry

Conduct a contextual inquiry that includes projective and metaphor-elicitation techniques, ethnographic studies, interviews, and observations to gather evidence you will use to base your audience profile on.

4. Analyze Your Data to Map the Significant Behavioral Patterns

Analyze your data to map the significant behavioral patterns. Clusters of subjects who map over a particular set of key variables might represent a significant audience segment. For instance, audiences for a VR gaming experience might be spread somewhere on the spectrum between must-have narrative fans and players who do not need strong stories to enjoy games. Since your XR experience cannot equally please everyone, you might want to decide where your primary audience sits on this issue and design the game accordingly. In the case of a medical app, you will want to determine variables such as the experience level, in both medical terms and comfort with immersive technology, of the person using the application.

5. Outline Top Emerging Audience Segments

Based on your analysis, outline top emerging audience segments. Develop up to three possible segments to the point of understanding the implications of each one for your project outcome. Pick the most promising core audience that most closely aligns with your vision. Core audience identification might become a pivot-inducing moment when your team discovers a unique audience segment that might actually make you rethink your original project vision. Your original vision is one of many provisional statements or hypotheses to be tested. Discovering the reality of your potential audience through research will serve as a hypothesis test for that particular assumption.

6. Review and Prioritize Segments

Review and prioritize the segments that emerge from your research. The most compelling media experiences, services, and applications are created with one primary

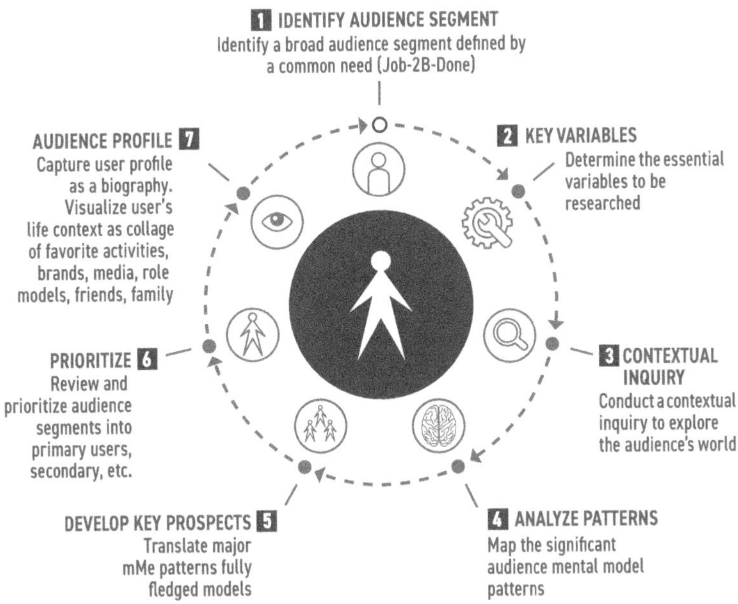

Figure 2.1 The audience profile modeling process

Source: Concept and illustration by © Peter (Zak) Zakrzewski.

persona in mind. In media design, the reality of scarce resources means that those resources have to be narrowly focused on strongly appealing to a well-defined audience. Focus on one specific audience will not only assure the right allocation of your scarce resources, but it might enhance the experience's desirability for other audience segments.

Too many audience segments dilute focus and resources and bloat projects with feature creeps. A typical design project will identify up to three personas: a primary audience, or those you truly need to appeal to; a secondary audience, which might be either less committed users or users whose needs are only partially addressed by your solution; and a tertiary audience, who might be nonusers somehow influencing the primary users. For instance, a primary audience for an educational app would be the students themselves, the secondary audience might be the teachers working with the students. The tertiary audience might include the parents or the school administrators purchasing the XR app. Once you settle on your primary audience, your team will have clarity about who they are working for at each step of the process.

7. Primary Audience Profile

Capture the findings from your mental model exploration (mMe) as the primary audience profile expressed as a media rich portrait, a photo album, and a journal that includes the significant details of the core audience's lived experience, unmet needs, goals, challenges, hopes, and dreams. Add more emotive quality to the description by including ethnographic images of their environments, prized objects, tools, brands present in their daily lives, hobbies, media they consume, etc.

The best material for the audience portrait comes from the ethnographic data gathered during the research process. Resist the temptation to use meaningless and potentially misleading decorations such as posed stock images or illustrations. These will add little value in terms of building empathy and understanding. Try writing the profile as if it were your own journal. Write about 300 to 500 words. Remember, you are not writing a novel but a design document. Treat the narrative as a brief character sketch. Begin by introducing the participant's relevant personal details.

You can add relevant demographic information related to each audience cluster, such as identity, age range, occupation, etc. To help make your mMe profiles more realistic, add relevant context you have gleaned from your interviews, such as the user's environment, social relationships, schedules, frustrations, current technologies used, and other contextual details. Make sure to base all of the context details on real observed data. Do not engage in fiction to make your mMes more exciting. The only fictional details we add to our audience profiles might be their names.

The most important part of the entire process is the identification of key user goals that logically relate to the experience you are designing. The compiled audience portrait will help your team develop a more empathic understanding of the audience and will allow you to relate to them on a humane level.

Learning About Your Audience

XR audience research can be broadly divided into pragmatics (the assessment of behavior) and semantics (the assessment of communication and meaning). The early stages of media projects addressing the core audience and value hypothesis are usually more concerned with finding resonant meanings, while the latter stages dealing with the usability hypothesis tend to focus on assessing the behavior observed during the user–system interactions.

Both of these categories can be approached by the application of subjective or objective measures. Subjective assessment can be obtained by conducting interviews with subjects representing the core audience. Objective measures can be obtained from biometric data and digital breadcrumbs left behind by audiences interacting over digital platforms. Design researchers and theorists Wina Smeenk, Oscar Tomico, and Koen van Turnhout urge designers to triangulate their audience research by blending three research views comprising the first-, second-, and third-person perspectives.

Within their systematic analysis of the mixed-perspective framework, intuitive understanding is based on subjective or interpretive research that allows the design team to develop and trust their informed creative intuition. Objective, empirically based, scholarly, data-driven, and secondary research is framed as a third-person perspective. The first-person and the third-person perspectives are complemented by the important empathic view referred to as the second-person way of understanding others (Smeenk et al., 2016).

Many of the audience exploration methods we cover in this chapter will require creative interpretation, empathic understanding, and informed intuition, in contrast to the more empirically oriented usability studies conducted during the later hypothesis-testing stages of XR projects. The techniques we present here will help the design team to explore the audience's mental models of the world, core drives, goals, and motivations through meaningful metaphors, ethnographic observations, and creative self-reporting. We will start by covering the vital but often misunderstood and most definitely misnamed user research tool known as user personas.

Exploring the Audience Goals and Motivations

Throughout the book, we refer to the 5Es framework of usability proposed by the co-founder of the Center for Civic Design, Whitney Quesenbery. The framework offers five lenses of effectiveness, efficiency, engagement, error tolerance, and ease of learning for building and evaluating sociotechnical systems (Quesenbery, 2004).

The first and, in our opinion, the most important E of successful XR experience design refers to the effectiveness of the proposed design solution. How useful is the proposed solution in meeting the users' needs and goals? Discovering your audience means understanding what does the audience want to gain from engaging in the experience? Do they want to be entertained? Do they want to explore a subject and learn something new? Do they want to meet new people and be thrilled by an exhilarating social experience? Do they need help with performing a particular task or job in their life? The XR design team needs to have a clear answer to these

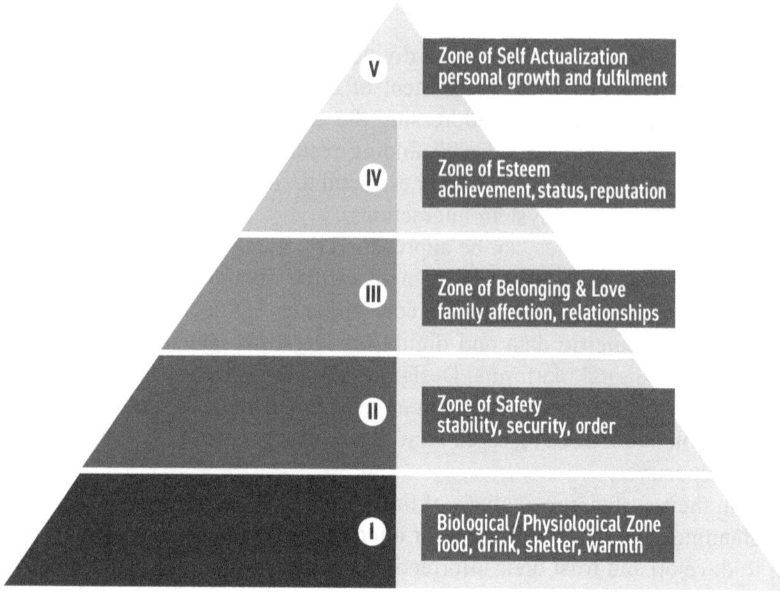

Figure 2.2 Audience needs and goal zones

Source: Illustration by © Peter (Zak) Zakrzewski, based on Abraham Maslow's Hierarchy of Needs.

questions before they can set out to create a truly resonating media project. We will cover the 5Es in more detail in Chapter 9.

In experience design, user goals can be divided into two types: (1) performance goals that are based on needs and (2) experience goals that are based on the aesthetics of sensory inputs. We like to think of peak performance moments as the Aha! moments and the peak experience ones as the Wow! moments.

One commonly used and useful framework for identifying user motivations is Abraham Maslow's well-known hierarchy of needs. Maslow, a humanistic psychologist, listed five core zones of human needs stacked up in order of importance to survival and a sense of well-being: (I) physiological needs, (II) safety, (III) love and belonging, (IV) esteem, and (V) self-actualization (Figure 2.2). Many of the jobs to be done within the audience members' life project can overlap two or more zones. The five zones can serve as an excellent guideline for turning core needs into immersive experiences.

Eight Drives That Motivate Engagement

Yu-Kai Chou, the developer of the Octalysis gamification framework, argues that systems created from the engineering perspective end up being function focused, meaning that they are optimized to complete various explicitly stated tasks as quickly and as efficiently as possible. However, XR audiences might not be primarily driven by the desire to rush through the experience and might even resent being compelled to do so. XR experience design is based on the understanding that participants are motivated by their own idiosyncratic reasons, emotions, and goals.

Chou suggests that only the experiences that engage their audiences' core drives are capable of eliciting strong motivation, resulting in meaningful user behavior. Chou's Octalysis framework of eight core categories of human motivation is now widely used in gamification projects in areas as wide-ranging as education, fitness, healthcare, marketing, and corporate training to enhance user engagement (Chou, 2024).

1. The Epic Meaning and Calling drive is activated when the participants believe they are engaged in doing something greater than themselves and were "chosen" to take some special and meaningful action. Participants motivated by this deeply seated need do not heed to receive any extrinsic reward aside from contributing to serve a greater goal. The implementation of this drive is not limited to purely altruistic behavior. It can be applied in many situations. Many world-changing events have been precipitated by people who believed they were participating in something greater than themselves.
2. The Development and Accomplishment need drives people towards accomplishing something and achieving a sense of growth. Badges and trophies without a perceived challenge are meaningless for people motivated by this drive. Once the challenge is introduced into the equation, rewards such as points, badges, and leaderboards (PBL) can become very effective motivators.
3. Empowerment of Creativity and Feedback constitute the third core drive. This drive is activated when players become engaged in a creative process, creatively challenged to figure out new solutions repeatedly. People motivated by this need seek ways to express their creativity, see the results of their creativity, and receive feedback. Playing such games as *Minecraft* offers a perfect challenge to those driven by the empowerment of creativity. Properly integrated game mechanics that empower users to be creative become self-sustaining.
4. The need for Ownership and Possession offers participants the feeling of owning or controlling something. People who experience ownership over something frequently want to increase and enhance what they own. People motivated by this drive often overvalue their possessions.
5. The need for Social Influence and Relatedness forms the basis for core drive 5. This motivation overlaps with Maslow's social needs. The motivating factors can include social acceptance, friendship, and even competition. As Alex Pentland observed, social network incentives are a more powerful method of affecting idea flows and change than individual incentives.
6. Scarcity and Impatience addresses people's propensity to want things simply because they are rare, exclusive, or unattainable. Games using this motivator utilize such mechanics as "appointment dynamics" and "torture breaks" to force players to return later to claim their rewards. Lack of immediate availability compels some people to double their effort to obtain the object of

desire. Rolex is one of the prime examples of a brand following this motivating strategy.

7. Fear of missing out (FoMo) is behind core drive number 7. Unpredictability motivates participants to be constantly engaged because they do not know what might happen next. Our brains find patterns and recognize when things do not fall into our regular pattern-recognition cycles. In cases such as gambling, players stay in higher alert modes, paying attention to deal with the unpredictability of the situation. Any games of chance and lottery programs use the same core drive to motivate people to stay engaged and keep coming back for more.
8. The very human propensity for Loss Avoidance drives people to actively try to avoid negative things happening. On a small scale, it could be to avoid losing previous work or changing one's behavior. Opportunities that can seemingly fade can make people act in the spur of the moment. Some people might also be unlikely to walk away from things they have already invested in, no matter how hopeless the chances of success.

Strategic Empathy

Before we can create immersive worlds our audiences find compelling, we as media designers have to empathically enter the worlds of our audience. The application of empathy to the process of designing experiences is one of the distinguishing characteristics of design thinking methodology.

Empathic design is a design approach developed originally by Patricia Moore and championed by Jane Fulton Suri and her colleagues at IDEO. Empathic design offers XR experience design teams a unique second-person perspective to gain information that is not accessible through traditional observation-oriented user research methods. The approach involves deep immersion into the environment and life context of the audience that gives media designers a unique window onto the challenges and motivations of the core audience.

Empathic design practitioners have developed a number of research methods that go beyond purely cognitive understanding of the user's needs. Fulton Suri argues that designers creating user experiences of any kind need to start the process by learning about people's needs, aspirations, rituals, and values within their personal, social, cultural context. She suggests that experience designers must integrate the intersubjective perspective to reflect the users' thoughts and feelings.

The inclusion of the media makers' ability to make empathic inferences from observation is an essential ingredient of discovering what matters to audience members. An empathic design approach requires designers to hone their ability to creatively interpret what people think, feel, and dream in order to envision future possibilities. Incorporation of empathic design methods has led to improved rates of adoption of many digital and physical product experiences, ranging from the Oxo Good Grip line to the Airbnb app.

Design researchers Merlijn Kouprie and Froukje Sleeswijk Visser have created an explicit framework for designers to engage in the empathic discovery of the

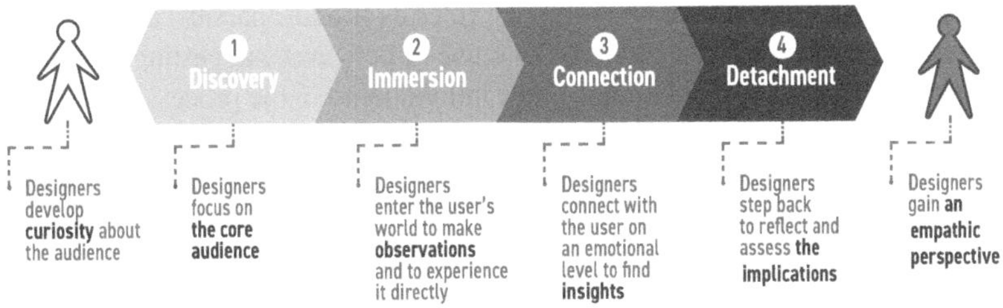

Figure 2.3 The empathic discovery process

Source: Illustration by © 2022 Peter (Zak) Zakrzewski, adapted from Merlijn Kouprie & Froukje Sleeswijk Visser, A Framework for Empathy in Design.

users' world. They refer to their process as "stepping in and out of the user's life" (Kouprie & Visser, 2009, p. 1). The method guides designers through four stages of gaining an empathic perspective: (1) discovery, (2) immersion, (3) connection, and (4) detachment.

The process begins with the designer's focus on a particular audience segment and a willingness to explore the context of the user's life. In the Immersion phase, designers leave their studio to take an active role in wandering around the user's world. In the Connection phase, designers make an emotional connection through their own embodied experience with the user's life context. At this phase, designers engage on both the affective and cognitive level to achieve an understanding of both the user's feelings and meanings surrounding the situation under investigation.

In the last phase, designers detach from the emotional connection in order to enter a problem-solving "helpful mode," with a new, increased understanding. By stepping back to reflect, the designer can deploy the newfound insights in the ideation phase of the project. Microinsights from the empathic discovery should be captured and summarized through the user persona (mental model exploration) method.

User Personas: A Powerful Audience Research Tool With a Misleading Name

The empathic exploration of the audience insights has to be captured and summarized in a way that gives actionable direction to the entire XR experience design team. User personas have been adopted by the UX design community as a de facto standard for summarizing the exploration of user needs and modeling audience segments.

The word "persona" comes from a Latin term referring to a theatrical mask. In literature, "persona" refers to a fictional character. In sociology and psychology, the term has been developed by such scholars as Robert Park, Carl Jung, and Erving Goffman to imply a strategic mask of identity or a public image people adopt in the course of social performance. We wear our social masks to impress others and to make ourselves appear more socially desirable. We can adopt multiple personas depending on the social situation, which can include such roles as friends, parents, and co-workers. Carl Jung actually cautioned about the overuse of personas, suggesting that the practice may lead one to suppress or lose one's individuality in the process.

Judging by this description, personas actually seem to represent the opposite of what designers need to be looking for in an effort to understand their audience and to uncover their unmet needs. While according to this definition of the term, personas represent a mask or a surface structure potentially obscuring an individual's mental model of the world, what experience designers are really after is a glimpse into the deep structure of that model.

"Mental model exploration (mMe)" is a term Peter has developed in the course of his UX design career for the process of gaining a strategically empathic understanding of the audience members that is as unique as a fingerprint regarding their worldview and mindset.

Aside from the unfortunate name, user persona generation is a powerful usability tool that allows design teams to get an empathic glimpse at very focused audience segments. It represents an outcome of a discovery process aimed at identifying and understanding a specific small group of users who share a common need and usability preferences. Once defined, mental model explorations deliver three significant benefits to the design project. They provide (1) a specific design focus for conceptualizing early prototypes, serve as (2) a collaborative communication tool that guides the combined efforts of the entire design team, and provide (3) criteria for evaluation of design objectives and prototypes.

Mental model explorations provide insights into unique user motivations and goals, sets of specific patterns of user behaviors, and the life context in which the experience needs to function. The three ingredients necessary for effective user modeling include (1) an open-minded, curious, and caring attitude, (2) a commitment to conduct valid research and analysis, and (3) the ability to synthesize relevant insights into actionable implications.

mMe: Mental Model Explorations

Alan Cooper and his *About Face* co-authors remind us that all models are useful abstractions applied to represent a more workable version of complex phenomena in the natural and social sciences. Good models do away with nonessential details to emphasize the most relevant features of the structures and relationships between key parts of the system under investigation (Cooper et al., 2014).

Modeling audience mental models, like all good models, needs to focus on capturing the essential features of the users' core needs, beliefs, attitudes, and aptitudes as they relate to the immersive experience. One of the important insights regarding VR and MR headsets being used to watch movies is that they tend to make people feel isolated. Audiences seeking social movie experiences are unlikely to choose options involving VR goggles. Relatedly, social stigma about privacy was one of the main reasons behind the low adoption of Google Glass.

In the case of VR games, the design team might be looking to understand whether the prospective audience members are casual gamers or people who live and breathe digital game experiences. Some gamers care about the narrative dimension of games, while others prefer more interactive experiences. For those who love a good story, what narrative themes resonate with them? Does your audience enjoy social, multiplayer games, or do they prefer to play on their own? Comfort level with AR technology might be an important factor for users of an AR shopping app, etc.

In a book titled *Science and Sanity*, Alfred Korzybski, the developer of general semantics, compared human cognitive processing to a map representing a territory of external reality (Korzybski, 2005). His unique insight was that due to limited resources, human cognition abstracts the world by generalizing, deleting, and distorting sensory inputs to represent only those features that are most important to functioning and survival.

The human brain is a prediction machine that interprets what it perceives based on prior knowledge and experience. To deal with the massive task of coping with external reality, each one of us creates our own unique map of shared reality by reducing the world's complexity by abstracting and simplifying it. We then use these maps to navigate all of our lived experiences. While a map is not the territory it represents, a map can be a very effective navigational tool if it accurately reflects the structure of the territory it represents.

The author of *The Fifth Discipline*, Peter Senge, described mental models as deeply ingrained assumptions and generalizations that often unconsciously influence how we understand the world and take action (Senge, 2006). Philosopher and psychologist Philip Johnson-Laird argues that any form of reasoning is a simulation of the world fleshed out with experiential knowledge. We form and use mental models to derive conclusions about how to take action and solve problems (Johnson-Laird, 1995). Taking the time to understand the deep structure of how audience members see the world goes a long way toward being able to empathically create experiences that users find compelling and meaningful.

Audience mental model explorations will vary from project to project, but a comprehensive mMe should contain such key variables as (1) autobiographical details that define the common traits of the audience's perceived identity, (2) resonant themes and goals (jobs-to-be-done) and motivations that can span the gamut of biological/physiological, safety, social, belonging/love, identity/esteem, and self-actualization needs, (3) core beliefs, (4) related habits, (5) interaction needs, (6) visceral, behavioral, and reflective experiential needs; (7) aptitudes (novice, intermediate, advanced), and (8) context of use (at work, at home, while commuting, etc.).

Like standard personas, mMes are compiled out of direct contact with real potential audience members. The resulting deep dive into the mental landscapes of the audience should give the team members an empathic understanding of what desirability means for this particular group. The outcome of the mMe process should be visually translated with a photo album of the audience, including the activities and places they enjoy, media they watch, etc. Journals and diaries can also be prepared by the members of the design team to capture the observed insights. The resulting multimodal presentation should capture enough detail to make the design team feel that they can relate to the audience members as if they were close acquaintances or even friends.

We believe that the empathic exploration of the audience's mental models is not a paint-by-numbers endeavor. Much insight can be gathered from algorithmic explorations of behavioral user data, but when looking for unexplored and unmet needs, design teams will need to face the audience in a face-to-face conversation. Next, we share a few of the methods we have found not just useful but often eye-opening.

Projective Techniques: Metaphor Elicitation

As much as 90% of the meaning extracted from the messages we exchange with each other is nonverbal. Much of the meaning obtained from verbal language is determined by nonverbal cues such as nods and smiles. Visual communication is processed differently than verbal communication and is usually not subject to the same critical scrutiny as explicitly stated verbal messages. Information we infer from watching people and looking at images is internalized in a way that profoundly affects our attitudes and behavior, the extent of which might be hard to assess.

The pioneers of metaphoric communication, George Lakoff and Mark Johnson, argued that metaphors are central to how humans understand and interpret the world. Metaphors shape not only language but also thought and perception. They hypothesized that commonly used metaphors such as "things are looking up" and "argument is war" originated from the embodied experience that preceded spoken language (Lakoff & Johnson, 2003).

Metaphors allow us to understand one kind of experience in terms of another. By combining examples from seemingly unrelated domains, we can represent unfamiliar ideas. In his famous Sonnet 18, playwright William Shakespeare conveys the feelings he has regarding a Fair Youth he presumably is well acquainted with to a summer's day in order to help us develop an understanding about a person we do not personally know. Comparing someone to a summer's day gives the reader familiar with summer days a starting point. Summer days are associated with warmth and comfort. This common but also very personal understanding allows the reader to not just read the poem but experience it. The metaphor then allows Shakespeare to argue that the person in question is "more lovely and more temperate."

Metaphor exploration can be a powerful way to understand how prospective audience members think and feel and why they are attracted to certain types of content. Marketing professor and consultant Gerald Zaltman views metaphors as an essential tool for understanding the ways in which people perceive things, make sense of what they encounter, and guide their actions. Zaltman has identified seven grand metaphoric themes he has encountered throughout his research career. These are balance, transformation, journey, container, connection, resource, and control (Zaltman & Zaltman, 2008). He developed a unique methodology XR experience designers can apply to explore the users' world through meaningful metaphors.

Zaltman argues that since most of what truly influences people's behavior occurs below the level of awareness, researchers interested in discovering what motivates audiences need techniques that can explore the hidden knowledge of what people do not know or cannot tell they know. He discovered the power of imagery in ethnographic research while traveling in Nepal in 1990. In the impromptu project, Zaltman gave local residents disposable cameras with a goal of chronicling the trip from their perspective. He asked his informants to take pictures that would explain the experience of living in their villages.

The developed photographs intrigued Zaltman to the point of prompting him to return to the village to ask residents to explain the meaning conveyed in the pictures. He was amazed by the explanations villagers gave in response to his probes. They revealed ideas hidden deeply in the imagery that would have been very difficult and perhaps even completely unacceptable if they were expressed solely in words. For instance, the villagers framed their pictures to avoid showing people's feet in the photographs because in the Nepalese culture, bare feet signified poverty. Zaltman eventually developed a comprehensive research technique based on his experience. The method is known as the Zaltman metaphor elicitation technique (ZMET).

The ZMET technique is a three-step process. In the first step, participants are either assisted in selecting or asked to collect pictures representing their beliefs, thoughts, and feelings about the topic of interest. As in Zaltman's Nepalese study, images form a critical starting point for exploring deeply held, nonverbal thoughts and feelings.

In the second step, the researchers interviewed the subjects to uncover the deep structures of the participant's mental models that guide their thinking about a topic. The deep structures represent the unconscious orienting frames of the audiences' belief system that affect how they are likely to respond, process, and react to information and stimuli offered by the media makers. Once brought to the surface, these deep insights are relatively easy to understand and document as they manifest themselves in surface-level metaphors used in conversational language.

In the third step, the researchers group key images and ideas to identify higher-level frames and structures participants use to understand a given topic. The design team can use the idea clusters and frames to help them articulate resonant themes, topics, characters, and stories.

Projective Techniques: Collages, Camera Journals

Direct questioning such as surveys might be good sources of information consciously available to the audience, but the unconscious nature of mental models means that they cannot be elicited by direct questioning. The deeper structures of human cognition need to be teased out through indirect means such as metaphor elicitation and projective techniques, ethnographic observations, deep interviews, and language analysis.

Projective techniques aim to uncover the deep surface of the subjects' subconscious values, motivations, and responses. Projective techniques are well suited to allowing respondents to project their subjective viewpoints onto places, objects, or even other people. Design consultancy IDEO has used a research technique that asks the users to keep a camera journal of their experiences. Images for a camera journal can come from various sources such as smartphones, screen grabs, internet searches, etc.

In his design research practice, Peter has found a lot of fascinating user insight by asking test subjects to use their phone cameras to answer a series of questions, including: This is who I am; this is important; this makes my life difficult; this would make my life easier; and this looks beautiful. There are many other questions that are most relevant to the type of user information required that media makers can ask their prospective audience members to creatively learn about their particular life context.

Seeing the creative interpretation of the camera journal challenge demonstrates that people hold vast storages of unexamined, unconscious information about their needs and aspirations that can be rich sources of insights for media designers. Encouraging informants to communicate in metaphors has the potential to uncover a lot of otherwise inaccessible insights. Metaphors are a form of figurative thought that allows people to link ideas from seemingly unrelated domains to express otherwise elusive concepts. People use metaphors in everyday speech to express such deeply held and sometimes suppressed emotions as feeling a "heavy burden" or having a "clear picture" of something.

In the collage method, participants are asked to create a collage from a set of images either chosen by the participants themselves or provided by the research team. They are subsequently asked by the researchers to explain the significance of the images, their placement, relationships, and hierarchy. The method helps participants to verbalize complex and difficult-to-articulate concepts and feelings.

Projective Techniques: Self-Documentation

Asking prospective audience members to participate early in the media-creation process has several benefits. It allows for a long-term ethnographic study when the researchers cannot be present in person. Additionally, it allows the research team to see how the participants view their own life, challenges, and community connections through their own eyes. In a self-documentation study, we ask subjects to use cameras and journaling to follow specific instructions in answering questions about the context of their life most relevant to the media project. For instance, in a study of gamers, we might ask our subjects to turn on a webcam and record themselves at a time when they access their online community to play video games.

Self-documentation has been extensively used and promoted as a core design research method by product creation teams at IDEO. In a technique IDEO calls "Camera Journal," designers ask potential audience members to keep a visual diary of their impressions, circumstances, and activities related to a relevant experience. In the era of smartphones, this can easily be done by asking participants to use their phone camera or by giving them an inexpensive phone or a digital camera. The generated images can be combined with notation or interviews to prompt users to further reveal their subjective points of view, motivations, interests, pain points, and patterns of thought and behavior.

In a related technique called "A Day in the Life," designers explore the audience's life context by asking them to catalog activities related to a chosen topic as they go through a typical day. This method allows the design team to reveal any hard-to-anticipate issues inherent in the routines and circumstances audience members experience in the course of daily life.

Recruiting the right participants is critically important to the success of self-documentation studies. Our subjects must not only have the potential to reveal the insights we seek, they must also be willing and able to follow the instructions in taking pictures or recording journal entries.

As in the other types of user research, the chosen sample must include as full a spectrum of potential audience members as possible. This might include a special case of highly self-aware audience members known in the design community as extreme users. This special category of participants can help designers to discover key unarticulated needs, desires, and behaviors relevant to the rest of the audience segment because of their awareness of their needs and their ability and willingness to communicate about their experience. Effective recruitment of research participants might require access to the community insiders and other leads who can help the team gain access to the user community.

Self-documentation studies may take from a day up to a few months to complete, depending on your research focus. The study starts with an introduction meeting at which the researcher briefs the subjects. This is followed by a period of self-documentation, where the subjects record their experiences in their natural environments. It is important to give the subjects specific instructions in order to capture honest answers, which will lead to important insights regarding difficult-to-articulate problems, needs, and opportunities.

In the final stage, all the data gathered in the research is analyzed. Self-documentation studies, like ethnographic studies, generate insights from the bottom up. Key insights emerge from user observations and statements. Capturing bottom-up insights from the qualitative data gathered in self-documentation studies is best done using techniques such as affinity diagramming, which is outlined later in this chapter.

Ethnographic Observation

Ethnography is a form of observation method originally used in anthropology. Ethnography has increasingly gained popularity in user research in design and marketing. The immersive study of human cultures allows for an open-ended exploration of the rich context of people's natural environments.

While interviews are ideal tools for exploring meanings and attitudes, ethnographic research is an excellent tool for behavioral observation. In ethnographic studies, participants are observed in their homes, offices, or other places where they use products or perform relevant tasks. Ethnographic studies give design teams the potential to provide a deeper and more empathetic understanding of the users' lifestyles, cultures, habits, problems, processes, and workarounds.

Design researchers at IDEO have created a methodology that starts the design team exploring the area of opportunity by reviewing existing literature and secondary research. After this initial phase, designers engage with the users directly by observing them, asking them to help the team understand the pain points, and finally, walking in the users' shoes by trying things out for themselves.

An ideal ethnographic team includes a moderator, who maintains a dialog with the subject, a note taker, who records in written form all of the key observations, and a photographer or videographer, who records the session. Video is usually preferable, as relevant still frames can be extracted later from the footage.

As in all forms of a participatory type of research, recruiting the right participants for ethnographic studies is key to getting the right results. The informants not only possess the insights sought by the design team, but they must also be willing and able to share them. A good user sample must include a spectrum of potential users who can help to discover key unarticulated needs, desires, and behaviors valid for the rest of the audience. Successful recruitment requires access to the best candidates. Accessing good community contacts is key to finding the best candidates.

A typical ethnographic research session takes 2 to 3 hours. It starts with (1) an introduction and warm-up activities. The goal of this step is to put the subject at ease and to create rapport. Stage 2 entails pure observation, in which the subjects are asked to perform tasks in their natural environments without any interruptions from the researchers. The goal here is to capture honest signals, leading to important insights. The session ends with (3) a subject interview, which allows probing deeper using verbal questioning. The team meets immediately after the session for (4) a debrief to share what they have learned while all impressions are still fresh in their minds.

Using Language to Probe the Deep Structures of Human Experience

We hold many conversations in the course of our busy lives, but by focusing on what needs to be done, we often miss a lot of subtle details that are presented to us by people we converse with. Listening deeply to people can uncover a whole new world of understanding. To a trained researcher, words people use can serve as honest signals that represent their deep internal mental states. By asking good questions and carefully analyzing the answers, we can deepen our understanding of another person's hidden internal world. Interviews with audience members are frequently utilized by UX designers in the discovery phase of projects.

Researcher Sue Jones highlights a few critical issues involved in verbal ethnographic research. First and foremost, Jones cautions against the inclination to see the subjects as "cultural dopes," or simple organisms responding to external stimuli. She argues that all meaningful user interviews begin when the researchers recognize their informants as individuals who construct meaning to find significance in their life experience. Jones argues that people view events through a filter of a complex personal framework of beliefs and values that they have developed over a lifetime of living. The mental models people use in the process of surviving and thriving in the world consists of categorizing and explaining as well as attempting to predict events.

To Jones, trying to understand why people act the way they do is predicated on our ability to understand the meaning and significance people give to the actions they take in the world they have constructed. In order to comprehend another person's constructed reality, researchers need to not only ask well-thought-out questions but also ask them in a way that allows people to answer on their own terms. Asking people to express their mental models on their own terms also means to seek the depth of understanding that explores the rich context and substance of those internal mental structures.

Jones cautions against prestructuring the direction of an enquiry, which keeps the interview locked within the frame of reference of the interviewer rather than the person being interviewed. Getting good insights from interviews requires open-mindedness and flexibility to move with the interviewee where they need to go in order to explore in depth their psychic and social worlds.

The human mind excels at pattern-making and meaning-making. We are very comfortable piecing together meaning from fragments of information we receive. In our everyday conversations, we often sacrifice accuracy for speed while trying to cope with time constraints. Two particular tradeoffs of this need for speed are deep listening and fullness of communication. In a conversation, people recreate in their mind the experiences of the person they listen to by filling in information beyond the spoken surface structure.

For example, in the phrase "a woman bought a car," an experienced speaker of English will intuitively perceive two messages. The surface one will be identical to the uttered sentence, while in the second, deep structure, the listener will fill in the missing data to expand the statement. The deep structure will include the information that is not explicitly communicated about a woman buying a car from someone for some money. Buying something necessarily involves at least two parties and an exchange of money. Neither of these details was explicitly mentioned in the original sentence.

The application of formal linguistics allows researchers to examine the deep structure of any sentence and find out what pieces of critical information about the user's experience might be missing. In the context of user interviews, researchers do not want to imagine any missing information, as they might do in a conversation. Skilled interviewers identify such gaps as opportunities to ask interviewees to fill in any missing information in their own words. This can be best done by gently challenging the surface structures with questions that probe the deeper structure of the informant's actual life experience.

As we mentioned before, because of the complexity and sheer volume of incoming sensory data, people do not operate directly on the physical world around them but rather create representations, mental models or maps of the world, which they use to generate behavior. Language itself is one type of representational system that attempts to map the human experience. A language is a system of symbols a human being learns as a child growing up within a culture. In time, each of us develops our unique way of representing our particular individual experience in the world through culturally acquired language.

In their book *The Structure of Magic*, communication strategists Richard Bandler and John Grinder outlined a framework for analysis of verbal communication based on the principles of transformational grammar. They argued that since the same nervous system is responsible for producing both language and every other mental model of the world, the analysis of verbal communication using formal linguistic principles can provide an explicit framework for understanding human representation of reality (Bandler & Grinder, 2005).

The developer of transformational grammar theory, Noam Chomsky, proposed that natural human language operates on two levels. One of these represents an observable structure of a sentence, while the other represents meaning contained in the sentence. Two sentences, "John loves his dog" and "The dog is loved by John," differ in their surface form but express the same thought and, as a consequence, share a common deep structure of human experience.

After observing how master therapists Virginia Satir and Milton Erickson work with their clients, Bandler and Grinder identified three universal strategies people commonly use to model and present their version of reality. On the positive side, these universal mental shortcuts allow us to function in a massively complex world. Their downside is that they can occasionally impoverish our view of it.

The first universal is generalization, a mental process by which certain salient characteristics of a person's perception of reality become detached from the initial experience to become representative of an entire category of concepts, of which the original experience was just one example. Generalization is a useful mental process by which people learn from experience as in the "once burnt, twice shy" life lesson about touching stoves. Generalizations, however, can also distort and impoverish a person's model of reality by removing much detail from the original sensory experience. Not all stoves are hot all the time. Not all dogs bite, etc.

Deletion is a process by which people pay selective attention to certain portions of their experience and exclude others. Deletion is a useful tool, which allows people to focus, for instance, on navigating pedestrian traffic while walking on a busy sidewalk to exclude all kinds of other visual data such as colors of passing cars, billboard ads, or the clothing of people around them. Deletion reduces the world to dimensions we feel capable of handling. On the negative side, deletion can lead to people missing valuable portions of their experience and, as a consequence, misinterpreting important cues and interactions.

Finally, distortion is a mental process by which people make shifts in their perception of the original sensory data. We distort present reality in order to imagine events before they happen or rehearse actions before actually performing them. Distortion is the process that allows all types of imaginative creativity to occur. Einstein's famous thought experiments were mental distortions of his present reality. Authors, painters, and design and media professionals use distortion toward creative ends. But by using the same process, people can limit the fullness of their actual lived experience when they misrepresent words, actions, and events.

All of these processes shape our audience members' mental models and, at the same time, can become obstacles blocking our access to those models via surface level verbal questioning.

Applying Linguistic Intuitions to Enter the Users' World

Three aspects of linguistic theory can help designers to uncover the deep structures in which their audiences experience the world. These include well-formedness, constituent structure, and logical semantic relations.

Well-formedness allows experienced speakers of any language to make quick intuitive judgments about whether a group of words constitutes a meaningful sentence. Syntactic well-formedness refers to a correct order of words in a sentence. Semantic well-formedness refers to the way a group of words conveys a meaning. Noticing deficiencies in syntactic or semantic well-formedness offers an opportunity to ask the subject clarifying questions that bring the deep structure of their lived experience to the surface.

Constituent structure is a linguistic judgment experienced speakers make about which words in any given sentence go together as a unit. For instance, in the phrase "John heard a bell ring," the words "bell" and "ring" work together as a unit. Analyzing verbal accounts for constituent structure opens up possibilities for clarifying the details of the informant's reported experiences.

Logical semantic relations have five components:

1. Completeness describes the way a native speaker, when presented with a verb, will be able to determine the number of things involved in the action. For instance, the verb "hit" implies the existence of a person or a thing doing the hitting, a person or a thing being hit, and possibly an object used for hitting.
2. Ambiguity describes the way native speakers recognize a sentence as having two possible meanings. For instance, the phrase "Tom took Jim's hat off" can either mean that Tom took off a hat from Jim's head or that he took Jim's hat off his own head.
3. Synonymy refers to the ability of listeners to recognize two differently ordered sentences as having the same meaning. For instance, "John put on a hat" and "John put a hat on," while having a different word order, essentially mean the same thing.
4. Referential Indices describe the ability of native speakers to determine whether a word or phrase describes a specific object in their experience, such as "Tom's baseball hat," or a class of objects, such as hats in general.
5. Presuppositions refer to the ability of native speakers to deduce information about the speaker's experience from their statements. For instance, upon hearing the phrase "I damaged my car," the listener can infer that the speaker owns a car.

There are other linguistic transformations researchers can train themselves to identify. Synonymy means that a deep structure can be expressed by a number of different surface structures as in two synonymous phrases, "a woman bought a cat" and "a cat was bought by a woman." They both mean the same thing.

Ambiguity is the opposite of synonymy; it means that the same surface structure may be connected with different deep structures to represent more than one semantic meaning. For example, the phrase "chasing dogs can be dangerous" can either mean that the dogs who chase can pose danger to those being chased or that attempting to chase dogs can be a dangerous thing to do.

Indefinite noun phrase deletion can point to another instance of missing detail. In conversation, people frequently omit the indefinite noun phrase, such as "someone," from the sentence. Instead of saying "John talked to someone a great deal," a speaker might say "John talked a great deal." A deep structure–probing response in this case would be to inquire: Who did John talk to a great deal?

A complete logical semantic representation deletion can simplify a phrase, such as "John went to the movies and Bob went to the movies too" into "John went to the movies and Bob went too." The meanings of these appear the same even though an element of the deep structure regarding Bob going to the movies was deleted.

Nominalization is a particularly interesting form of linguistic transformation. It is a process by which a verb or action words such as "to decide" are changed in the surface structure into an event word like "decision." Unnoticed, nominalizations can be very misleading because they convert an active process into a passive event. There is a great deal of difference between hearing someone say "I decided" and "a decision was made." In the case of the former, we learn that the person we are speaking with decided on some course of action. We also realize that since action words represent fluid processes, the person's model of the situation includes the possibility of deciding to change that course of action. In the latter case, the surface structure (the verbal representation) presented to us does not tell us who made the decision or how this decision might be changed.

The referential index connects to a specific part of the model of the speaker's experience. Notice the difference between the phrase "John laughed" and "John laughed at Bob." "Bob" is the referential index that gives the listener a complete picture of whom or what John was laughing at. Deleting a referential index from the surface structure results in an incomplete representation of what actually happened.

Another useful concept of presuppositions describes statements that imply that certain other sentences must be true in order for them to make sense. For the sentence "John did not realize he had a flat tire" to be true, not only John's realization must be true, but John having a flat tire must be true as well. It is understandable that such details are frequently omitted in our everyday conversations, but in the case of attempting to accurately understand another person's lived experience, such omissions can give us an entirely false perspective. Listening deeply is a matter of practice.

The techniques we presented in this section will take time to understand and use in audience research, but applying them to in-depth interviews can give design researchers a more accurate ability to determine whether the participant's expressed surface structure is a complete and accurate representation of their life context.

Qualitative Data Analysis and Affinity Diagramming

The great strength of qualitative research is its flexible nature and its capacity to generate deeply intuitive but testable hypotheses from data. On the flip side, qualitative studies tend to produce a great deal of data with plenty of subtle detail. Finding actionable insights from qualitative data sources can be a time-consuming task.

Trustworthy data analysis needs to be well organized, rigorous, systematic, and iterative in nature. Researchers using qualitative approaches need to follow analysis strategies that balance the purpose of their study with appropriate data organization, conceptual concerns, and the inductive nature of the qualitative approach (Bingham, 2023).

Andrea Bingham suggests that qualitative data analysis comprises a five-step process of (1) organizing the data, (2) sorting the data, (3) understanding the data, (4) interpreting the data, and (5) explaining the data. In the first step, researchers need to develop codes that identify such attributes as data type, source, time, and location. In the second, they sort the data into categories that are most relevant to the research questions. The third step is dedicated to identifying the emerging themes. In the fourth, the team identifies the broader patterns emerging from the themes and begins to draft a statement of findings. In the last phase, researchers apply the theoretical framework to their findings and use it to explain the findings and draw inferences.

One of the most accessible and useful ways to process qualitative data in design research is known as affinity diagramming, which was proposed by cultural anthropologist Jiro Kawakita, who viewed the method as an alternative to Western quantitative methods in ethnography. In this four-step method, the research team captures their key observations by writing down single data units or "factoids" of data onto sticky Post-it notes. After capturing all observed "factoids," the team groups them into clusters of related concepts. Each cluster is labeled in a way that captures the essence of the affinity between the factoids

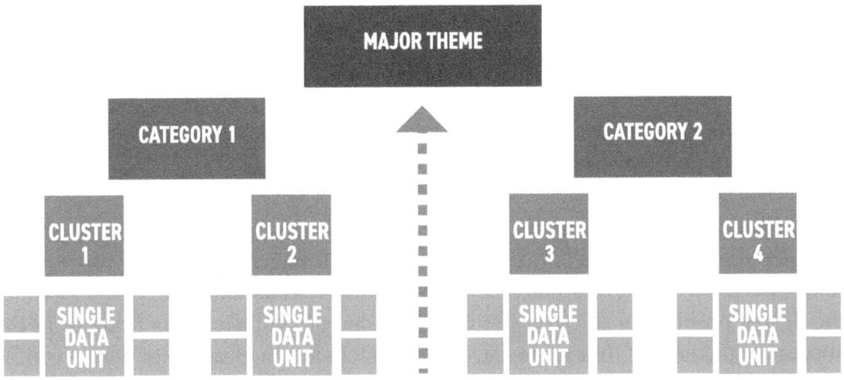

Figure 2.4 Affinity mapping

Source: Illustration by © 2024 Peter (Zak) Zakrzewski, based on Jiro Kawakita's Affinity Map method.

contained within it. In the third step, similar clusters are grouped into broader categories, which are labeled with different-colored Post-it notes. In the fourth stage, all categories are organized into highest-level classes with yet another color of notes. The emerging clusters, categories, and classes should explicitly articulate actionable insights.

Building Your Audience mMe Portraits

It is important to remember that audience profiles are abstractions not fictions. While they abstractly represent the entire audience segment, they are built on a firm foundation of user research. After the deep dive into the world of our audience, we are now ready to come to some actionable conclusions expressed as detailed audience mMe profiles (aka personas). There are a number of approaches to building user persona profiles. The chosen model will depend on the type of project the team is working on and the budget and time available for research.

Still, having a realistic glimpse into the world of the audience is always preferable to making wild guesses and relying on assumptions. Excellent coverage of this topic can be found on the Interaction Design Foundation and Nielsen Norman Group websites. The entire persona creation process has been also outlined by Robert Reimann, Kim Goodwin, and Lane Halley at Cooper in their seminal UX book *About Face* (Cooper et al., 2014). There are at least seven common categories of information a good audience profile should contain:

A. Biographical data, such as name, age, gender identification, aspirations (show images including life context and important metaphors)
B. Key insight, clearly articulating what the audience segment needs or considers resonant, expressed as a quote that might give the design team more insight into the audience segment's mindset
C. Unmet needs, goals, motivations, and concerns they would have in regard to using the application
D. Habits and known behavioral patterns
E. Experiential and interaction needs (visceral, behavioral, or reflective)
F. Relevant aptitudes (novice, intermediate, advanced), skills, and expectations the audience segment has in the area of the app the team is developing
G. Relevant context of use (when and where)

All of these variables can be combined into a single layout for every audience segment the team creates for the project. As mentioned before, your team might find it necessary to develop more than one persona to address different aspects of the design problem or to cover the diverse range of characteristics needed for the success of the project. In Figure 2.5, we have included an example of what a profile layout could look like when filled with the research summary. Even though we have omitted actual photographs for privacy reasons, we recommend using photographic images of actual users to add reality to the portrait.

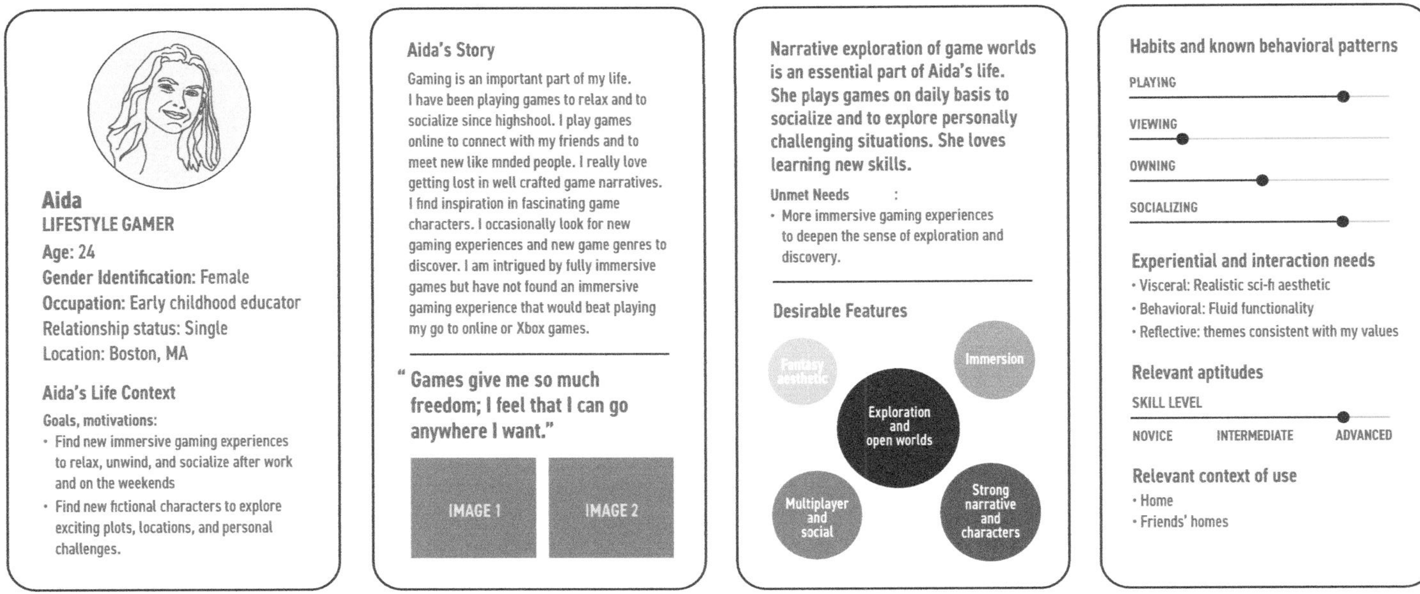

Figure 2.5 Primary audience mMe profile

Source: Concept and illustration by © 2024 Peter (Zak) Zakrzewski.

Summary

In this chapter, we explored the critical importance of understanding the audience's needs, behaviors, and life contexts when creating immersive media experiences. We began by emphasizing that media consumption is an active, cognitive process in which audiences reconstruct media elements into coherent experiences, highlighting the necessity to consider audience participation as a core aspect of immersive experience design.

We argue for a rhetorical approach to media creation, advocating for design that respects audience agency and choice. We propose various qualitative research methods adapted from psychology, anthropology, and design to help you understand your audience. We then introduced "mental model explorations" (mMe) as a technique for empathizing with and understanding the audience's psychological and contextual realities.

Leveraging frameworks like Maslow's hierarchy of needs and Yu-Kai Chou's Octalysis framework (2024) for understanding human motivation in interactive systems results in more effective identification of audience goals and motivations. We introduced the concept of "strategic empathy," which is a distinguishing feature of design thinking that allows design teams to gain deeper insights into audience needs, aspirations, and values.

We highlighted empathy-building research methods, including metaphor elicitation, self-documentation, ethnographic observation, and language analysis, each offering a unique lens for understanding the audience. We presented affinity mapping as a useful method to organize and interpret qualitative data.

We also outlined the construction of audience mMes as a six-stage process that includes selecting target user categories, determining behavioral variables, and synthesizing research into comprehensive profiles. We shared these design research tools to advocate for a rigorous, empathetic approach to understanding audiences in immersive media design that let design teams move beyond superficial audience models to grasp the complex, nuanced realities of how people interact with media to create experiences that resonate deeply, align with their motivations, and fulfill their unmet media consumption needs.

With this clear vision and understanding of your audience, we are ready to craft your value hypothesis, which is the next chapter's topic.

References

Bandler, R., & Grinder, J. (2005). *The structure of magic: A book about language and therapy*. Science and Behavior Books.

Bingham, A. J. (2023). From data management to actionable findings: A five-phase process of qualitative data analysis. *International Journal of Qualitative Methods*, 22, 16094069231183620. https://doi.org/10.1177/16094069231183620

Chou, Y.-K. (2024). Octalysis framework. *The Octalysis Group*. https://octalysisgroup.com/framework/

Cooper, A., Reimann, R., Cronin, D., & Cooper, A. (2014). *About face: The essentials of interaction design* (4th ed.). John Wiley and Sons.

Johnson-Laird, P. N. (1995). *Mental models: Towards a cognitive science of language, inference, and consciousness* (6th print). Harvard University Press.

Korzybski, A. (2005). *Science and sanity: An introduction to non-Aristotelian systems and general semantics* (5th ed., 3 p.). Institute of General Semantics.

Kouprie, M., & Visser, F. S. (2009). A framework for empathy in design: Stepping into and out of the user's life. *Journal of Engineering Design*, 20(5), 437–448. https://doi.org/10.1080/09544820902875033

Lakoff, G., & Johnson, M. (2003). *Metaphors we live by*. University of Chicago Press.

Magliano, J. P., & Zacks, J. M. (2011). The impact of continuity editing in narrative film on event segmentation. *Cognitive Science*, 35(8), 1489–1517. https://doi.org/10.1111/j.1551-6709.2011.01202.x

Quesenbery, W. (2004). Balancing the 5Es: Usability. *Cutter IT Journal*, 17(2), 4–11. www.wqusability.com/articles/5es-citj0204.pdf

Senge, P. M. (2006). *The fifth discipline: The art and practice of the learning organization* (Rev. and updated ed.). Random House Business Books.

Smeenk, W., Tomico Plasencia, O., & van Turnhout, K. (2016). A systematic analysis of mixed perspectives in empathic design: Not one perspective encompasses all. *International Journal of Design*, 10(2), 31–48.

Zakrzewski, P. (2022). *Designing XR: A rhetorical design perspective for the ecology of human+computer systems*. Emerald Publishing Limited. http://doi.org/10.1108/9781802623659

Zaltman, G., & Zaltman, L. H. (2008). *Marketing metaphoria: What deep metaphors reveal about the minds of consumers*. Harvard Business Review Press.

3 Developing Your Value Hypothesis

The Beautiful Problem

Particle physicist Savas Dimopoulus observed that one of the key characteristics of great creative minds is "purely an artistic ability to discern what is a good idea" from "what is a beautiful idea," a problem "worth spending time on" (Alayón, 2018). He argues that the essence of creative success in any endeavor lies in finding "a problem that is sufficiently interesting, yet sufficiently difficult" to the point that "it has not yet been solved, but the time for solving it has come now" (2018).

The emerging immersive media landscape is full of beautiful problems to be solved. What will differentiate media makers entering this field will be the creative ability to discern what is merely a good experience from what is a beautiful and meaningful immersive experience that is truly worth working on. Most importantly, it will be an ability to recognize sufficiently resonant and sufficiently challenging audience needs that have not yet been addressed, but the time for addressing them has come now.

> *What differentiates media designers is a creative ability to discern between what is a good idea and what is a beautiful idea worth spending time on. It is about finding an unmet audience need that is sufficiently resonant yet sufficiently difficult that it has not yet been addressed, but the time for addressing it has come now.*
>
> BASED ON: PROFESSOR SAVAS DIMOPOULOS

We believe that choosing the right value proposition to deliver to one's audience is one of the most important steps of the digital product or media creation process.

DOI: 10.4324/9781003442462-7

What a Value Hypothesis Is and Why You Need One

Eric Ries, a proponent of the lean start-up methodology, suggests that the best strategy to deal with the uncertainty inherent in creating innovative products and experiences is to think of the entire process as a series of experiments. The philosophy behind this iterative design process suggests that one audience test is better than a hundred "expert" assumptions. Ries refers to all significant assumptions a creative team makes as "leap-of-faith" hypotheses (Ries, 2014).

The four main hypotheses involved in the creation of new value comprise (1) the audience hypothesis (who is the audience? Can they relate to your vision?); (2) the value hypothesis (Does your project's theme resonate with the audience? Will the experience perform a job-2B-done in the life of your audience?); (3) the usability/feasibility hypothesis (How will the system perform? How immersive will it be?); and (4) the growth hypothesis (How will it spread? How big can it scale?).

The process starts with the design team's hypothesis about a particular audience with an unmet audience need. An effective hypothesis has to be specific, measurable, and falsifiable. As discussed before, all leap-of-faith hypotheses are validated by building prototypes, conducting experiments, and getting real-world feedback. A compelling value hypothesis is essential for any media project hoping to resonate deeply with the audience. A well-stated value proposition can not only unite and guide the design team but also become the starting point in inviting the audience to participate in the media franchise.

In creating media projects, we can learn a lot from start-ups who use the agile approach to address resonant human needs to build value for all stakeholders. Start-ups go through the process of articulating a testable value hypothesis, validating the hypothesis, and, finally, using the validated value hypothesis to build compelling experiences that, if done right, spread like wildfire.

Jobs to Be Done: Why Focusing on Value Is the Foundation of Immersion

A strange thing happened in March 2020. Amid the COVID-19 pandemic lockdowns, Netflix, the tech start-up that took over Hollywood, signed up 16 million new subscribers instead of the projected 7 million (Kafka, 2020). All 9 million of the additional global subscribers showed up at the exact moment when governments around the world required their citizens to stay home. The new viewers did not just sign up for the streaming service; they proceeded to engage in marathon viewing sessions known as binge-watching.

The desire to binge-watch our favorite content can be described as viewing multiple television program episodes in rapid succession. It is typically accomplished by streaming services like Netflix, Hulu, or Amazon Prime. Binge-watching has given audiences a higher degree of viewing autonomy than traditional media's scheduled television show format. The trend started with the advent of DVDs, which gave consumers more independence in the choice of shows and more control over their media consumption (Rahman & Arif, 2021). Audiences who binge-watch consume between two and six episodes of the same show in one sitting. So why would a reasonably mature streaming company like Netflix explode its viewership during one of the most significant pandemic outbreaks of the last 100 years?

As media makers, we sometimes might forget that deep down, all hit movies, shows, and theater performances succeed by satisfying real psychological needs that often provide intangible value to their audiences. The originator of the disruptive innovation concept, Clayton Christensen, coined a term for this value creation concept. He referred to the value created for the user in the innovation process as the "job-to-be-done" (Christensen et al., 2016).

Every successful product or service does some important job in the life of a real flesh-and-blood person (in contrast to quantitatively derived demographic market segments and other abstractions that obscure the real insights needed for successful experience design). Christensen argued that as needs arise in our lives, we seek products and services and hire them to do those tasks. We are not suggesting that XR content can be equated with consumption of bagels, milkshakes, and chocolate bars, but it is important to remember that at the core, all media experiences have to provide some kind of value to their audiences.

What unique value did Netflix provide to find it at the center of such a massive behavioral shift? Researchers found that as early as January 2020, during the onset of the COVID-19 pandemic, general anxiety and health anxiety, in particular, emerged as a major theme across social media platforms such as Reddit. Many people's personal challenges significantly increased during that time. Posts related to such issues as economic stress, isolation, boredom, and loneliness more than doubled during the pandemic (Low et al., 2020).

Media audiences engage in binge-watching to satisfy their need for cognition and as a means of seeking sensation, enjoyment, control, and experiencing a sense of fandom. Binge-watching can also be a highly social activity, as individuals are especially drawn to binge-watch shows that are recommended by others in their network (Shim & Kim, 2018). During the pandemic, Netflix was one place where audiences could escape to be immersed in gripping narratives, explore exotic locations, and be part of vibrant social interactions.

It is usually easier to assess the success of a value proposition in hindsight than it is to forecast it into the future, but the Netflix example shows how there is a direct connection between the fulfillment of deep psychological needs and the success of media experiences. Being clear on the value hypothesis from the start of the project allows the team to integrate all of their efforts and resources with a single focus on doing the job that the core audience needs to be done. Crafting a value hypothesis that deeply resonates with the core audience is the necessary next step toward the wide adoption of a proposed immersive experience.

Articulating a Compelling Value Proposition

On October 23, 2001, Apple's CEO Steve Jobs stood on a stage in Cupertino and announced a new digital device that could store most people's entire music library right in their pocket. Most people born after 2001 might have a difficult time imagining the impact of the value proposition of a media device that could hold 1,000 songs and be carried around in one's jeans pocket at a time when most people listened to their music on CDs that held 15 songs each and required a backpack to carry.

Apple's iPod was a technological breakthrough that used ultra-thin hard drives, MP3 compression, and high-capacity batteries to deliver an experience of walking around with a choice of a personal soundtrack. To music fans around the world, the iPod was magic. For immersive media experiences, as it was for the iPod, that magic can be found by offering a compelling value proposition that bridges the gap between the feasibility of the current disruptive media technology and the audience's unmet needs.

For an AR application designed to help with retail shopping, for example, the value proposition cannot merely focus on the novelty of AR technology. Instead, it needs to simplify and increase the engagement of the shopping experience by allowing customers to find and review products in real time or to visualize them in their own homes before making a purchase decision. Such a proposition speaks directly to the common problem of uncertainty in online shopping, making the technology immediately relevant to the user's actual shopping needs and not just the engineering prowess of the design team.

Demonstrating the value becomes crucial to move beyond the initial novelty to offer tangible or intangible benefits. An immersive VR training program for surgeons, for instance, provides more than an innovative training method; it significantly reduces the risk of errors in surgeries, which could potentially improve patient outcomes and save healthcare institutions millions in malpractice costs. By demonstrating how trainees can reduce error rates, the value proposition becomes quantifiable and grounded in concrete benefits, transforming the technology from a novel idea to a must-have tool.

Value proposition differentiation is especially vital in the crowded spaces of new applications and experiences. Consider an immersive VR museum experience that not only allows users to explore art collections from their homes but also integrates interactive elements, such as curated tours by artists and historians, and the ability to create personal art galleries. Such a unique blend of education, interaction, and personalization can set the experience apart from many other virtual museum tours, highlighting the unique selling proposition of the platform. It's not just about viewing art; it's about engaging with it in a way that was previously impossible.

By focusing on solving specific user problems, providing clear, quantifiable benefits, and highlighting unique features, your team will be able to craft value propositions that not only demonstrate the value of your offerings but also establish deep, lasting connections with your audience.

Here are six brief steps we recommend for the value proposition brainstorming session: (1) Name your project. The name you choose needs to be unique and easy to pronounce and remember. (2) State the unmet audience need. What is your audience currently missing in their media offerings? (3) Identify the audience segment. (4) Without getting bogged down in details, describe your solution and key user benefits. (5) Articulate how your solution will help the audience achieve their goals

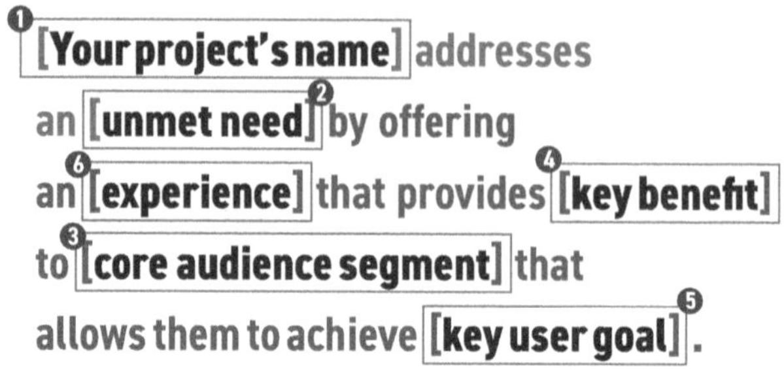

Figure 3.1 Anatomy of a value proposition statement

Source: Concept and illustration by © 2022 Peter (Zak) Zakrzewski.

(job-2B-done). (6) Describe how compelling and engaging your experience will be for the audience. Edit the entire statement as a short (one minute or less) elevator pitch you can share with your funders and partners and in your promotional materials.

In crafting your value proposition statement, make sure the value you offer to your audience stems directly from your project's vision. Does your vision address a hair-on-fire problem or a hard-fact problem, or are you proposing a vision of the future? As we suggested in the vision chapter, the components of the value proposition for each of these will be quite different.

In the case of a hair-on-fire problem, your audience members are aware of their need and are actively looking for solutions while comparing existing ones. A hair-on-fire problem value proposition has to tout a best-in-class solution in order to be noticed. The application and its experience cannot just be faster or cheaper; it must be truly differentiated.

In the case of a hard-fact problem, you must convince the potential audience that the problem is solvable and that adopting it is worth changing their current behaviors. The value proposition must be compelling enough to be worth making a change.

If your team is offering not just a product or an experience but a vision of the future, your value proposition must overcome not just a force of habit but a disbelief. Your value proposition must paint a picture of a ready-to-use, engaging, and complete ecosystem. Vision-of-the-future products succeed only when they make the new paradigm irresistible. This long-term value proposition presents a grand challenge and an opportunity to shape the future and to reshape the world.

Testing Your Value Hypothesis

Alex Pentland argues that the main task of being engaged in scientific endeavors starts with developing a compelling story about the world and then making a concentrated effort to test it against reality. A common misstep in digital product innovation is assuming that you know what your users want without actually interacting with them. This common pitfall is referred to as a solution in search of a problem.

Real-world testing is indispensable for grounding your hypotheses in actual user experiences and needs. Without this reality check, even the most brilliant ideas are likely to fall flat.

Take Dropbox as an example. The problem of file synchronization was accepted as a hard fact of life, and the proposed solution was so complex that the founders initially struggled to explain the concept to potential funders. The breakthrough came when the team posted a simple demo video showing Dropbox in action on YouTube. The Easter egg–filled video was actually a pitch for tech-savvy beta testers that could prove the value hypothesis of the new app. The video generated a beta tester wait list of 75,000 almost overnight. This value hypothesis validation was a crucial proof of concept that showed the market viability to potential funders.

By testing the value hypothesis, you seek the alignment between the minimum viable XR experience being built by the design team and the core audience segment you identified for the project. XR media makers can test their value hypotheses by building proofs of concept and using the hypothesis-testing loops we discussed in the vision chapter. The lean start-up approach leverages the scientific hypothesis-testing process that recognizes the value of abductive reasoning, a form of logical inference that seeks novel conclusions from a set of observations. Leap-of-faith assumptions need to be vigorously tested as early and frequently as possible in order to ensure that what we are creating resonates with the audience and helps them achieve a "job to be done."

The process of discovering and building a compelling and humane immersive experience is an outcome of many experiments. Iterative testing cycles, combined with a deep understanding of design thinking and strategic business principles, can guide media makers through the "pivot or persevere" decisions that define start-up growth. This disciplined approach to innovation—validated learning through direct user engagement—ensures that when you do decide to pivot, as Twitter famously did in its early days of being podcast focused, it is based on solid evidence of what the users truly value.

By uniting a creative team around a clear vision and value proposition, you can turn innovative ideas into experiences that users not only try but integrate into their lives, providing continuous feedback that fuels further refinement and value creation. This loop of building, measuring, learning, and iterating is the backbone of transforming challenging problems into opportunities for growth and impact.

Think of testing your ideas like going through a superhero checklist to make sure your gadgets are ready for action. Here are the essential questions you need to ask, made simple. (1) What problem does our product solve? (2) Does our audience really want or need this? (3) Is it fun, engaging, or interesting to use? (4) Can we deliver this experience as envisioned? (5) Do we have the tools and skills to build it? (6) Is it something new, different, and differentiated in a way our audience can relate to?

The important decision involves committing to a specific, testable value hypothesis. With this in hand, the team needs to keep asking: How do we know if our new immersive media app is as awesome as we think it is? For each of these questions, your team will need to think carefully and try out your ideas to see if they actually work in practice. This level of discipline will help to make sure that your cool new idea is not just fun to dream up but can actually become something real that everyone loves to use.

Summary

In this chapter, we delved into developing your value hypothesis within the context of your overarching project vision and with a solid understanding of your audience. We recommend an iterative and experimental design approach, underscoring the significance of creating a value proposition that is deeply resonant with your audience, which can be created through a series of methodical experiments designed to test and refine the core assumptions of your design team.

Each significant assumption made by your creative team is considered a "leap-of-faith" hypothesis. The process involves developing hypotheses about the XR experience, which must be specific, measurable, and falsifiable, and then testing these through prototypes and real-world feedback to ensure resonance with the audience. To help you navigate the value proposition process, the chapter outlined a structured approach based on the lean start-up methodology.

We also reintroduced Clayton Christensen's concept of the job-to-be-done that highlights that every successful product, service, or media experience fulfills a significant "job" in a real person's life, addressing their needs beyond mere demographic segments.

A compelling value proposition serves as another unifying guide for the design team and an invitation for the audience to engage with your media experience. You will undergo the stages of articulating a testable value hypothesis, validating it, and leveraging the validated hypothesis to generate a compelling value proposition.

Now that you have a working draft of your vision, an audience profile, and a solid value proposition, you and your team are ready to create a context scenario that outlines the specific details of the participant's experience.

References

Alayón, D. (2018, January 24). Quantum computing. *Medium*. https://medium.com/future-today/quantum-computing-913edcaca90

Christensen, C. M., Hall, T., Dillon, K., & Duncan, D. S. (2016). *Competing against luck: The story of innovation and customer choice*. HarperBusiness.

Kafka, P. (2020). The pandemic has been great for Netflix The streaming service added another 10 million customers in the second quarter of the year. *Vox*. https://www.vox.com/recode/2020/7/16/21327451/netflix-covid-earnings-subscribers-q2

Low, D. M., Rumker, L., Talkar, T., Torous, J., Cecchi, G., & Ghosh, S. S. (2020). Natural language processing reveals vulnerable mental health support groups and heightened health anxiety on Reddit during COVID-19: Observational study. *Journal of Medical Internet Research*, 22(10), e22635. https://doi.org/10.2196/22635

Rahman, K. T., & Arif, Md. Z. U. (2021). Impacts of binge-watching on Netflix during the COVID-19 pandemic. *South Asian Journal of Marketing*, 2(1), 97–112. https://doi.org/10.1108/SAJM-05-2021-0070

Ries, E. (2014). *The lean startup: How today's entrepreneurs use continuous innovation to create radically successful businesses*. Crown Business.

Shim, H., & Kim, K. J. (2018). An exploration of the motivations for binge-watching and the role of individual differences. *Computers in Human Behavior*, 82, 94–100. https://doi.org/10.1016/j.chb.2017.12.032

Section IV

Ideation and Inscription

4 Creating the Context Scenario

Computational Worldmaking: How Imaginative Scenarios Can Inspire the Future

The legendary personal computing pioneer Alan C. Kay told this insightful story from his XEROX Palo Alto Research Center (PARC) days. The parent company executives who were visiting the lab were apprehensive about the impact of computer technology on the future of their copier business. They kept asking Kay: How can our company be affected by the personal computers? One day, exasperated, Kay told them to stop worrying about the future and start inventing it. He argued that the time of rapid technological changes was the best time in the history of human civilization to be proactive about the future (Hiltzik, 1999).

We agree wholeheartedly with Kay's sentiment, with one exception. We believe the best way to predict the future is to design it. The design approach to innovation goes beyond technological invention to consider all aspects of disruption, including its psychological, social, and economic impact. Design as a methodology offers immersive media makers an entire toolbox of methods for imagining and building the future.

One of the most impactful of these is the context scenario that leverages the power of narrative as a design tool. People who have made the greatest contributions to the future of things have not been inventors but writers who inspired the inventors. The term and concept of "genetic engineering" was first introduced by the science fiction author Jack Williamson in his 1951 novel *Dragon's Island*. The book was published two years before James Watson and Francis Crick showed a DNA molecule's double-helix structure.

The idea of the cell phone first entered the imagination of its inventor, Martin Cooper, while he was watching *Star Trek*. When he saw Captain Kirk talking on his communicator without being wired into a wall and without dialing, Cooper did not see it as a fantasy, although it was to the rest of the world. To him, it was an objective.

Research conducted by archeologist Michael Brian Schiffer shows how one media technology—the portable radio, introduced in the late 1930s—was not a product of straightforward technological inventions but a phenomenon that emerged out of "a constellation of interconnected social and technological factors"

DOI: 10.4324/9781003442462-9

(Schiffer, 1991, p. xv). This constellation included the writings of such technomancers as the father of American science fiction, Hugo Gernsback, who, in the early 1900s, founded a string of magazines that fired the imaginations of American children and inspired them to seek technical careers. Gernsback coined the English term "television" in *Modern Electrics* as far back as 1909. The masthead of his seminal science fiction magazine, *Amazing Stories*, boldly proclaimed that its "Extravagant Fiction Today" would become "Cold Fact Tomorrow" (p. 36). Throughout the 1930s Great Depression era, magazines such as *Popular Mechanics* and *Popular Science Monthly* stoked the fire of electronic hobbyists' imaginations with illustrations and descriptions of people driving, cycling, and camping while listening to the radio.

Finally, in the fall of 1938, American electronics company Philco opened the floodgates by introducing the first portable radio aimed at mass-market consumers. Based on Sylvania's new battery tubes, the Philco 71T was a small suitcase covered with airplane cloth. It was a far cry from the handheld radios kids carried to the beach in their pockets in the 1950s, but its wide adoption proved the use case for the entire product segment. Many engineering geeks have been moved to action by extravagant fiction and visions of the future presented in books and movies.

One of the visionaries most closely associated with XR is American writer Neal Town Stephenson, whose work of speculative fiction titled *Snow Crash* introduced the term "metaverse" as a name for a virtual reality–based Internet resembling a massively multiplayer online game (MMO). Stephenson's Metaverse was populated by user-controlled avatars and daemons. Social status in the Metaverse could be attained by the sophistication and the lifelikeness of one's avatar and by having access to restricted areas such as the exclusive Metaverse club known as Black Sun.

Avi Bar-Zeev, who designed Google Earth, was so taken with Neal Stephenson's ideas that he tried to arrange a meeting with the author while working on the Keyhole app suite, which was later turned into Google's mapping technology. Stephenson, who apparently considered his metaverse to be a vision of dystopia, did not oblige.

The secret to creating compelling future projections lies not in technically detailed descriptions of the underlying technologies but in portraying vivid interactions that empower the protagonists to do extraordinary things. Software designer Alan Cooper, the pioneering proponent of context scenarios, suggests treating most enabling elements in the scenario as "magic" in order to capture the true transformative vision of the compelling user experience first.

Scenarios of Possible Futures: The Apple Knowledge Navigator

For anyone thinking that Apple's iPhone and iPad were the company's first forays into smart personal devices, it might be surprising to learn that the company has been envisioning such products since the 1980s. Rich context scenarios are among the most inspiring tools for envisioning the future and designing compelling user experiences.

There might be no better examples of such envisioning than Apple's 1987 *Knowledge Navigator*. The companywide project used the transformative power of a context scenario to propose a future of personal computing that ultimately came to life in the form of an iPad. This far-enough-into-the-future but feasible-enough-to-build scenario took a team of experts from all company areas. The project was sponsored by the director of Apple's higher education marketing group, Bud Colligan. Hugh Dubberly and Doris Mitsch of Apple Creative Services wrote and creatively developed the resulting video presentations. Technical and conceptual input was provided by the members of Apple's Advanced Technologies Group, including Mike Liebhold and the former Xerox PARC lead Alan Kay. Gavin Ivester and Adam Grosser at Apple Design created the industrial design prototype of the *Knowledge Navigator* (Colligan, 2011; Dubberly, 2007).

The resulting concept video features a knowledge worker—a college professor—going through several work- and personal life-related tasks, which include preparing for a lecture, remotely collaborating with a colleague, and planning a birthday party. Despite looking very realistic decades later, all user interactions were simulated with Hollywood-level advanced special effects. The context scenario accurately predicted such future technologies as touch screens, web browsers, remote collaboration tools, and AI assistants.

At their core, context scenarios require a technologically savvy group of creative individuals with a vision, imagination, and no more than a pen and paper. Scenarios can be easily communicated with written stories, storyboards, photos, or videos. What is critically important is that scenarios can consequently be put through a "breakdown" process by various subject matter experts, such as industrial designers, developers, and 3D artists, to determine the necessary information architecture and digital assets needed to execute the actual product. We recommend starting the ideation phase of immersive media projects with a narrative description of the proposed experience.

The Expanded Chain of Persuasion: The Six Stages of Immersive Transportation

Marie-Laure Ryan (2015) draws a comparison between the old tradition of immersion we experience while reading stories presented in books and the mediated immersion we experience in virtual worlds. Both the printed books and XR worlds are products of imagination and technology.

However, the comparison mostly ends there. Reading a book is purely a mental phenomenon. A book might transport us to exotic locations while our sensory channels remain situated within our physical space. Reading a story is a product of our own imagination based on visual input of black type printed on a page or a screen.

Immersive media, on the other hand, have the capacity to tell stories through eight different sensory channels. Within XR environments, our entire bodies can interact with sensory inputs provided by the media makers. Despite the difference in the complexity of the underlying technologies, we find Ryan's application of Richard Gerrig's folk theory of narrative immersion very useful in describing the levels of immersive engagement audiences can go through in the process of immersive experiences.

Ryan outlines Gerrig's folk theory of immersion as a six-step process.

1. In the first step, the participant is transported into a textual world that temporarily reshapes their identity.
2. This narrative sleight of hand is accomplished by some means of "transportation."
3. The immersion into the story world is a result of the participant performing the act of turning textual directions into an internal representation of the imagined world.
4. In step four, the participant travels some distance from their world of origin and must adapt to the laws of the new narrative world, which differ from the laws of their native reality.
5. The idiosyncratic laws of the textual world overrule the principles of the participant's native reality.
6. In the final step, the participant returns to the world of origin changed by the journey (2015).

We add Gerrig's and Ryan's theoretical frameworks to augment the chain of persuasion introduced in Chapter 1. The expanded model of immersive persuasion now addresses the XR technology as the vehicle by which the entry into the immersive world is accomplished. It also highlights the critical step when the participant internalizes the rules of the XR world in the process of navigating and mastering its environment.

The revised process starts when:

1. The protagonist is attracted by the relatability and resonance of the XR experience.
2. The protagonist enters into the immersive world by means of a well-designed XR "magic bus"—the XR system whose technology forms the supportive structure for the immersive experience. This step asserts the critical importance of the feasibility of technology, which we cover in the next chapter.
3. Immersive presence is a result of the protagonist performing the act of turning multimodal inputs provided by the system into an internal representation of a virtual world as conceived by the media makers but co-created by audience participation.
4. In the next step, the participants travel some distance from their world of origin and start to adapt to the laws of the new world, which differ from the laws of their native reality. The virtual world temporarily begins to reshape their

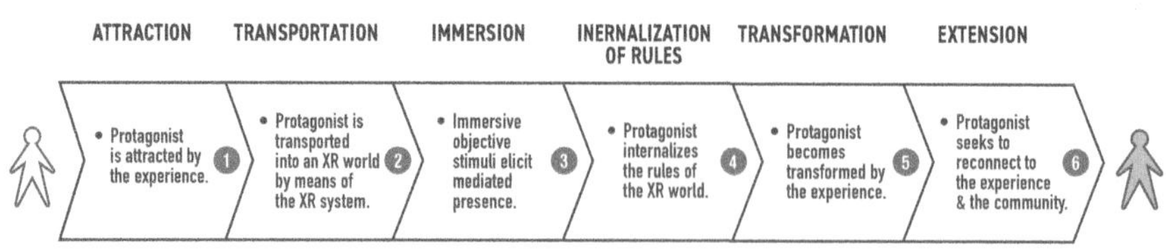

Figure 4.1 The chain of immersive persuasion

Source: Concept and illustration by © 2024 Peter (Zak) Zakrzewski, partially based on Richard Gerrig's Folk Theory of Narrative Immersion.

identity and worldview as they internalize its rules. Internalization of rules is an important principle of game design, which we cover in Chapter 7 when we discuss XR environments as complex systems.

5. As the participants internalize the rules of the virtual world, these idiosyncratic laws (system rules) override the principles of the participants' native reality to the point that they become transformed by the experience.
6. In the last step, the participant returns to the world of origin changed by the journey. At this stage, they seek to reconnect to the experience and to the community. Truly immersive experiences turn audience members first into participants and then into protagonists in their own journey of discovery.

The Four Degrees of Immersive Engagement

Another critical factor to consider when designing immersive experiences is the optimal level of engagement designers aim to elicit from the participant. Marie-Laure Ryan (2015) provides another useful framework that makes a distinction between four degrees of immersive engagement progressing from (1) distracted audience remaining vulnerable to the distracting stimuli of external reality to (2) imaginatively involved audience who are part of the immersive world but remain able to contemplate it with aesthetic or epistemological detachment to (3) entranced audience so completely caught up in the immersive world that they lose sight of anything external to it and finally to (4) the audience so addicted to the immersive experience that they lose the capacity to distinguish the immersive world from the actual one.

As in the case with all forms of subjective experience, the influence the design team can exert is limited to the provision of external stimuli. The impact of various immersive experiences on different audience members will vary greatly, and hence, it is not predictable. But the design team can use the described framework to calibrate the optimal experience appropriately. With the exception of the addicted extreme, which is not a desirable emotional state, the other degrees of immersion need to be strategically targeted based on the type of experience being designed.

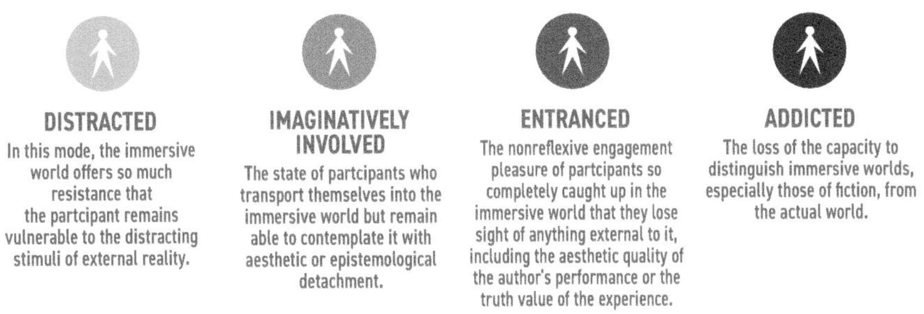

Figure 4.2 Four degrees of immersive engagement

Source: Illustration by © Peter (Zak) Zakrzewski, based on Marie-Laure Ryan.

For AR applications, for instance, that require the user to stay present within their physical environment, imaginative involvement and even a distracted sense of immersion might be perfectly appropriate. Teams working on VR or MR games and entertainment experiences would very likely need to aim for maximum mediated presence and entrancement.

Context Scenarios

After investing the time and resources to understand your audience, you will need to bridge the research–design gap to make sure that the immersive experience you are creating is as rich and rewarding as possible for your chosen audience. Scenario writing is different from other forms of writing, such as fiction, nonfiction, technical, and advertising copywriting because as part of design planning, it must fulfill a number of strict criteria.

First and foremost, the scenario must be strategically consistent with the project's vision. It must be practically actionable, clear, and concise. When writing XRXD context scenarios, you explore the user's life context, needs, and behaviors as the major motivation for all of the key touchpoints at which the users interact with the system. In your context scenario, you will explore a number of questions from the audience perspective that include the participants' ability to know where they are, what they need to know at any given moment, and what options for action are available to them.

Cooper et al. suggest that the bridging process between the audience discovery and building stages of the project has four distinct phases. In the first phase, the design team develops a story that imagines one entire path through the desired audience experience. In the second phase, the design team "breaks down" the scenario to extract the design requirements, much like the various departments within a film production would want to extract the details about the costumes, props, and locations from the film script. In the third phase, the team defines the scope of the necessary technology, digital assets, and fundamental interaction framework, which together will form the structure of the user experience. We will cover the second and third phases of this process in the next chapter. In the fourth phase, the framework created by the scenario is iterated and completed with ever-increasing amounts of design detail (Cooper et al., 2014).

The XRI Omniverse

The mixing of the previously separate domains of XR, AI, IoT, and robotics is an outcome of a steady path marked by an evolution away from personal computers actively operated by users cognitively situated within the physical realities of their homes and offices and toward mixed reality spaces that immerse users in advanced computational media. The combined visions of Milgram, Kishino, Weiser, and such robotics pioneers as George Devol, Joseph Engelberger, and Rodney Brooks have laid the foundation for a new and unknown territory of the hybrid world of smart objects and robots seamlessly blended with the virtual world of digital twins.

Intelligence augmentation is emerging as its own hybrid domain comprising XR, machine learning (ML), robotics, and the distributed computing of the Internet of Things (IoT). Extended reality (XR) is emerging as an entry point of an omniverse that operates within the overlap between the physical and virtual worlds where four possible interaction patterns can take place: (1) physical-to-physical, (2) physical-to-virtual, (3) virtual-to-physical, and (4) virtual-to-virtual.

The XRI world will allow people equipped with extended and/or avatarized bodies and even just physical bodies to roam freely between the physical world populated by information-rich and connected enchanted objects (IoT devices) and extended virtual environments populated with avatars, digital twins, and synthetic AI agents (actants). The human in the loop equipped with an XRI extended body will be able to interact with both the enchanted objects within their physical environment and the digital twins and synthetic AI agents (actants) in the virtual world.

Enchanted objects, such as smart thermostats, coffee makers, and home and industrial robots, will be able to act independently or as directed by the human users in the physical world. Enchanted objects, digital twins, and synthetic agents (actants) will be able to interact directly with each other or with the virtual environment with or without any human intervention. XR will not just be an interface into this new world. It will actually be a door that will allow users to step into the omniverse of information and augmentation.

Human in the Loop: The Body in the XRI Space

There are many ways in which the human body can enter the XRI space. When we think of XR, we often think of the one extreme of the continuum where we find such well-documented forms of XR embodiment as virtual avatars commonly used within the context of VR applications such as digital games where users are immersed in virtual environments. The fully avatarized participants differ from body accessorization and partial avatarization made possible in the context of augmented reality (AR), where users experience virtual content embedded in physical environments through technologies such as head-mounted displays (HUDs).

Wearable AR devices such as glasses, smart watches, fitness trackers, and haptic gloves can extend the body's natural senses to connect the user to digital information in mixed and augmented reality. At the other end of the continuum, participants can engage with tangible enchanted information-rich IoT objects embedded within the physical environment using only their physical bodies.

Using our own bodies seems so natural that it appears to be a product of an almost transparent process. As we discuss in other chapters of the book, this seemingly effortless (for most of us) ability conceals complex unconscious processes that can only be understood in the context of research in neuroscience, psychology, and somatic therapy. Our senses offer the multimodal channels that connect us to the physical world inside and outside our bodies, but our senses have their limits. Our eyes can only perceive the visible spectrum of light. Our ears can only process frequencies in a range from about 20 Hz to 20 kHz.

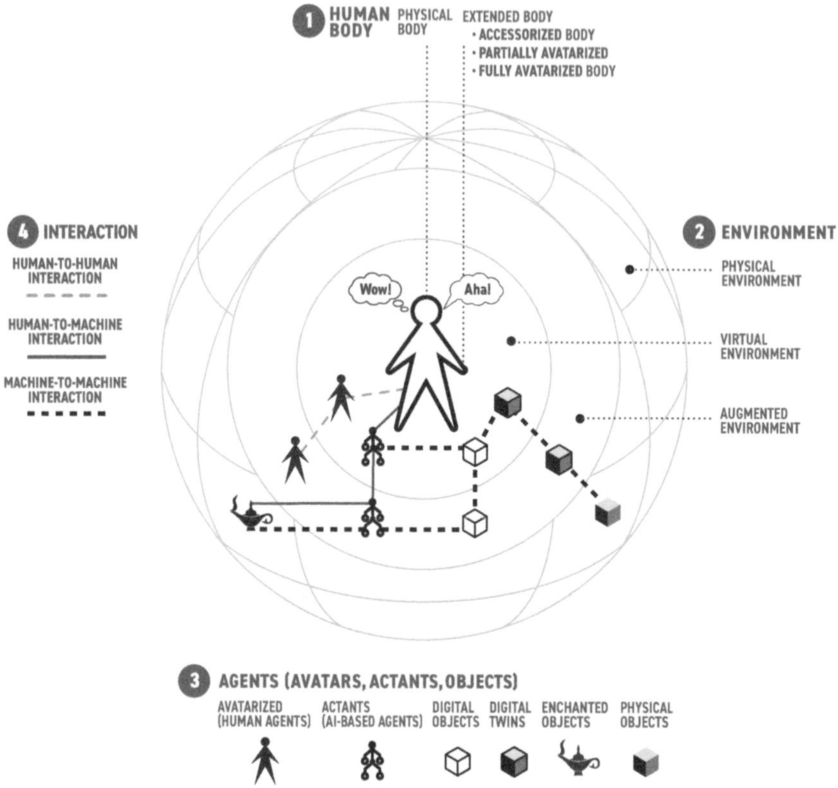

Figure 4.3 The human body in the XRI continuum

Source: Concept by Peter (Zak) Zakrzewski with David Tamés, illustration by © 2024 Peter (Zak) Zakrzewski.

Sensory augmentation is not really anything new. Hearing aids, glasses, binoculars, and microscopes are all examples of mechanical devices capable of augmenting the human senses. Identifying with augmentation technology is nothing new either. The rubber hand illusion (RHI) experiment described by Botvinick and Cohen (1998) is the most well-known illustration of the insight that the right sensory input presented at the right time can lead to the illusion of extending the body.

The extended body is the physical body augmented with assistive technologies defined by Huber et al. as media technologies enabling "user interfaces and interactions that seamlessly integrate with a user's mind, body and behavior" to provide "enhanced physical, sensorial and cognitive capabilities" (Huber et al., 2018, p. 1). Assistive augmentation can assist all kinds of users in extending their innate senses, adding relevant information, and developing advanced skills.

We do not use technology only to augment our senses. We have been extending our natural cognitive capacity with such mechanical tools as pen and paper to aid our memory through writing things down and cognitive processing when we use them to perform calculations and accounting tasks or to draw pictures to visualize information. The library card system is a mechanical technology aimed at augmenting our ability to keep track of books. Wearable computing can be traced back to the invention of a watch and a portable abacus.

One early instance of an AR device is a mechanical gunsight patented in 1901 by Irish telescope manufacturer Sir Howard Grubb. The inventor overlaid the near-view targeting reticle over the distant target using a series of optical elements, solving the problem of the inability of the human eye to focus on more than one depth of field and creating an augmented rifle with much greater accuracy in the process.

Mathematicians Edward O. Thorp and Claude Shannon created computerized timing devices in the 1960s to help them win at roulette. In the early 1980s, MIT student Steve Mann built a backpack-mounted wearable multimedia computer to pave the way for the 2007 introduction of the iPhone. By the mid-2020s, wearable computing devices such as smart watches and fitness trackers incorporating sensors such as accelerometers, heart rate monitors, and electrocardiogram (ECG) and blood oxygen saturation (SpO2) monitors are used by billions of people around the globe. Google Glass was the first notable mass market head-mounted display, controlled by gestures, that attempted to lay computer generated information over the physical world.

Genay et al., who created the body avatarization continuum framework, argue that users embodied in augmented reality (AR) can also leverage the concept of the avatars to enable them to see themselves acting in the real world but from a perspective of a body that is altered in shape or capability. They argue that mixed-reality (MR) technologies using coherent sensory feedback can produce the illusions of virtual ownership similar to those seen in VR environments.

The psychological effects of embodiment documented in VR may also occur in AR in the form of change in behavior and perception, which could be explored particularly in medical treatments and therapies (Genay et al., 2022). Within the body avatarization continuum framework, extended body augmentations can occur at different levels, ranging from the simple addition of wearable computing accessories to a complete change of identity, appearance, and morphology (2022). The body avatarization continuum describes the scale of the virtual user representation with an aim to contribute to the evolution of phenomenology of embodiment experiences.

The continuum starts on the extreme left, with the user only using their physical body without any augmentation. At the opposite end of the continuum lies full avatarization, with the user embodying and controlling a virtual body. The fully avatarized body can perceive the environment in the first person from the egocentric perspective through the eyes of an avatar or in the third person viewing the scene from a more allocentric point of view. The space between the two extrema of the continuum is occupied by diverse degrees of virtual body augmentation.

Starting from the left, Genay et al. propose the category of body accessorization. Body accessorization describes the physical body extended into XRI space through superficial elements such as virtual clothes. One of the best examples of body accessorization can be found in augmented reality retail, where customers can overlay products such as clothing, eyewear, makeup, and jewelry on their bodies.

Moving further toward the right side of the continuum, Genay et al. place partial avatarization which involves modification of bodily characteristics through the addition of virtual body parts or changes to the real ones. Partial avatarization of the body can include limbs, skin color, face shape, and other features of the human anatomy (Genay et al., 2022).

Partial avatarization has been applied in the medical field with systems that train amputees to use virtual prostheses before real prostheses are implanted, and to help stroke patients to recover motor functions during rehabilitation (2022). The augmented prosthetic procedure not only helped to accelerate patient adaptation but also reduced phantom limb pain through the development of long-lasting ownership illusions.

Virtual embodiment illusions applied to patients with paralyzed limbs eased their recovery by generating cognitive stimulation that helped the brain to remap body ownership, agency, and kinesthetic perception through mediated visual feedback (2022). Partial avatarization also shows great promise in modifying a body structure and extending possibilities of hybrid interactions involving virtual limbs connected to the robotic systems that act on real objects in physical environments (2022).

Full avatarization within the XRI continuum extends the creative potential to use the avatar-based embodiment to engage participants in identity exploration through different appearance, shape, size, and physical capability or to allow them to observe themselves in a given situation through virtual out-of-body experiences. Full avatarization is predominantly the domain of virtual reality, but it can also be applied to AR.

For instance, Anderegg et al. explored the control of full-bodied avatars in AR mobile games. In the proposed interaction metaphor, the players were allowed to control their avatars like a puppets by using intuitive gestures (2022).

Hoang et al. Johnson and Sun used AR based full avatarization in health education and health awareness by providing users with on-person anatomical displays laid over their own physical bodies. The results of the studies show that the subjects demonstrated stronger levels of "connectedness" and ownership of the projected content than with other forms of mediation, particularly when the presented data was consistent with a participant's physiological state such as synchronized heartbeat.

Extended Environments

When Mark Weiser envisioned, back in 1991, an interaction framework of computer technologies that disappeared into the background while augmenting and enriching the real world with digital functionality, the concept of networks of smart devices was already nearly ten years old. In 1982, researchers at Carnegie Mellon University modified a Coke vending machine to report its inventory level and to give status updates on whether the newly loaded drinks were cold enough to dispense to customers.

In 1985, Peter T. Lewis, speaking to the Congressional Black Caucus Foundation Annual Legislative Weekend in Washington, D.C., described the feasibility of an integrated network he referred to as "the Internet of Things," consisting of people, processes, and connectable sensor-enabled devices that would allow users to remotely monitor status, evaluate, and manipulate entire sociotechnical systems (Lakhwani et al., 2020). Weiser advocated for a world of augmented intelligence where human–computer interactions would be as pleasant and natural as "taking a walk in the woods" (p. 8). He envisioned thermostats, music streamers, and ovens interconnected within computer networks and embedded in extended physical environments such as homes and offices.

Extended environments populated by enchanted objects do not necessarily require users to interact with computational media and smart appliances through the familiar graphical user interface (GUI) consisting of windows, icons, menus, and pointing devices such as a mouse or a trackpad, humorously referred to by the tangible user interface (TUI) researchers as WIMP. From the 1990s onward, TUI proponents proposed an interface approach that relied on users' knowledge and skills of interaction with objects in the physical world (Shaer, 2009).

The consumer version of the extended environment of IoT was first developed in the form of a Polish speech synthesizer inspired by Stanley Kubrick's *2001: A Space Odyssey* named Ivona, which became the precursor to Amazon's Alexa. The development of the app's interface was further inspired by the computer voice conversational system used on board the starship *Enterprise* in the Star Trek TV and film franchise. Alexa can process natural language (NLP) and other forms of weak AI to perform such tasks as streaming podcasts and music, creating to-do lists, setting alarms, and providing weather, traffic, and news updates. It can also control smart devices as part of an integrated home automation system. The app's capabilities may be extended by installing skills developed by third-party vendors.

Objects within extended environments can also interact with extended bodies. One example from the mobile computing world is geolocation—the use of location technologies such as GPS and device IP addresses to identify and recognize the users of connected devices such as smartphones. Geolocation has numerous applications across various consumer and enterprise applications. Geolocation technology has been leveraged by emergency service providers such as police, fire, and ambulance to locate individuals in need of assistance. It is frequently used by people who seek directions while traveling, look for nearby restaurants, or want to find a ride-sharing service. Retailers use geolocation data to provide more personalized shopping experiences for their customers. Geolocation is also widely used for fleet management to allow companies to track the location of their vehicles and to optimize routes.

Enchanted Objects

The term "enchanted objects" was proposed by David Rose in his book by the same title, but it goes back to the tangible user interface research pioneered in the 1990s by HCI researchers George Fitzmaurice, Hiroshi Ishii, and William Buxton. The research agenda of "graspable interfaces" was focused on such common physical features as handles that people could use to manipulate digital objects. Ishii and his colleagues presented tangible user interfaces that turned the world of physical objects into an interface by connecting various tangible objects and surfaces to computational media. Their vision was to offer users an embodied multisensory experience of digital information by changing "painted bits" into "tangible bits" that would take full advantage of the multimodality of the human nervous system in its interactions with the natural environment (Ishii, 1997, p. 7).

The proposed TUIs, unlike their digital twins, did not rely on pure semantics but leveraged the expectations of users' everyday experience to communicate the affordances of functionality and interaction. Early TUI experiments harnessed the multimodal human perceptions by mapping digital data onto tactile physical form, using objects such as marbles, plastic cards, and blocks. Various novel input devices that drew on the concept of TUIs were widely adopted in such gaming systems as the Wii home video game console released by Nintendo in 2006. Wii featured a remote controller which, when used as a handheld pointing device, could detect players' movement in three dimensions. Wii is yet another example of gaming paving the way for the wider adoption of immersive media.

Rose defined tangible interfaces as providing interactions between humans and computational media that did not require a screen but instead relied on gesture, tactility, sound, light patterns, and haptics (Rose, 2014). He argues that enchanted objects start as ordinary things such as light bulbs, shoes, and wallets that gain their remarkable power by being augmented through such technologies as sensors, actuators, embedded processing, and wireless connections. In the process, they become useful, more informative, and more delightful to use (2014).

One of the best examples of enchanted objects is a set of tangible albeit modified instruments connected to the digital game *Guitar Hero*. Rose, who collaborated on the project, argues the game's massive appeal stems from its ability to resonate with a common dream of becoming a rock star and from providing a deeply satisfying experience of playing music without the frustrating learning process.

Mirror Worlds, Digital Twins, and Synthetic Agents

Digital twins are the enchanted objects' virtual counterparts. They are computational representations of people, physical objects, or processes recreated in a virtual environment. What they lack in corporeal solidity, they make up by their enhanced synthetic intelligence, fluid embeddedness within networks of digital information, and ease of participation in the simulation of complex real-world systems. Digital twins can be used to simulate entire environments for optimizing system workflows.

The concept of digital twins was anticipated in 1991 by a pioneering computer scientist David Gelernter, who preempted by 20 years venture capitalist Marc Andreessen's Silicon Valley manifesto titled "Why Software Is Eating the World" (Andreessen, 2011) with his own *Mirror Worlds: or the Day Software Puts the Universe in a Shoebox* (Gelernter, 1992). Gelernter argued that digital mirror worlds would allow humans to explore and control the world with never-before-seen depth and detail without ever changing out of our pajamas.

He imagined digital duplicates populated by virtual twins representing entire cities complete with patterns of economic development, infrastructure, and moving traffic or visualizations of operations of massive multinational conglomerates all neatly presented on a computer screen. In the world mirrored by the computer, humans would gain control over the real world by gradually duplicating reality with software imitation. The concept was embraced by the engineering community, who saw that digital twins built according to exact real-world conditions would allow them to test product concepts and improve performance before the products were even built.

The first example of the digital twin as a conceptual model underlying product lifecycle management was presented by Michael Grieves in 2002 at a Society of Manufacturing Engineers conference in Troy, Michigan. Despite a few different versions, such as virtual twin, the name "digital twin" seems to have stuck after the publication of John Vickers NASA Roadmap Report in 2010.

The digital twin concept as defined by the engineering community consists of three distinct parts. The first component is the original physical object or process and its physical environment. The second is the digital representation of the object or process and the environment under investigation. The third part consists of the communication channel, or the digital thread linking the physical and virtual environments. The digital thread includes information flows and sensor-based data flows between the physical and virtual objects and their environments.

Gelernter's proposal advanced another speculative vision exemplified by a use case of an engaged citizen who could leverage the power of mirror worlds to explore the performance of the local schools by virtually chatting with teachers and other parents. Gelernter imagined this concerned citizen being able to plant software agents to report back on topics of importance and, upon deciding to run for the local school board, be able to conduct much of the campaign itself within the mirror world. This vision has come much closer to reality with the mass adoption of such generative-AI platforms as ChatGPT and the rapid developments in the agentic architecture in which groups of AI agents can take autonomous responsibility for determining both the intent of a task request and planning for its response.

Computation has moved from applications written as sets of instructions to assemblies of AI-based actants capable of independently accomplishing complex tasks on behalf of the user. The XRI world that straddles both the real world and its digital image will be increasingly populated with synthetic agents representing the interests of individuals such as passive investors, legal claimants, researchers, and travelers who want to leverage the power of AI without active conscious engagement. Synthetic agents will be able to query databases, call APIs, and connect to other automated services on behalf of their human clients.

What makes AI agents different from all other software programs autonomously completing tasks determined by the software developers is that AI agents can "evaluate" decisions based on their inputs and data to optimize performance and results by "sensing" their environments with physical or software interfaces. AI agents are the next generation of software programs that can interact with their environment, collect data, and use the data to perform self-determined tasks to meet any predetermined user goals. In a collaborative human+computer (H+C) architecture, humans set goals but allow the AI agents to independently choose the best course of action to achieve those goals.

Even though self-driving cars are still far from being ready for wide adoption on public roads, they serve as a great example of a potential offered by autonomous agents. Commuters using autonomous vehicles decide where they want to travel while leaving the responsibility for all driving activities, such as controlling the vehicle, monitoring important systems within the car, and navigating the environment to the AI system. Self-driving cars accomplish these tasks based on algorithms defining their overall performance and on data inputs from multiple sensors.

Banks use AI systems to simulate familiar interactions with real customer service representatives to make such interactions are more effective and efficient while still being engaging. As individuals or coordinated groups of agents, synthetic agents can carry on complex or repetitive tasks in mirror worlds while we sleep.

The Scenario Ingredients Framework

User scenarios give the entire immersive experience design process clear direction. As the case of Apple's Navigator example shows us, they can become the glue strong enough to hold together a common vision shared by the entire design team, from researchers to information architects, software and hardware engineers, visual designers, and storytellers for decades. Your scenario will be built based on the ingredients shown in Figure 4.3. These include the four major considerations of (1) the participant's embodiment, (2) the features of the environment, (3) the active objects in the environment, and (4) the possible interactions. We will cover all of these in greater detail in the next chapter, which is concerned with feasibility.

1. In a nutshell, the participant's body can be conceptualized as being
   a. completely free of augmentation, as in the case of ubiquitous computing or tangible user interfaces (TUI).
   b. It can be accessorized with XR devices that connect it to the immersive world. This can take a form of a phone or tablet used for mobile AR.
   c. The participant's body can be partially avatarized to allow them to be seen in the VR environment. It can also be fully avatarized to give the participant a greater sense of agency and presence in the virtual world.
2. The scenario needs to describe the context of the immersive experience by discussing the features of the physical, visual, and/or augmented world the audience will navigate.
3. The environment will be populated by a variety of objects and entities including
   a. other human actors,
   b. physical objects,
   c. enchanted objects present in the physical world,
   d. digital twins,
   e. AI-based agents (actants), and
   f. digital objects.

4. The interactions within the XR environment will fall into three categories:

   a. human-to-human,
   b. human-to-machine, and
   c. machine-to-machine exchanges.

   Here are five quintessential questions your context scenario will need to answer.

- **Who** is the participant, and what goals are they pursuing?
- **How** will the application (XR system) interact with and respond to its audience?
- **When** will the application (XR system) be experienced?
- **Where** and under what circumstances will the participants engage with the experience?
- **What** information and resources does a user need at any given point in the experience?

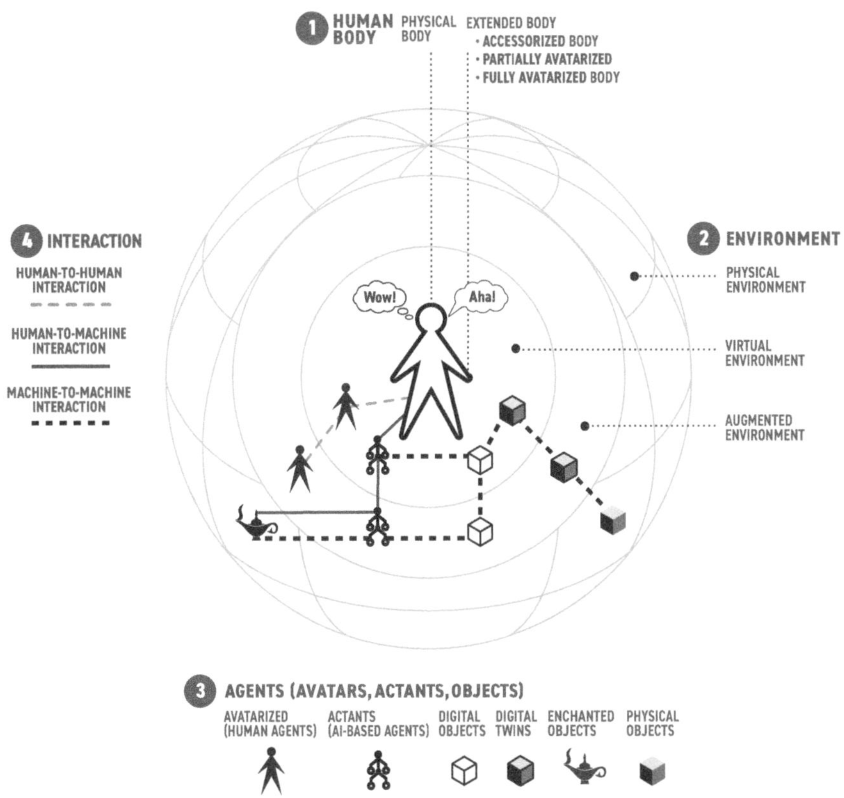

Figure 4.4 The human body in the XRI world

Source: Concept by Peter (Zak) Zakrzewski with David Tamés, illustration by © 2024 Peter (Zak) Zakrzewski.

Ideating Immersive Spaces

Immersive media making involves creating a variety of environmental constraints within which participants can move freely to arrive at their own goals and interpretations. Usability means that the constraints defining the immersive space are easy to learn and to internalize. Systems that are too complex or too difficult to navigate cause user frustration. Presence is replaced with frustration when participants either cannot understand what actions to take or are incapable of taking those actions.

During the scenario-writing process, it is important to keep in mind that audiences bring with them their own powerful constraints: their mental models of the world and their personal goals. Audiences reach their goals and subgoals by putting the system in a desirable configuration. The entire context scenario is a choreography for a dance between the immersive system and the participant who engages in an exchange of communication behaviors. Anticipatory play is enabled by the system designers, allowing the participants to form predictive models of the system's behavior.

Immersive media design focuses on building systems that respond to participants in ways that encourage the formation of meaningful interpretations based on constantly inviting the audience to make interpretive moves on their path through the experiences. Each dramatic beat of an encounter with the immersive system is initiated by the various touchpoints and resolved by an interpretive move made by the participant.

All touchpoints have both a sensory and a semiotic dimension. On the sensory level, touchpoints offer the embodied and aesthetic experience. On the semiotic level, they can either advance a narrative or move the participant closer to achieving their goal. Potential plot twists woven into context scenarios are cruxes that can alter the audience's understanding of the action system takes in a narrative meaning-making sense. Cruxes that trigger deeper understanding in the audience can be considered epiphanies.

Scenarios allow the media design team to strategically sort out the sensory, narrative, and challenge-based components of the immersive experience. The main goal of creating context scenarios is to conceptualize an immersive space that offers participants access to a constant flow of opportunities to discover something new and exciting without ever straying too far from the comfort of what is safe and familiar. Based on the participants' goals, the path through the experience can lie on the continuum from utter efficiency to creatively interesting and even convoluted in ways that evoke a sense of discovery and fun.

Scenario generation is an act of sketching out a form of an aesthetic and semiotic playground that prompts the participant to actively generate interpretive moves that bring on the discovery of new understanding. At any given time, the audience is faced with a horizon of possibility—all possible actions that can be taken by the participant in the pursuit of their goal.

The best immersive experiences offer their audiences six core characteristics. Choice represents a range of all possible actions. For participants to feel like they are making a natural choice, their actions must make a verifiable change in the system's state. Variety offers participants constantly emerging new horizons and opportunities to engage. Consequences mean that all user actions have meaningful outcomes. The reliability rule dictates that the outcomes can be understood and anticipated. The uncertainty rule ensures that outcomes are not too predictable or predetermined. Satisfaction assures that the participants attain desirable outcomes (Upton, 2015).

Immersive Scenario Writing Principle 1: Addressing the Full Spectrum of the Embodied Sensory Experience

Cognitive scientist Margaret Wilson argues that our cognition is "deeply rooted in the body's interaction with the world" (Wilson, 2002, p. 625). The embodied nature of cognition leads us to employ a number of rationalities evolved over time that include seeking utility, making aesthetic judgements, cooperating with others, and making meaning through personal reflection and communal storytelling. Our bodies are central to our perception and experience of the world. And if the goal of immersive media design is to help the participants organize the meaningful blobs of sensory input into content that moves them to experience something extraordinary, we have to start the process by considering the embodied nature of XR presence.

Wilson articulated six claims modern cognitive science holds about how our cognitive processes help us navigate the world.

1. All cognitive activity takes place in the context of real-world environments and is directly linked to perception and action.
2. Cognitive performance is sensitive to time pressure in the context of our real-time interactions with the environment.
3. Because of limits on our information-processing abilities, such as limits on working memory and attention, we offload difficult cognitive tasks onto our environment. For instance, we have invented such tools as abacuses and calculators so we do not have to perform mathematical calculations in our heads. We use reference books, Google searches, appointment calendars, and phone apps to help us manage cognitive tasks throughout the day.
4. Because of the dense and continuous nature of the flow of information between our cognitive system and our environment, the environment we find ourselves in can be considered a part of our expanded cognitive system.
5. The main function of cognition is to guide action. All cognitive mechanisms such as perception and memory need to be understood in terms of their contribution to situation-appropriate behavior.
6. Even when we decouple mental activity from our immediate environment, our cognition is grounded in mechanisms of sensory processing and motor control. For instance, in the case of counting on our fingers, we run a simulation of an aspect of the physical world as a means of representing information that helps us to draw useful inferences.

The ability to mediate all of the five external senses and to affect the three internal sensory channels gives XR media makers a unique opportunity to create a fully embodied immersive presence. We recommend starting the XRXD process by first addressing the participant's body. The mediation options here include two major choices. Designers can envision the experience as requiring the participant to either go through it with nothing more than their physical body or by extending the body with XR gear such as headsets, headphones, or mobile devices (Figure 4.4).

Traditional media do not require the audience to wear any special gear. People can curl up on a couch to stream their favorite shows. This, in many ways, is the golden standard for natural user interfaces (NUI). In the immersive space, this type of unadorned-by-technology experience takes a lot more work to accomplish.

For the more immersive mediation of the visual channel, you can use your Android or iPhone smartphone and the Polycam app to create high-quality 3D models from photos, generate LiDAR scans, and capture immersive 360-degree images and export them in a variety of industry-standard formats to map any room to free an audience from having to use flat screens or wear headsets to bring panoramic or 360° content that can be shared by groups of people. The development of spatial audio has contributed to creating more immersive XR environments by providing more realistic auditory cues that align with the visual components of the XR environment, significantly enhancing the sense of presence in XR environments.

Tangible user interfaces, sometimes referred to as enchanted objects, can also offer tactile access to computing that does not require the users to wear any XR equipment. Adding any form of haptic feedback can provide tactile sensations that simulate the feeling of touching and interacting with real objects.

XR media makers can leverage computer vision and advanced computation to allow users to interact with digital content using gestures that mimic real-world actions. Leveraging natural gestures as an interaction mode can create experiences that feel natural, engaging, and deeply immersive, giving the audience a heightened sense of presence and agency within virtual worlds.

The extended body option includes accessorizing the participant's body with various types of XR gear ranging from handheld mobile devices to full body suits. Within fully virtual environments, the extended body can include partial or full avatarization that allows participants to see each other and to see their own virtual twins, respectively.

Because sensory perception is the gateway to all embodied human experience, creating immersive media content can be considered a form of multimodal composition in which compelling experiences are created by designers applying objective sensory stimuli to elicit subjective states in the audience (Zakrzewski, 2022). As we discuss in detail in Chapter 5, the holy grail of XRXD—immersive presence—relies entirely on the alignment between the virtual world and all of the participant's senses. Context scenarios need to be written based on a solid understanding of XR technology available to the media makers.

Many philosophers and neuroscientists argue that our perception of the world is an elaborate mental construct built on top of sensory input. Neuroscientist Heather

Berlin likens what our brains go through in the process of sentient experience to living inside a dark perception box from which the brain orients itself by receiving vital environmental signals from the five sensory channels. This is not to suggest that the things in the world do not actually exist but that our experience of them is a construction of our brain. Our perceptions are formed when the expectations based on past experiences meet the signals coming into the nervous system from the outside.

Since each of us is neurologically unique, we experience even the same situations slightly differently. No two people will experience the same blue in an image of the blue Caribbean ocean. Some may relate it to a family holiday, others to a childhood memory, others yet to a fear of sharks. Also, individually, we can experience similar situations differently based on our ever-changing moods.

We can learn a lot about the true impact of sensory processing on cognition from the instances when such sensations are impaired. Researchers have found that symptoms of poor sensory processing can occur within a broad spectrum of severity and can be experienced by all of us. This can entail a variety of behavioral problems and emotional issues, including depression, anxiety, and difficulties performing everyday tasks.

The neuroscientist, educational psychologist, and occupational therapist Jean Ayres explored the idea that sensory information processing and integration are vital substrates for adaptive behavior. Consequently, she developed the sensory integration (SI) theory to explore the relationship between the fundamental neural processes that allow individuals to receive, modulate, and integrate sensory input and the resulting behavior. By directly affecting how the nervous system receives messages about the inner and outer environments, XR designers can profoundly affect the participant's ability to take appropriate action and make motor and behavioral responses (Zakrzewski, 2022).

Our central nervous system is a parallel processor that is constantly integrating sense data, sensations, and memories. For example, if we hear a car horn while crossing the street, air molecules impinging on our eardrums cause electrical signals to be sent to the auditory cortex. This brain activity causes a cascade of other neural events that recruit various related groups of neurons that, together, keep us safe. Neurons in the auditory cortex evaluate the pitch of the sound so that we can tell the difference between the car horn and the sound of an ice cream truck or a guitar at a rock concert. Our brain activates another group of neurons to determine where the sound came from.

These neural activities invoke a visual orienting response that makes us turn in the direction the sound came from to establish what made it. Our prior experience with car horns helps us understand the context of the sense data we are processing. If we decide that it is necessary to jump out of the way as a result of activity from the neurons in the motor cortex, this activity will be prompted by the neurons in the amygdala (our emotional center) that told our nervous system that we are in danger (Levitin, 2006).

Our minds come to understand the world around us through the interplay between low-level and high-level processing. Low-level processing in our brain hears disparate sounds and sees flickering pixels on the screen. It might even be able to put patterns of sounds or pixels together to recognize a basic source or shape in our auditory or visual vocabulary, but it is the high-level processing that puts it all together to generate meaningful mental images.

As we discussed in Section II, the best way to integrate the senses is to start by choosing a desired emotional state for each dramatic beat, interaction touchpoint, or entire zone of the XR experience and to strategically support the elicitation of that state by a well-chosen set of sensory stimuli. Figure 4.5 shows this five step process in detail.

Another tool we find useful for ensuring sensory integration is the chart cross-referencing the diachronic (unfolding in time) view of the XR experience with the submodalities of each sensory channel (Figure 4.6). Mediated presence will be enriched by more channels involved in the experience and the level of their synchronization.

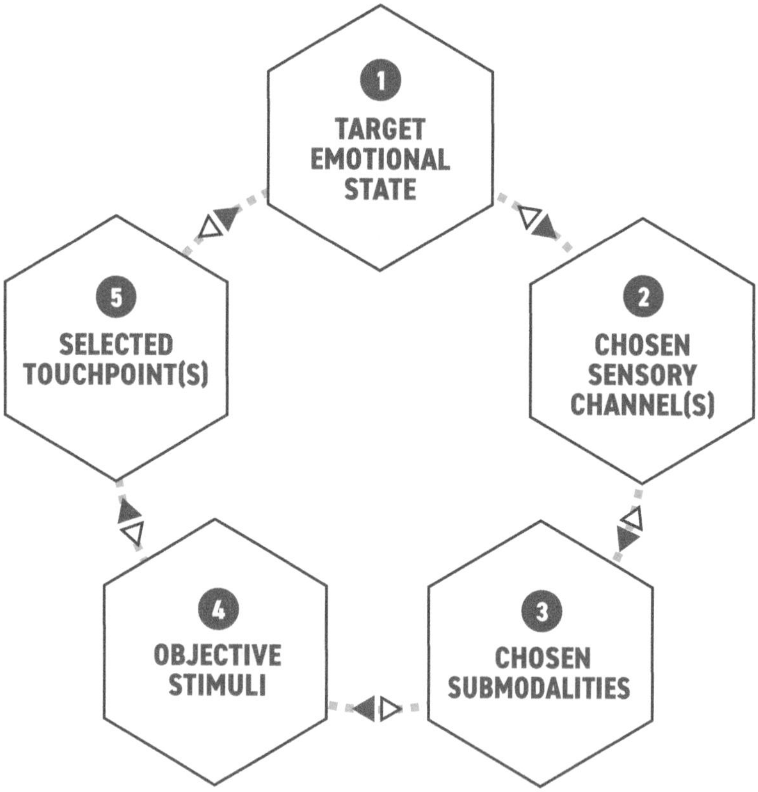

Figure 4.5 The presence elicitation model

Source: Concept and illustration by © 2024 Peter (Zak) Zakrzewski.

	OBJECTIVE STIMULI								SUBJECTIVE STATE
	SEE	HEAR	TOUCH	SMELL	TASTE	VESTIBULAR ORIENTATION	PROPRIOCEPTIVE COORDINATION	INTEROCEPTIVE STATE	TARGET SUBJECTIVE STATE
ATTRACTION									
ENTRY									
IMMERSION									
TRANSFORMATION									
EXTENSION									

Figure 4.6 Targeting subjective emotional states

Source: Concept and illustration by © 2024 Peter (Zak) Zakrzewski.

Immersive Scenario Writing Principle 2: Setting the Stage With a Synchronic View of the XR Environment

Family therapists and communication theorists Paul Watzlawick, Janet Beavin Bavelas, and Don Jackson articulated an important insight about the relationship between a living organism and its environment. They argued that because existence depends on the vital relationship between the organism and its environment, we experience our environments as a set of instructions necessary for our survival and thriving in the world. With the advent of media, the vital messages coming from our environment have undergone a shift from the purely biological to the more psychological realm (Watzlawick et al., 1967).

Within the emerging domain of immersive media, existence and communication with the virtual environment continue to be inseparable concepts. We consider the design of the immersive environment to be second in importance only to the consideration of the participants' embodied experience in the virtual world. Immersive media experiences unfold over time, but they unfold within three dimensions of physical and virtual environments. XR designers can exercise nearly complete control of the virtual spaces. In the case of mixed and augmented reality, they need to understand the environmental context and the extent to which the physical and the virtual worlds can be aligned.

We believe that during the ideation (inscribed) phase of the XRXD process, design teams will find it helpful to conceptually separate the three-dimensional spatial design from the fourth dimension of time. This separation between the synchronic (3D) and the diachronic (4D) aspect of immersive experiences is obviously artificial. Events occur over time and occur on stages with other people, agents, and objects, be it inside a theater, VR game, retail space, or restaurant. The separation we propose here is a matter of convenience that allows the design team to manage the inscription process effectively.

The synchronic view of media making during the inscription or production phase has a well-established tradition in the concept of mise-en-scène in cinema. This term was popularized by French film critics, including André Bazin and François Truffaut, to refer to the art of arranging critical elements within a cinema scene or theater stage setting. It encompasses the visual components presented to the audience, including actors, sets, props, lighting, etc. Bazin emphasized that a primary function of mise-en-scène is to convey the characters' emotions, motivations, and psychological states through visual cues. This can be achieved by carefully selecting set dressing, objects, props, colors, lighting schemes, and spatial arrangements to imbue the narrative with distinct meanings.

Throughout cinematic history, directors have developed unique approaches to mise-en-scène. For example, in *The Cabinet of Dr. Caligari* (1920), director Robert Wiene chose a jagged, expressionist architecture and exaggerated sets with altered perspectives to convey psychological horror and used stark lighting with deep shadows to symbolize the dual nature of sanity and madness and emphasize the horror encoded in the narrative. The film demonstrates how skillful crafting of the mise-en-scène can be used to externalize human emotions and desires. The film epitomizes the German expressionist movement and continues to influence directors working in horror and noir genres to the present day.

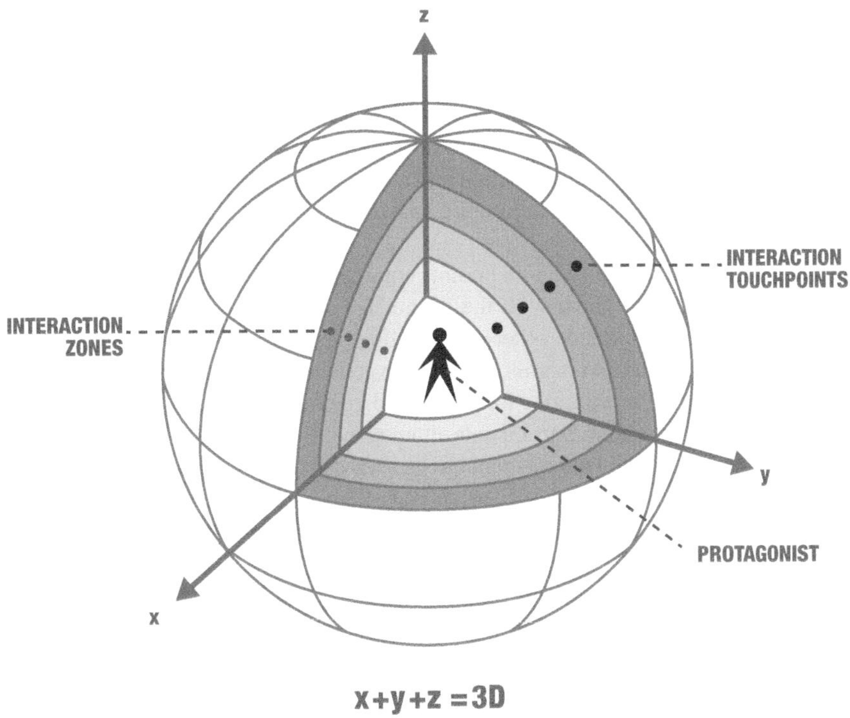

Figure 4.7 The synchronic snapshot of the XR system

Source: Concept and illustration by © 2024 Peter (Zak) Zakrzewski.

In *Blade Runner* (1982), director Ridley Scott transports viewers into a dystopian story world where technology and hyperindustrialization have overwhelmed the natural environment. Neon signs illuminate streets cloaked in ever-present darkness drenched in rain and fog to create a mood of perpetual night, contributing to the film's noir cyberpunk atmosphere. Although set in a future Los Angeles, the mise-en-scène is immediately familiar, in large part due to the way Syd Mead—in his role as concept artist and visual futurist—defined the look and atmosphere of *Blade Runner* using objects with familiar characteristics (e.g., flying cars, video phones, the Voight-Kampff machine used to measured emotional responses, etc.) recontextualized to create the future world. Mead explains that "The difficult part of designing for a film is that you have to use audience in-head memory. People will believe something is real only if it is done right and compared to what they *think* it should be" (Lightman & Patterson, 2020). Mead's contributions crystallized the cyberpunk aesthetic and influenced science fiction films like *Ghost in the Shell* (Mamoru Oshii, 1995) and *The Matrix* (Lana Wachowski & Lilly Wachowski, 1999).

In the model shown in Figure 4.7, the synchronic view of XR experiences comprises two major aspects of immersion: interaction zones and interaction touchpoints. Those who have played digital games will be familiar with interaction zones as being equivalent to game levels. Game designers restrict access to various interaction zones in order to strategically control the quality of the gaming experience. Access to some zones has to be earned through successful navigation of various challenges. In terms of cinematic experiences, interaction zones can be thought of as locations or scenes. The application of interaction zones to immersive media design allows media makers to define and control virtual spaces the way architects design rooms within homes and public spaces such as museums, shopping malls, and transportation hubs.

The concept of interaction touchpoints comes from the field of brand experience design. Participants interact with the immersive system through carefully placed touchpoints that can include other participants, physical objects and props, digital objects and props, and AI agents. Interaction touchpoints bring immersive spaces to life by creating opportunities for action and communication. In a familiar social media platform example, participants interact with other social media users and their content through the aid of such touchpoints as "like" buttons and status notifications. In the gaming world, players interact with such touchpoints as other players and with digital objects such as magic rings, sports cars, and sabers.

Immersive Scenario Writing Principle 3: Affordances: Experience Happens at Interaction Touchpoints

The concept of user touchpoints originated in the field of brand design. User touchpoints are an integral part of the strategic curation, creation, and placement of interactive objects, communication channels, and experiences that, together, create a single perception of a coherent media experience. Touchpoints are the vehicle and not the destination. They act as gates to the perception that can build a powerful immersive transportation through a consistent storyworld.

The *Star Wars* saga is a perfect example of a media franchise with multiple touchpoints, all adding to one coherent narrative. The many individual contact points between the audience and the media franchise are where the powerful narrative and identity are born. In the context of the XRXD model, touchpoints are affordances through which the immersive environment communicates with the audience.

The concept of affordances was first proposed by the American psychologist James J. Gibson in his 1966 book, *The Senses Considered as Perceptual Systems*. It was later adapted to UI/UX and popularized by the cognitive scientist Don Norman. Gibson defined affordances as signals objects in the environment offer living organisms to help them identify whether an object might offer utility or pose danger (Gibson, 1983).

In modern psychology, affordances refer to the action potential of objects present in the environment. In design, products are created to deliberately suggest their intended function to users. For instance, door handles suggest that a door needs to be pulled, while a metal plate installed on the surface of the door would suggest that the door needs to be pushed. As affordances, interaction touchpoints can create the foundational aspects of the environment that spawn the bond that emerges between the audience and the XR media application.

Immersive experiences are not just about what we see or hear but, in large part, about what we can do within the XR environment. Besides serving as functional affordances, touchpoints can become props that help to define the participants' immersive journey.

Psychologist Jerome Bruner considered particularity of places and objects to be an essential feature of narrative transportation. Narratives are built out of ostensive references to particular happenings in particular locations, carried out with particular props. Narrative particulars are tokens of broader story types (Bruner, 1991). A romantic script genre will very likely call for some kind of gift that enables the process of gift-giving. The gift might take the form of jewelry, flowers, or a book of poems, but regardless, each of these will add depth and verisimilitude to the narrative.

Media makers give touchpoints their emblematic status by embedding them in a story. *Star Wars* would not be the same without such objects as the Death Star and the lightsaber. *Star Trek* would not be the same without the Federation starships, the communicator, the tricorder, tractor beams, and phasers. Bruner argued that narratives cannot function without particular embodiment. Narrative worlds draw their most suggestive quality from the emblematic nature of their particulars.

The ability to interact with virtual objects and characters will add an extra layer of depth to the experience, making it more lifelike and engaging for users. The richer, more lifelike, and more intuitive immersive interactions are, the easier it becomes for users to feel truly present in the virtual world. Interactivity is an essential factor in achieving immersion and presence within virtual reality experiences, as it enables users to engage with their digital environment in a natural and intuitive manner. By designing interactions that mirror real-world behaviors and responses, developers can create experiences that feel more lifelike and engaging for users.

Mixing Humans and Nonhumans: Actants as System Touchpoints

For audiences of hundreds of millions playing simulation video games such as the *Sims* series, fun and meaning can be found in managing virtual cities populated by virtual humans and other playable and nonplayable synthetic characters. The creator of *Sims*, Will Wright, described the game as an "interactive doll house" (Seabrook, 2006). The designers behind the interactive dolls within virtual doll houses can now utilize generative AI to dramatically advance the simulation genre by giving the characters more lifelike emergent social behavior powered by large language models (LLMs).

For instance, researchers from Stanford and Google created a dedicated LLM architecture that can store a complete record of an AI-based agent's experience to synthesize those memories into higher-level synthetic "reflections," which can be dynamically retrieved to plan future actions (Park et al., 2023). In one experimental simulation, a group of 25 Sims-like agents, working within a pixel art sandbox world, independently spread the word about a Valentine's Day party proposed by one AI-based agent. In the process, they struck up "friendships," asked each other out on "dates," and showed up for the party on time two days later. The underlying LLMs trained on the building blocks of human communication and behavior scraped from the social web showed the potential to create incredibly lifelike and human-like emergent behavior.

The AI agents powered by the corpus of the social web can not only create more immersive gameplay, they can serve as a basis for next-generation virtual *Truman Shows* that endlessly entertain audiences in a way not possible with pre-scripted TV shows and movies (Parvini, 2024). It seems inevitable that over time, we will go beyond games and entertainment to consider AI agents to be more than just toys but potential friends, confidants, personal assistants, coworkers, teachers, advisors, and perhaps even lovers.

We humans seem to be predisposed to enter into deep and complex relationships with our tools. Our ability to leverage props and aids to accomplish our goals is the distinctive feature of human intelligence (Clark, 2003). As our technologies build up and become more complex, they increasingly go beyond functioning as simple mechanical extensions of the body and inject themselves into the workings of the human psyche to reshape our social life (Zakrzewski, 2022).

Bruno Latour was one of the first social scientists to discuss the role of technologies, even as simple as door stoppers, to enhance or promote specific activities and behaviors while discouraging others. Latour argued that once put in place, technological sociotechnical reinforcers develop their own dynamic potential to function independently of people who installed them and contribute to the emergence of wider system behavior. He suggested that one of the most valuable insights we can glean from applying social sciences to technology was that "agencies over which we have no control" can "make us do things" (Latour, 2007, p. 50). Within sociotechnical systems, emergent action is distributed among multiple agents, few of whom actually "look like humans" (p. 50).

Latour and his science and technology studies (STS) colleague Michel Callon, along with sociologist John Law, developed the actor–network theory (ANT) as a theoretical framework capable of analyzing the complex interactions between the heterogeneous social and technological components contributing to the development of sociotechnical networks. One of the main ANT arguments suggested that material objects and objects of technology in particular could play a significant role in amplifying social trends and making the constantly shifting weak ties of social interactions more durable. Social systems, according to Latour, could be made to "last longer and expand further" by the power of "entities that don't sleep" and "associations that don't break down" (Latour, 2007, p. 70).

Latour and his colleagues used Julien Greimas's literary concept of the "actants" to represent all active human and nonhuman agents in the network as potential forces within sociotechnical systems. Latour argued that a literary term such as an "actant" gave system analysts more freedom than any existing social science concepts, because in fiction, the actants could be made to act through the agency of magical powers (2007). Greimas proposed his Actantial Model of narrative interaction based on the theories of the Soviet folklorist and scholar Vladimir Propp, who suggested that actants were not types of persons in the stories but rather common patterns of behavior or expressions of dramatic potential.

The concept of an actant defined by a fluid capacity for configuration fitted well with ANT's assumption that social groups were constantly made and remade and formed and reformed. Latour defined actants as unconfigured agents that modified a state of social affairs by making a difference in the outcome of the system. The nonhuman actants can affect or even determine human action by authorizing or disallowing, encouraging, rewarding, suggesting, influencing, blocking, or enabling various moves allowable within the rules of the sociotechnical system (Zakrzewski, 2022). The core question asked by ANT practitioners that can be applied to immersive media is: Does this agent make a difference in the course of other agent's actions or not (Latour, 2007)?

As discussed, the concept of the actant generalizes the notion of the agent within a network to something akin to a dynamic force within a system. Actants as touchpoints offer immersive system participants opportunities to interact with the system in more complex and rewarding ways. Design of synthetic agents (actants) and entire agentic architectures will be an important area of immersive media design. All AI-based agents operate from the base of their unique architecture, which usually consists of a software program.

The architecture for magical objects and robotic AI agents can be a combination of software programs and a physical structure. Robotic agents can be created out of an assemblage of such hardware as actuators, motors, sensors, and robotic arms. The software architecture for an AI agent may include voice activation, text prompts, APIs, and databases that enable autonomous operations. Actants are designed to break down the goal into several smaller actionable tasks based on the specific instructions or goals received from the user. Data collection and translation into actions that support the agent's objective are based on the agent's intended function.

Designing the agent function involves considerations regarding the actant's AI capabilities, its knowledge base, types of information to be processed, feedback mechanism, and other technologies required by the actant to accomplish its assigned tasks. The implementation of the actant's function is determined by its program, which involves developing, training, and deploying the actant on the designated architecture aligned to its overall function, technical requirements, and interaction and performance elements.

The emerging advanced agentic architecture makes it more and more feasible for AI actants to independently implement a variety of tasks. In the process of performing assigned tasks, AI actants will need information based on such sources as system performance data, user data, and internet searches. Actants can also be designed to access or exchange information by interacting with other actants or machine learning models in a machine-to-machine interaction pattern.

The Actantial Model can be a powerful tool for adding nonplayer characters (NPC) to immersive applications. Within immersive media experiences, actants, who can be people, animals, or inanimate objects, are all entities that perform actions and have roles in a participant's journey. Actants can change roles or functions, showing that the way characters, objects, and even ideas interact in a story can be very dynamic and complex.

Even objects and concepts can play critical roles in how actions unfold within these networks. The ANT framework, which views social groups as ever-changing networks, encourages us to ask: Does this actant change the course of action for others? For instance, let us imagine a virtual reality game in which the participant role-plays a detective solving a mystery. The Actantial Model could be used to design the story so that objects in the game world (a locked diary, a hidden key, or a mysterious letter) act as dramatic forces causing the protagonist to follow certain paths within the story. These objects would not just be passive decorations; they would influence the player's decisions and actions, effectively driving the narrative forward and making the experience more engaging.

In educational AR applications, various elements like historical figures, artifacts, or even concepts (like gravity or evolution) could be presented as actants. For example, when learning about the solar system, planets can become actants that students are able to interact with. By manipulating the elements of the solar system, students could experience firsthand how changes in one planet's orbit might affect others, making the educational experience more interactive, dynamic, and memorable.

In an immersive installation that tells the story of a forest ecosystem's changes over time, nonhuman actants such as trees, animals, and even the weather can play significant roles as narrative forces that interact directly with the audience. By interacting with these elements, visitors could gain a better understanding of the interconnectedness of the ecosystem and the impact of human actions on environmental balance. By considering both human agents and nonhuman actants in the design process, immersive media makers can develop more nuanced and engaging experiences that reflect the dynamic nature of stories and XR systems.

Actors and Actants: Mixing Humans and Nonhumans

Designing compelling and humane immersive experiences requires distinguishing between human actors and nonhuman actants. This important distinction can help to avoid confusion between human psychology and embodied, conscious experience of the world and programmed behaviors of even the most advanced objects of technology.

Unlike the term "actor," which is commonly used in humanities and social sciences, "actant" originates from literary theory, reflecting the creation of artificial worlds. Human actors possess intrinsic agency, allowing them to make choices based on personal beliefs and experiences. Their actions are influenced by both unique genetic makeup and personal and cultural development.

Methods from such disciplines as psychology, cognitive science, and social sciences can help designers to better understand behavioral patterns of human participants through the application of computational social science, ethnography, deep interviews, and projective techniques we discussed in Chapter 2. Media designers can attempt to elicit emotional states in the audience through the application of objective stimuli such as tone and pitch of music or color and shape of object, but we cannot directly control these subjective states of the audience.

The agency of the nonhuman actants, on the other hand, is extrinsic in nature. Their action potential can be designed and assigned by the XR system creators. AI agents can engage in puzzling and impressive feats of generative synthetic intelligence, but they, according to our definition, do not possess conscious choice.

Their potential for behavior can be defined and modified by the system designers to achieve desired system outcomes. The interplay between human actors and nonhuman actants in XR systems aims not only to enhance the audience experience but can also enhance the overall system performance.

XR technology can amplify human agency by facilitating outcomes that humans alone cannot achieve, but in and of themselves, digital objects, no matter how advanced, lack conscious human judgment. We believe that the clear taxonomy of actor and actant roles will enable designers to strategically approach the development of humane XR systems that strike a balance between understanding human subjective experiences and designing and adjusting nonhuman elements that achieve an overall system function. This balance is essential for immersive experiences that are both effective in terms of user goals and respectful of their human agency.

Nonplayer Characters, Bots, and AI Agents: Designing Digital Objects as Actants Within a Network

Digital games have evolved from packaged entertainment to social networks designed for immersion and interaction. Open-world games such as *Red Dead Redemption 2* (RDR2) immerse users in their virtual worlds by populating them with computer-controlled or nonplayer characters (NPC) hanging out on the streets of sleepy Wild West towns, at vaudeville theaters, and in saloons. These synthetic characters bring immersive virtual game worlds to life by offering participants exciting opportunities for interaction. AI assistants and bots are poised to transform all types of immersive media experiences by enabling interactions through intuitive interfaces of gesture and natural language.

AI agent research is still a developing field without clear definitions. But what the AI models and algorithms behind agentic architecture all have in common is that they can autonomously make decisions and take actions in a dynamic and complex world on behalf of their human clients. The idea behind AI agents is the creation of a system that can execute a wide range of tasks such as tutoring a child in math or booking a perfectly planned vacation. Well-trained AI agents will remember if we prefer to save money on our hotels or optimize an experience of staying at the most interesting places. They will not only be able to suggest the best itinerary for the trip, the best flights, and hotels, but they will also be able to go ahead and book them for us. They will be able to make us a list of things to pack based on the travel plan and the weather forecast for the locations we will be visiting. They could suggest places to dine and visit. They might ask if we want to share our itinerary with friends and colleagues who might be at the same location at the time we will be there.

AI agents, like Google's Astra, will be multimodal. They will be able to process natural language, images, audio, and video, allowing us to simply point our smart devices at things and ask questions. They will be able to respond to our queries with text, audio, and video outputs.

The term "agentic" has been redefined in the LLM era. Kapoor et al. (2024) suggest that the term "agent" can be defined based on the level of sophistication an AI agent can demonstrate on three spectra of ability that include the complexity of the environment and goals the agent can navigate, the interface the agent offers and the supervision needed, and the complexity of the agent's system.

AI systems capable of operating in more complex environments and those that can pursue complex goals without specific instructions are considered to be more agentic. AI systems that can be instructed by the user in natural language and can also act autonomously on the user's behalf will be placed higher on the agentic spectrum. AI systems that use more sophisticated design patterns such as web search and programming or planning, reflection, and subgoal decomposition are also more agentic. The AI-driven enhancements can significantly improve user engagement, usability, and learning outcomes in immersive media applications.

Post-ChatGPT start-ups use generative AI to create characters that can converse freely without relying on prewritten scripts and even seem to possess agency and personality of their own. These generative AI-based characters can continue to play the game while the human players are not around to watch. Character creation has been a foundational pillar of role-playing games (RPGs) since the original Dungeons & Dragons (D&D). The most immersive RPGs allow participants to choose from the existing preset options to customize their key features of their identity such as appearance, gender, and class.

By combining LLMs with text-to-image diffusion models such as Midjourney or Stable Diffusion, media makers can go beyond offering presets to generate a unique character for each player and playthrough. For instance, Spellbrush's Anime gacha RPG Arrowmancer allows participants to generate their own party of unique characters that include art and combat abilities. The game's personalization is part of its monetization system, since the players can import their AI-generated characters into custom gacha banners where they can pull duplicate characters to augment their party.

Interactive digital objects and digital twins have been applied to such fields as medical training, industrial simulation, customer service, and virtual museum guides leading visitors on tours of museum exhibits. AI assistants can improve simulated medical procedures, offering real-time feedback and guidance in a conversational manner.

For example, an AI-powered VR training module for surgeons can adapt the complexity of a simulated surgery based on the learner's responses and questions, making the learning experience personalized and effective. Researchers at Anglia Ruskin University developed a clinical training scenario in VR to provide medical students the opportunity to repeatedly practice surgical procedures while eliminating harmful consequences to patients. The system orchestrates simple suturing practice through an immersive environment and AI algorithm. The system detects the relative movement of the users hands and provides feedback through the AI system that collects data from the student and compares it with a database to assess the player's performance accordingly. Real-time feedback enabled the student to correct mistakes.

These implementations demonstrate the seamless integration of AI with VR and AR to create dynamic, responsive environments. AI assistants in VR medical training enhance the acquisition of complex skills through adaptive learning, while in museums, AI bots transform static exhibits into interactive, personalized learning journeys that adapt to individual user needs, providing customized experiences that can be more engaging.

Participants encountering the immersive environment decide which aspects of the immersive space can be classified as being useful to be incorporated, something irrelevant to be ignored, or something actively countered as hostile and threatening. The Actantial Model can be a very useful framework to analyze the role of various agents within the XR system. While being obviously very applicable to game and narrative types of immersive experiences, with creative application, the framework can add a lot of depth to even such seemingly mundane user goals as shopping.

We have adapted Greimas's model to immersive media design (Figure 4.8) by breaking narrative action into discrete classes of entities starting with (1) the protagonist who seeks a particular goal even if that goal amounts to no more than free and open exploration. During their journey, the protagonist will interact with various (2) enchanted objects that possess various "magical" powers to accomplish tasks and advance the action. The protagonist might also encounter other human agents who also seek their own goals. The human agents cannot be controlled by either the protagonist or the design team, as they can become (3) collaborators in a work-related task or (4) competitors in a game. The protagonist can also encounter AI-based actants who can become (5) helpers in their quest or (6) opponents in a game-like scenario.

The categories of the actantial analysis framework were originally divided into three relational and motivational patterns, which could be used as the axes of the dynamic potential within the system. Agents within the narrative network could be motivated by desire, power, or knowledge. Peter has modified this framework to adapt it to the design of digital applications and experiences. The revised framework in the new model is divided into five relational and motivational patterns based on Maslow's hierarchy of needs, which can be used as the axes of the dynamic potential within the XR system.

1. The axis of physiological needs addresses such basic needs as food, drink, shelter, warmth, and general physical comfort. This category might, at first glance, seem trivial until we imagine staging a state-of-the-art survival experience at a remote island or in a remote forest.
2. Within the psychological safety zone, the design team focuses on providing the participants with a sense of security, order, or economic stability. These essential needs can be highly applicable to health- or investment-oriented applications.
3. In the axis of social interaction, various touchpoints can address such needs as belonging, love, family, affection, and other forms of relationships. This area is the domain heavily explored by social media apps and platforms. Within the XR media, social VR users have been among the earliest adopters of immersivity.
4. The axis of esteem and social capital builds on the previous axes to provide participants with a sense of personal empowerment, achievement, status, and reputation. The last and highest order of needs is represented by the axis of meaning and self-actualization. Within this axis, the participants seek personal growth, fulfillment, knowledge, deeper understanding, and meaning.

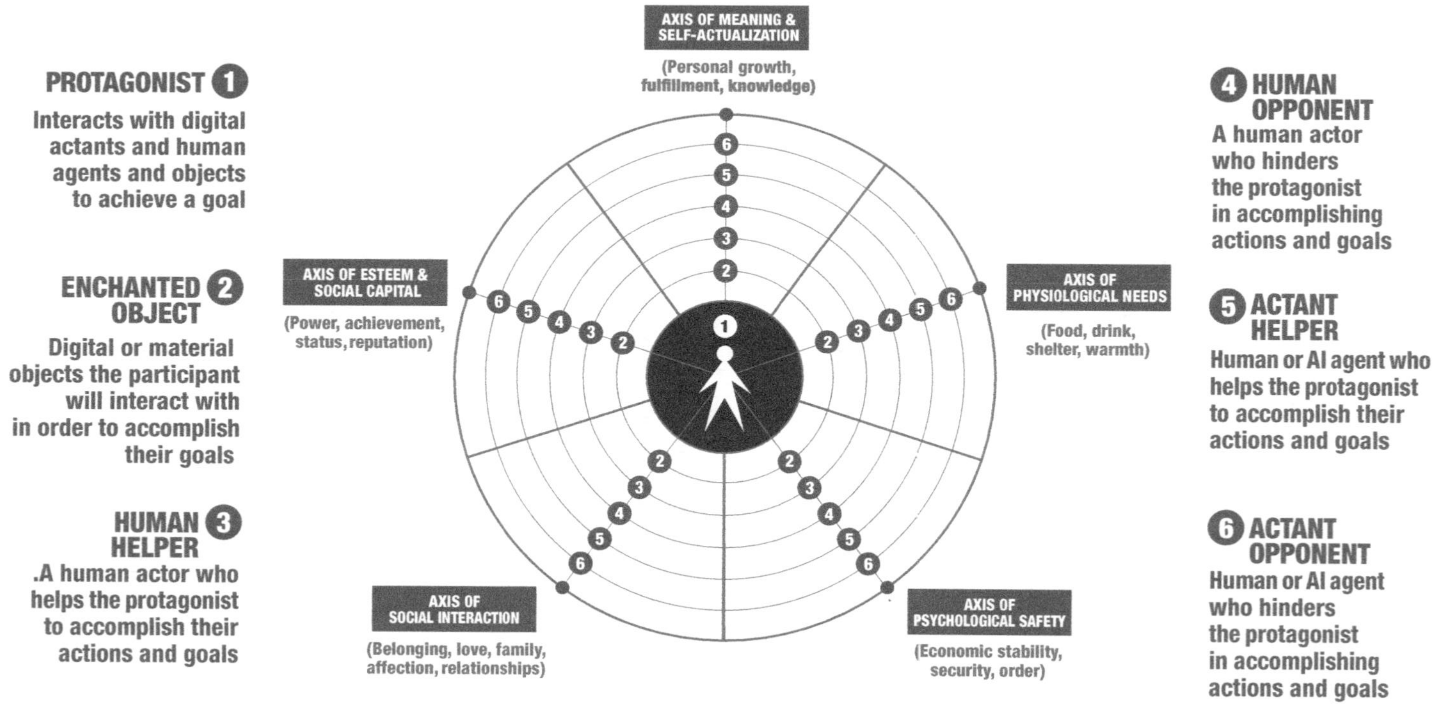

Figure 4.8 Actantial interaction touchpoints within the XR system

Source: Concept and illustration by © 2024 Peter (Zak) Zakrzewski, based on The Actantial Model by Julien Greimas & Maslow's Hierarchy of Needs.

Let's take a look at some immersive media examples using the Actantial Model. *The Invisible Hours* (Tequila Works, 2017) immerses players in a complex murder mystery within a grand mansion, allowing players (protagonist) to act as invisible observers to solve a crime (a goal). The game's narrative pushes the player to uncover clues (objects), while various hints and tools (helpers) assist in solving puzzles. The red herrings and traps (opponents) add challenge to the immersive experience. Success in solving the crime benefits the player, who gains satisfaction and possibly in-game rewards that satisfy the need for esteem, power, achievement, status, and reputation.

Unlike traditional games in which the player directly controls a character, here, players fulfill their desire to solve the mystery through observation and exploration, navigating the mansion's nonlinear narrative to piece together clues at their own pace. In terms of social interaction needs, the players interact with various helpers and opponents manifested through the game's characters, each harboring their own secrets and motives. While the player doesn't interact directly with these characters, understanding the social dynamics between them is crucial. Players observe how characters help or hinder each other's goals, affecting the narrative's unfolding. This observation allows players to discern allies from adversaries in the context of the murder mystery, thereby indirectly influencing the player's investigative strategy.

In terms of meaning, fulfillment, knowledge, and personal growth needs, because of the game's design, which allows events to unfold simultaneously in various mansion parts, players can decide where to focus their attention. By choosing where to be and when, players receive different pieces of information, influencing their understanding of the narrative. This setup encourages repeated playthroughs and explorations, as each provides new information and perspectives, deepening the player's overall understanding of the complex story. Through these axes, *The Invisible Hours* crafts a deeply immersive and intellectually engaging experience, leveraging VR technology to place players at the heart of a narrative puzzle that challenges their perception, inductive reasoning, and understanding of the story.

Immersive Scenario Writing Principle 4: Interactivity

We grow up in the world of social relations, which is full of voluntary and involuntary interpersonal relationships we have with individuals within and outside of our social networks of family, friends, school and work colleagues, customers, strangers on the subway, and baristas at the local coffee shop. As individuals, we see the world through the network of our relationships. As constituent parts of a social system, we can be best understood relative to other people we interact with and to our social ecosystem as a whole.

The fields of sociology and anthropology recognize five categories of observable interactions starting from the basic behaviors such as various physical movements of the body. Actions represent movements social actors perform with meaning and purpose. Social actions are social behaviors meant to directly or indirectly address other people with a goal of soliciting a response from them. Social contacts form the basis of social interactions involving symbols that help us to define social relationships. We recognize people as teachers, co-workers, bosses, and baristas by their cultural context and form relationships with them accordingly. Seeing social interactions as a form of symbolic communication can show us how social life depends on the ways people define themselves and others by the use of social titles, objects, and uniforms.

Personal computer pioneers collectively chose the term "interaction" to define the entire field dealing with the way humans interface with computers. The field of human–computer interaction (HCI) research goes beyond the strict purview of computer science to include design, media studies, and behavioral sciences. The name was popularized by Stuart K. Card, Allen Newell, and Thomas P. Moran in their 1983 book titled *The Psychology of Human–Computer Interaction.* The use of the term "interaction" introduced the notion of a dialogue between human users and their computers to distinguish the latter from other tools with specific but limited applications. The fact that computers could be applied to a variety of tasks made an open-ended dialogue between computers and their users an important design consideration (Card et al., 1986).

Just as one positive interaction with a witty, chatty, and cheerful barista preparing our morning espresso can define our mood for the rest of the day, the interactions we have within virtual spaces will shape the way we feel about the XR experience. Direct I-person interactions can dramatically change the impact of media content on the audience.

In a virtual reality project called *Tree* (Milica Zec & Winslow Porter, 2017), participants are given an opportunity to role-play living as a kapok tree in the Peruvian Amazon rainforest, from seedling to full-grown. The experience incorporates multisensory components, including haptic feedback, wind, scents, and temperature changes, providing a multisensory experience that includes birds landing on branches and a forest fire. Participants are given a natural kapok tree seed to plant before putting on the VR headset, symbolizing a personal commitment to the environment.

The immersion puts the audience in a position to directly experience the dangers of deforestation and climate change, encouraging visitors to reflect on environmental conservation through the emotional and visceral connection of living through "tree subjectivity." To enhance the storytelling, Xin Liu and Yedan Qian in the Fluid Interfaces Group at the MIT Media Lab collaborated with the directors to design and construct the tactile interactions, precisely controlling the physical elements of vibration, heat, wind, and body haptics, with the goal of enhancing the participant's experience of immersively "being" a tree.

Two of the most impactful interactions to include during the scenario writing process include the human-to-human and human-to-machine ones. XR spaces are de facto social media platforms, so the mediation of the human-to-human interaction aspect must be addressed with attention and care. As discussed earlier, developments in AI technology have laid the foundation for a lot more intricate and exciting human-to-machine interactions.

Actantial design can go a long way toward making AI-based agents an integral part of meaningful and innovative media content. The last class of interactions within the XR systems include the machine-to-machine communication among individual bits of code and agents that contributes to the overall user experience and system outcomes.

In the Introduction chapter, we referred to artificial intelligence as the collective unconscious for a reason. Our unconscious patterns of action and movement free us to "just walk" or "just play" without worrying about all of the steps required to put one foot in front of the other. In the same sense, machine-to-machine interactions within XR systems can add value by performing mundane tasks without the users' conscious awareness. They can allow the audience to freely choose between the lean-forward and lean-back modes of immersion depending on their mood and motivation behind the desired experience.

Immersive Scenario Writing Principle 5: A Diachronic View of Experiences That Flow in the Fourth Dimension of Time

Cognitive psychologist Jerome Bruner argued that as humans, we organize our experience of the world around us in the form of narrative stories (Bruner, 1991). Our personal narratives offer us a special version of reality whose acceptability is governed by verisimilitude rather than by empirical verification or logic (1991). Stories are instruments of mind that help us to construct an understandable map of the complex reality outside.

Bruner sketched out ten features that he considered essential to effective narrative meaning-making. He named narrative diachronicity as the chief among them, arguing that, first and foremost, narratives are accounts of events occurring over time. While we may characterize stories in such nontemporal terms as a tragedy or a comedy, all genres, regardless of the tone and content of their stories, describe what are quintessentially patterns of events occurring over time.

What is equally important is that narrative time is measured in subjectively experienced human time rather than abstract time measured by a clock. The significance we give to narrative time is defined by the meaning we assign to events within its compass. Linear sequencing in spoken or written stories is only one device available to the storytellers. Such narrative features as flashbacks, flash-forwards, and temporal synecdoche can all represent various diachronic orders of subjective cognitive experience. Even nonverbal media such as comics and cathedral windows use left-to-right and top-to-bottom conventions to represent narrative diachronicity.

A unique pattern of events over time is an essential part of how we understand the world. This is why we believe that context scenarios need to be considered as diachronic representations of immersive experiences occurring over time. In Figure 4.9, we have outlined three distinct XRXD stages linked by two very important transitions. We have aligned these to the narrative arc where stories draw their audience into the narrative world through exposition. Further entry into the story is facilitated by a rising action.

Full immersion occurs at the moment of a dramatic "crisis" when the audience experiences the Aha! and Wow! moments. Transformation aligns with resolution of the presented dramatic conflict.

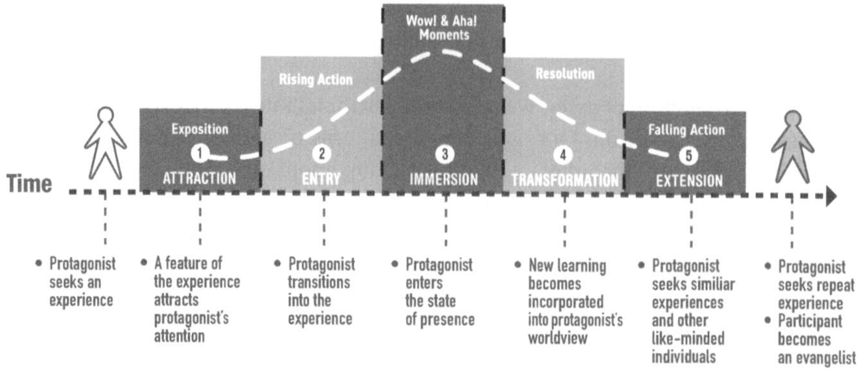

Figure 4.9 The diachronic view of XR experience

Source: Illustration by © 2024 Peter (Zak) Zakrzewski, adapted from Doblin Group's Compelling Experiences model.

Finally, the dramatic term "denouement" describes the final phase of a story in which the various strands of the plot are drawn together in a way that explains and resolves the entire plot. We align this dramatic stage with the extension phase of the immersive experience when the audience members become committed members of the media franchise community.

As discussed in Section II, the user journey map the diachronic model is based on was originally introduced by Larry Keeley of the design consultancy Doblin. The diachronic model can be used to lay out individual XR interactions or entire immersive user journeys, which need to begin with some sort of attraction or an entry point into the experience. An attraction can be a physical affordance like a door into a room or an appeal to the user to engage.

Entry represents the transition during which the audience is transported from their initial interest into the immersion stage of the experience.

Immersion is the stage within which the Aha! and the Wow! moments associated with presence can occur. Supporting presence in the scenario centers around letting the audience shut out the everyday and be fully immersed on the deepest level. Mediated presence is the pinnacle of the XR experience design process. Being immersively present can transform a first-time user into a fan or even a brand ambassador, or if it is disappointing, it can turn them away.

As the audience exits the immersive experiences during the Transformation phase, they rejoin their everyday world with an opportunity to reflect on the immersive experience. This phase offers an opportunity for a meaningful transformation during which the audience gains new knowledge and understanding of the world. The transition between mediated presence and repeat immersion takes a skilled process, culminating in the extension stage.

As we mentioned in Section II, the desired outcome of the extension stage is to support the audience members in staying connected beyond the initial experience and becoming dedicated long-term members of the media franchise community.

Extension is where individual users coalesce into dedicated communities of people who want to share the experience and connect with other like-minded users. Carefully planning for extension can make a difference between building and scaling an influential audience that not only comes back for more interactions with the brand but also recommends it to their near peers and others in their social networks.

Immersive Scenario Writing Principle 6: The 5Es of Usability

We would like to include one more essential scenario writing principle based on the framework proposed by the co-founder of the Center for Civic Design, Whitney Quesenbery. The 5Es framework offers a set of very useful lenses for building and evaluating artificially designed systems (Quesenbery, 2004).

Effectiveness, the first of the 5Es, addresses the question whether the solution is meeting the users' goals. Effectiveness of meeting user needs has to be a guiding question throughout the entire media design process, which starts with discovering the audience.

Efficiency probes the system's speed and performance in accomplishing the user goals. How fast does the system perform or respond to the users' inputs? How accurately does it perform the intended task?

Engagement focuses on the way the system is experienced by the users. How pleasant, satisfying, or interesting is the XR application to use?

Error tolerance refers to the system's ability to deal with any user errors. How well does the system prevent such errors? Does it help users recover from mistakes and misunderstandings?

Finally, ease of learning refers to the comprehensibility and learnability of the system and its rules. This E measure allows the design team to assess how well the system supports initial orientation and deeper learning of the system model. XR systems can be very complex environments with a variety of idiosyncratic rules that users will need to navigate. Helping the audience to learn and internalize the rules of the system is an important task that starts with the context scenario.

Summary

In this chapter, we have discussed how imaginative scenarios can shape the future. We underscored the power of narrative as an immersive media design tool. We covered the ingredients and the process of context scenarios. The chapter covered the importance of starting the ideation process by considering the participants' embodied experience before considering the immersive environment, the objects in the environment, and the types of interactions necessary to support the desired user experience. We also highlighted the power of utilizing the who, what, how, when, and where questions when imagining the audience's journey.

We discussed four context scenario–writing principles based on the body, environment, object, and interaction model, one based on the diachronic dimension of immersive experience, and one on the 5Es of usability framework. With the context scenario ready, the design team is well on the way toward considering the XR technology needed to support the immersive media experience, which is the topic of our next chapter.

References

Andreessen, M. (2011, August 19). Marc Andreessen on why software is eating the world. *Wall Street Journal*. http://online.wsj.com/article/SB10001424053111903480904576512250915629460.html

Botvinick, M., & Cohen, J. (1998). Rubber hands 'feel' touch that eyes see. *Nature*, 391(6669), 756–756.

Bruner, J. (1991). The narrative construction of reality. *Critical Inquiry*, 18, 1–21.

Card, S. K., Moran, T. P., & Newell, A. (1986). *The psychology of human-computer interaction*. Lawrence Erlbaum Associates.

Clark, A. (2003). *Natural-born cyborgs: Minds, technologies, and the future of human intelligence*. Oxford University Press.

Colligan, B. (2011, November 20). How the knowledge navigator video came about. *Dubberly Design Office*. www.dubberly.com/articles/how-the-knowledge-navigator-video-came-about.html

Cooper, A., Reimann, R., Cronin, D., & Cooper, A. (2014). *About face: The essentials of interaction design* (4th ed.). John Wiley and Sons.

Dubberly, H. (2007, March 30). The making of knowledge navigator. *Dubberly Design Office*. www.dubberly.com/articles/the-making-of-knowledge-navigator.html

Gelernter, D. H. (1992). *Mirror worlds or the day software puts the universe in a shoebox: How it will happen and what it will mean*. Oxford University Press.

Genay, A., Lecuyer, A., & Hachet, M. (2022). Being an avatar "for real": A survey on virtual embodiment in augmented reality. *IEEE Transactions on Visualization and Computer Graphics*, 28(12), 5071–5090. https://doi.org/10.1109/TVCG.2021.3099290

Gibson, J. J. (1983). *The senses considered as perceptual systems*. Greenwood Press.

Hiltzik, M. A. (1999). *Dealers of lightning: Xerox PARC and the dawn of the computer age*. HarperBusiness.

Huber, J., Shilkrot, R., Maes, P., & Nanayakkara, S. (Eds.). (2018). *Assistive augmentation*. Springer.

Ishii, H., & Ullmer, B. (1997). Tangible bits: Towards seamless interfaces between people, bits and atoms. *Proceedings of the ACM SIGCHI conference on human factors in computing systems*, 234–241. https://doi.org/10.1145/258549.258715

Kapoor, S., Stroebl, B., Siegel, Z. S., Nadgir, N., & Narayanan, A. (2024). AI agents that matter. *arXiv:2407.01502* [Cs.LG]. https://arxiv.org/abs/2407.01502

Lakhwani, K., Gianey, H. K., Wireko, J. K., & Hiran, K. K. (2020). *Internet of things (IoT): Principles, paradigms and applications of IoT*. BPB Publications.

Latour, B. (2007). *Reassembling the social: An introduction to actor-network-theory*. Oxford University Press.

Levitin, D. J. (2006). *This is your brain on music: The science of a human obsession*. Dutton.

Lightman, H. A., & Patterson, R. (2020, October 7). Discussing the set design of Blade Runner. *American Cinematographer*. https://theasc.com/articles/blade-runner-set-design

Park, J. S., O'Brien, J., Cai, C. J., Morris, M. R., Liang, P., & Bernstein, M. S. (2023). Generative agents: Interactive simulacra of human behavior. *Proceedings of the 36th annual ACM symposium on user interface software and technology*, 1–22. https://doi.org/10.1145/3586183.3606763

Parvini, S. (2024, September 25). Can AI make video games more immersive? Some studios turn to AI-fueled NPCs for more interaction. *AP News*. https://apnews.com/article/ai-artificial-intelligence-video-games-npc-c1327bb9130136d0a5f658f44176c5e7

Quesenbery, W. (2004). Balancing the 5Es: Usability. *Cutter IT Journal*, 17(2), 4–11. www.wqusability.com/articles/5es-citj0204.pdf
Rose, D. (2014). *Enchanted objects: Design, human desire, and the Internet of things*. Scribner.
Ryan, M.-L. (2015). *Narrative as virtual reality 2: Revisiting immersion and interactivity in literature and electronic media* (2nd ed.). Johns Hopkins University Press.
Schiffer, M. B. (1991). *The portable radio in American life*. University of Arizona Press.
Seabrook, J. (2006, October 29). Game master. *The New Yorker*. www.newyorker.com/magazine/2006/11/06/game-master
Shaer, O. (2009). Tangible user interfaces: Past, present, and future directions. *Foundations and Trends in Human–Computer Interaction*, 3(1–2), 1–137. https://doi.org/10.1561/1100000026
Upton, B. (2015). *The aesthetic of play*. The MIT Press.
Watzlawick, P., Bavelas, J. B., & Jackson, D. D. (1967). *Pragmatics of human communication: A study of interactional patterns, pathologies, and paradoxes*. Norton.
Wilson, M. (2002). Six views of embodied cognition. *Psychonomic Bulletin & Review*, 9(4), 625–636. https://doi.org/10.3758/BF03196322
Zakrzewski, P. (2022). *Designing XR: A rhetorical design perspective for the ecology of human+computer systems*. Emerald Publishing Limited. http://doi.org/10.1108/9781802623659

5 Engineering the Magic

The Feasibility of Immersive Experiences

Supporting the Wow! and Aha! Moments

Immersive applications that blur the boundaries between the physical and digital worlds to give us access to advanced computation and digital content are created with a wide array of still-developing XR technologies. Understanding them is essential to creating compelling immersive experiences that allow audience members to look past the technology and attain the Aha! and Wow! moments that signify a true presence. Based on a conversation with the founder of Magic Leap, Rony Abovitz, Kevin Kelly has provided one of the clearest explanations of the importance and the challenge of aligning the still-nascent XR technology with the desired immersive experience of mediated presence.

Shaped by eons of evolutionary adaptation, our brains have developed a set of particular perceptual strategies to make sense of the incoming sensory data. Media technologies from silent movies to superhero blockbusters and digital games full of computer-generated imagery (CGI) and animated visual effects give us a sense of realism by exploiting these perceptual system peculiarities. The perception of apparent motion is grounded in our brain's ability to integrate visual information across time and space. We interpret sequences of still images as continuous motion even though the images are static. The neuroscience of mechanisms for detecting direction-selective neurons and temporal integration enable the seamless experience of motion in everyday life (Deco & Roland, 2010).

Leveraging other perceptual peculiarities, visual effects artists can scan real-world objects such as cars or elephants from a variety of angles and render them as 3D volumetric images that can be displayed on screens or in VR headsets. They can go even a step further and show the audience a Godzilla instead of an elephant, and the audience will still subconsciously perceive the presence of the monster to be real. However, the gut-level illusion of reality will break if even one small detail of the image becomes misaligned. For instance, using a standard mobile phone screen in a head-mounted display will create a major alignment problem when placed right next to our eyeballs if the media makers want to generate the illusion of Godzilla charging at us from 50 feet away. For that illusion to work, our eyes need to be focused 50 feet away.

In a similar way, if the media makers want to let us look at a virtual Jiminy Cricket standing 6 inches from our face using VR goggles, the brain will telegraph the fact

DOI: 10.4324/9781003442462-10

that our eyes are not crossed as they should be in real life but are staring straight ahead instead. We might not be conscious of such optical mismatches, but over a period of time, the subconscious misalignment may contribute to break-in-presence or frequently reported discomfort and fatigue that take us out of the immersive experience.

While the engineers behind XR gear still have a lot of work ahead of them, there is a lot that media makers can do with the existing technology, provided we pay close attention to the chain of immersive verisimilitude. In this chapter, we will cover some of the most important technological building blocks of feasibility and methods of engineering immersive experiences.

Building the Immersive Substrate of Presence

With the context scenario, you and your team have defined the immersive experience you want to create for your audience. Now, it is time to support this vision with a feasible infrastructure. Technology is a fundamental building block of immersive media experience design (XRXD). The advanced media technologies behind all forms of XR distinguish our conscious experience of the physical world from the mediated XR presence experienced in virtual worlds.

Since the "message" of immersive media is the multimodal mediation of presence, when it comes to XR content, no medium quite literally means no message. XR media technologies are the necessary substrate of mediated presence. Breaks in presence occurring when the technology misaligns from the participant's embodied experience are XR content makers' greatest enemy. Immersive experience suffers when the mediated environment fails to consistently support its output message, such as in the case of virtual flight simulators that freeze or slow down when there are too many agents on the screen. Lombard and Ditton defined the experience of immersive presence as the "perceptual illusion of non-mediation" (Lombard & Ditton, 2007, p. 5).

Creating the effortless illusion of nonmediation is the secret to XR media making. The best XR experiences occur when participants forget the technology and immerse themselves in the magic of virtual worlds, not because the technology is not there but because the technology perceptually disappears. It becomes imperceptible.

Building a smooth alignment between the human nervous system, which is a product of phylogenetic and ontogenetic coevolution with a natural environment, and the emerging immersive media technologies requires a deep knowledge of both domains. The matching of human sensory perception with visual and audiovisual displays, haptic input devices, and other XR technologies is as much art as it is science. It is one of the most creative and demanding forms of engineering design because of the unique demands immersion places on its media infrastructure compared to traditional, computational, and mobile forms of media.

XR presents content makers with a constellation of new and unique affordances not previously available to creative professionals. Because of the technological complexity, the most compelling XR experiences can only be created by teams who thoroughly understand the limits and creative potential of currently available technologies.

As mentioned before, designing, prototyping, and developing immersive experiences requires a deep alignment between human sensory perception and the capabilities of emerging immersive technologies. This alignment requires a comprehensive understanding of both domains—how humans perceive and interact with the world and the technologies available to create immersive experiences. Matching sensory human perception with visual and audiovisual displays, haptic input devices, and other technologies is art and engineering.

XRXD demands a creative and technically adept approach to design, since immersive media places unique demands on the infrastructure required to deliver the experience compared to traditional media. Human sensory perception involves a complex interaction between various senses that we will detail later in this chapter. As you strive to create experiences that engage our senses in natural and convincing ways, you must have a fundamental understanding of the technological affordances available to your team and the performance standards required for prototyping and development.

For example, foveated rendering techniques implemented in MR headsets exploit the characteristics of human vision in that our peripheral vision presents us with far lower detail than where our gaze is directed by only rendering high-resolution images in the area of the headset display where the participant is directly looking, optimizing image-rendering performance without sacrificing the visual quality. The foveated rendering technique highlights the importance of aligning technological capabilities with the characteristics of human sensory processing to enhance immersion.

As XR media makers, we must also understand the psychological and physiological responses that the immersion technologies elicit. Creating a sense of immersion depends heavily on accurately synchronizing visual, auditory, and haptic feedback to match the user's actions and expectations (Slater, 2009). Haptic feedback offers one of the new and less explored channels of multimodal composition which can support a wider range of sensory input to enhance realism and engagement. Haptic devices, such as gloves in full VR or tactile feedback that simulates touch sensations when using a smartphone, can provide feedback that is similar to physical-world interactions.

Incorporating these unique-to-XR technologies can improve task performance for media experiences such as virtual surgery training or remote manipulation of robotic systems. Headset limitations such as field of view, resolution, motion-tracking accuracy, and latency have to be carefully considered for room-scale VR experiences such as training emergency room technicians to avoid issues of motion sickness or disorientation, which can severely undermine the training scenario.

Three Perspectives on Technology

In his seminal formulation of a theory of technology, economist and complexity theorist W. Brian Arthur defined technology as a means to fulfill a human purpose via a device, a method, or a process (Arthur, 2009). Technology refers to the application of scientific and engineering knowledge to achieve specific goals, solve problems, or manipulate the lived environment. It encompasses a wide range of tools, methods, processes, and systems designed to enhance human capabilities and improve efficiency in various endeavors. Media technologies include hardware, software, and tools used to compose, create, produce, deliver, and manage media content. This ranges from simple tools used like a pencil to sophisticated virtual reality headsets. Based on Arthur's foundational work, in the context of media design, we define technology in three different ways, depending on the purpose and perspective of the user.

Engineers and Technologists

To the engineer, technology represents the entire collection of devices, methods, processes, and engineering practices representing both the limits and the creative potential of feasibility of change at a given point in time (2009). From the perspective of engineers and technologists, technology is seen through the lens of methods, processes, and engineering practices. Engineers primarily focus on optimizing performance, overcoming technical challenges, understanding the constraints and possibilities of the system components they are working with, and ensuring the technological feasibility and sustainability of the system for the life of the service, product, or experience.

Engineers working on a new mixed reality headset might concentrate on perfecting spatial tracking and real-time rendering using the fastest processing components available to ensure that digital objects remain stable and responsive as users move around. Engineers working on an enterprise training program might focus on using off-the-shelf technologies that they are confident will be easily available and repairable for the expected life of the program.

Media Makers and Content Creators

To experience designers and media makers, technology represents an essential ingredient of creative disruption that amplifies human ability, desire, and behavior, enabling a fulfillment of human purpose, as a device, a method, or process (2009). XR technology, in particular, is the essential building block of the immersive substrate that supports and elicits presence. Media makers leverage media technologies such as cameras, audio gear, lighting equipment, computer workstations, and editing software to craft compelling narratives and build engaging environments. When using media technology, content makers emphasize desirability of the content for the audience and the fulfillment of the educational, entertainment, or task-specific purpose of the project.

Audience or End Users

Audiences of well-engineered digital products experience the underlying technology in a way described by science fiction writer Arthur C. Clarke as being indistinguishable from magic. Few people using their mobile phones throughout the day pause to think about all of the advanced technologies that allow them to effortlessly send and receive messages, play games, navigate to their next destination with real-time positioning updates, and check posts on their favorite social media platforms.

Our smart devices are useful because they just work. If movie audiences pondered the technologies behind filmmaking, they would miss the forest for the trees, the story for the technical gear. Perceiving technology as a form of magic allows us as audience members to focus on what matters to us—being entertained or effortlessly and seamlessly performing complex tasks in our lives. While we may not understand the underlying "under the hood" mechanisms, we appreciate the affordances that technology provides. People playing *Beat Saber* might marvel at the ability to wield a lightsaber with tactile sensations and spatial sound effects without the need to comprehend the sophisticated tracking and rendering technologies required to pull off the illusion.

Engineering Design: It's a Remix

Clarke's intriguing definition of advanced technology as magic positions novel application of disruptive technologies as a core ingredient of media innovation. Technologies can amplify certain kinds of behavior and suppress others (Clark, 1999). To an immersive media maker, technology represents an essential component of creative disruption when it is based on its power to address core human needs, abilities, and desires. Choosing the right technological ingredients depends on the team's ability to navigate the still-uncharted territory to mix and match the appropriate technologies to support an immersively mediated presence for the core audience. Immersive media designers construct experiences from the domains they know and understand.

Brian Arthur offers one of the most inspiring definitions of engineering when he argues that "Good design . . . is like good poetry" (Arthur, 2009). He suggested that (1) an "articulate utterance in technology requires deep knowledge of the domain in question"; (2) "a fluency in the vocabulary of components used"; (3) "a 'knowingness' of what is natural and accepted in the culture of that domain" that entails "a familiarity with standard modules, previous designs, standard materials" (p. 24). The other essential qualities include intuitive and imaginative forms of knowledge and "taste" that allow the designers to choose the right ingredients from among the many possible ones. Arthur has provided, in our opinion, one of the best definitions of the creative power and demands of media engineering design.

> *An articulate utterance [in technology] ... requires deep knowledge of the domain in question: a fluency in the vocabulary of components used; a familiarity with standard modules, previous designs, standard materials ... Intuitive knowledge, feeling, past use, taste—all these count ... Good design in fact is like good poetry ... in the sheer rightness of choice from the many possible...*
>
> W. BRIAN ARTHUR

From Interacting With Computers to Experiencing Information

The evolution of interaction design from the once groundbreaking human–computer interaction (HCI) to the emerging immersive media landscape marks a significant shift not only in how users will interact with advanced computer systems but, maybe even more importantly, how the system designers conceptualize experiential access to information. The HCI field developed during the era of personal computers, which gave an audience, outside of computer science, access to computation and information processing. Naturally, for decades since the 1980s, the focus of HCI design was on usability. In the era when information can be not just accessed but experienced through rich multimodal virtual environments, the conversation moves from providing usability to mediating presence.

Over the past six decades, HCI techniques have evolved significantly, driven by technological advancements and a deeper understanding of user behaviors, needs, and commitments. The earliest computer interfaces started as text-based command-line interfaces (CLI), which required users to memorize the machine language of commands and syntax. CLI was suitable for technically skilled individuals but challenging for the general public. The introduction of graphical user interfaces (GUIs), initially advanced at Xerox PARC in the 1970s and popularized by the introduction of the Apple Macintosh in 1984 and followed by Microsoft Windows in 1985, marked a significant leap in HCI. GUIs implemented visual representations of windows, icons, menus, and pointers (WIMP), making computers more accessible to a broader audience.

As multimedia applications and digital games became more prevalent, the limitations of single-modality interfaces became evident, leading to the call for richer multimodal interactions. The resulting imperative of natural user interfaces (NUI) involve interaction techniques that align with inherent human capabilities and behaviors, such as touching to sense objects, gesturing to express ideas, using voice to communicate meaning and emotion, and the natural way the human visual system helps us to interact with the world.

For instance, we overcome the limitations of our cognitive processing by selecting only what is important. The fovea, located at the center of the gaze, processes selected visual information with high-resolution clarity, leaving the less precise peripheral vision to track everything else. Foveated rendering is now a core part of XR technology. Advanced eye-tracking technologies also allow designers to make experiential access to information more seamless and intuitive. Technological advancements have created other opportunities to make human+computer collaborations more natural and intuitive.

While GUI interfaces made an enormous leap to allow individuals without a computer science background to interact with computers in a "usable" way, they still forced people to comply with the HCI way of doing things. Users had to stare at flat screens and operate computer mice while sitting, often for hours, in unnatural positions that caused strained eyes, wrists, necks, and backs. XR technologies offer us an opportunity to change this dynamic by leveraging interactions based on computer vision that can sense the user's body movements to create natural and intuitive access to computation that enhances the sense of immersive realism. Commercially available devices like Microsoft HoloLens, Apple Vision Pro, and Meta Quest enable XR experience designers to develop NUI-based interfaces that incorporate gesture recognition to allow users to interact with digital content in a way that mimics real-world interactions.

The Broad Landscape of Immersive Media Technology Architectures

As we mentioned previously, extended reality (XR) is an umbrella term that encompasses several technology architectures that may be broadly categorized along the virtuality continuum as AR, MR, or VR, broadly implemented as either mobile AR, projection-based AR, and MR or VR experiences implemented with head-mounted displays (HMDs) and tracking technologies. Each of these architectures offers unique advantages and limitations. Your choice will be driven by balancing technological affordances, the interaction and data input/output needs of your project, and budgetary concerns.

Understanding the affordances and limitations of each dominant architecture is crucial for choosing the right architecture for your project and selecting specific products for your first working prototype. While the technology is evolving rapidly, the following represents an outline of well-established architectures in terms of available technologies for both enterprise and consumer applications.

Mobile AR

Mobile AR utilizes smartphones or tablets to overlay digital content onto the real world. Users view the augmented content through the device's camera and screen. Mobile AR is highly portable and accessible, as it leverages the widespread availability of smartphones and tablets.

Smartphones are equipped with cameras, microphones, and various sensors to track movements, perceive depth, detect orientation, sense motion, determine when an object is close to the phone, and more. This provides XR apps with the data needed to implement the magic of AR apps. In this section, we'll describe the most commonly used sensors and capabilities that provide the technological foundation that enables the interaction and sensing functions of AR experiences.

Users can easily access AR experiences through apps without needing additional hardware. Mobile devices are equipped with cameras and sensors (e.g., GPS, accelerometer) that enable real-time tracking and interaction with the environment. Examples of mobile AR include IKEA Place, which allows users to visualize furniture in their homes using a smartphone or tablet, and the classic *Pokémon GO* mobile game that launched AR in the public imagination.

Projection-Based AR

Projection-based AR, also known as spatial augmented reality, uses projectors to cast digital content onto real-world surfaces and objects. This approach does not require users to wear or hold any devices. Since the content is projected directly onto physical surfaces, users can see the augmentations with their natural field of view. Users can interact with the augmented content without needing to hold or wear devices. Projection-based AR systems are typically stationary, with projectors and cameras set up in specific locations to track and augment the environment. Cameras and sensors track the user's movements and the position of objects, ensuring that the projected content remains accurately aligned. Users can interact with the projected content through gestures, touch, or tools, enhancing the immersive experience.

Application areas include manufacturing, where projection-based AR can guide workers through assembly processes by projecting step-by-step instructions onto work surfaces. Projection mapping has also been used in theater productions and live events to create dynamic visual effects, including projecting 3D art on buildings.

Projection AR overlays digital images and objects onto real-world surfaces via projectors. Unlike mobile AR, projection AR does not require users to wear a headset or glasses or use a smartphone or tablet; instead, the projection interacts directly with the real-world environment.

A widely acclaimed example of projection AR is TeamLab's *Borderless* at the Mori Art Museum in Tokyo. It was an interactive exhibit using 3D projection mapping, sensors, and interactive technologies to make the artwork respond to the movement and presence of visitors. Digital projections are mapped onto walls, floors, and ceilings, creating the illusion that digital content seamlessly interacts with the physical world. In *Universe of Water Particles,* a waterfall is projected onto a wall, and as visitors stand in front of it, the water reacts and flows around them. *Borderless* opened on June 21, 2018, and ran through August 2022. The exhibition reopened at Azabudai Hills in Tokyo on February 9, 2024, and is one of the most visited installations in art world history and serves as a powerful example of how 3D projection mapping combined with interactivity can create immersive experiences that blur the line between the digital and physical worlds.

Projection Mapping

Projection mapping involves projecting visuals onto physical-world surfaces, transforming everyday objects into interactive displays. Transformations adapt the projected content to the contours of the physical object the image is projected onto. To implement 3D projection mapping, you need a high-quality projector, a computer, and 3D content creation software. The type of projector required will vary depending on the size, scale, material, and ambient lighting conditions of the surface you want to project onto. 3D projection mapping manipulates the projected image to match the 3D surface and can create illusions and effects. Substance 3D from Adobe, while not specifically designed for AR, can be used to create detailed 3D models and textures for a projection AR project. You must ensure your 3D models and textures are optimized for real-time use.

Computer Vision

Computer vision algorithms are used in projection AR systems to (1) detect surfaces and perform 3D mapping to accurately project 3D models on complex surfaces; (2) recognize objects and track them for dynamic alignment of projections; (3) recognize user gestures and interactions; (4) determine pose and perform calibration; (5) adapt to the environment to account for changes in the physical spaces; and (6) enable seamless AR without the need for physical markers. These core functions enable digital content to blend effectively with the real-world environment and support interaction without devices.

Camera and Sensors

Cameras and sensors provide the environmental data required for immersive systems to place objects, track motion, project images, and more using computer vision and world-modeling algorithms to interpret the real world and the objects in it. Projection AR systems use cameras and depth sensors to capture the physical environment, allowing the system to project digital images directly onto surfaces, such as walls, tables, or irregularly shaped objects. In environments with irregular surfaces, the system relies on depth-sensing cameras and 3D mapping algorithms to ensure that projections conform to the physical contours of the room or object.

Mobile AR relies on the cameras and sensors in smartphones and tablets to interpret the physical environment. Data from the built-in camera is combined with data from sensors (e.g., accelerometers and gyroscopes) to detect movement and calculate the orientation of the device. LiDAR (light detection and ranging) scanners enhance depth perception and are widely integrated into devices like the iPhone, enabling AR applications to understand the spatial layout of the physical environment.

With MR applications, digital objects are overlaid onto the physical world and are used for interactions. MR headsets incorporate cameras, depth sensors, infrared scanners, accelerometers, and gyroscopes to collect the data needed to map the physical environment in real time and enable interaction and feedback.

In VR, cameras and sensors are used to track the user's movements and the position of the headset. While VR does not ordinarily depend on the physical environment, VR headsets may require external sensors (e.g., base stations) to implement outside-in tracking. On the other hand, inside-out tracking uses cameras in the headset to map the user's position in physical space without base stations. Accurate determination of position and movement is required to accurately position virtual hands or avatars and update the display in real-time to prevent VR sickness.

MR or VR Using Head-Mounted Displays (HMDs)

HMDs for MR and VR are wearable devices that provide immersive experiences by completely or partially overlaying digital content onto the user's field of vision. MR devices blend real and virtual worlds, while VR devices create entirely virtual environments. HMDs offer a high level of immersion by providing a wide field of view and blocking out the real world (VR) or seamlessly blending real and virtual

Table 5.1 Comparison table of immersive technology architectures

	Projection-Based AR (a.k.a. spatial augmented reality)	*Mobile AR*	*Head-Mounted Displays (HMDs) for MR*	*Head-Mounted Displays (HMDs) for VR*	*Cine VR (360 video)*
Description	Uses projectors to cast digital content onto real-world surfaces and objects. This approach does not require users to wear or hold any devices.	Utilizes smartphones or tablets to overlay digital content onto the real world. Users view the augmented content through the device's camera and screen.	Wearable devices that provide immersive experiences by overlaying digital content onto the user's field of vision. MR devices blend real and virtual worlds.	Wearable devices that provide immersive experiences by taking over the audiovisual sensorium completely. VR devices create entirely virtual environments.	Immerses viewers in a fully spherical video environment around them, allowing a sense of presence in the virtual scene; may be viewed with a headset or dome theatre installation.
Affordances	Natural field of view: Since the content is projected directly onto physical surfaces, users can see the augmentations with their natural field of view. Hands-free interaction: Users can interact with the augmented content without needing to hold or wear devices. Fixed installation: Projection-based AR systems are typically stationary, with projectors and cameras set up in specific locations to track and augment the environment.	Portability: Mobile AR is highly portable and accessible, as it leverages the widespread availability of smartphones and tablets. Ease of use: Users can easily access AR experiences through apps without needing additional hardware. Camera and sensors: Mobile devices are equipped with cameras and sensors (e.g., GPS, accelerometer) that enable real-time tracking and interaction with the environment.	Hybrid immersion: HMDs seamlessly blend real and virtual elements (MR). Mixing: Pass-through (complete recreation of real world, e.g., Apple Vision Pro) or Translucent Displays (e.g., Microsoft HoloLens) Advanced tracking: MR HMDs use advanced tracking technologies (usually inside-out tracking, to monitor the user's movements and interactions in real time).	Full immersion: HMDs offer a high level of immersion by providing a wide field of view and blocking out the real world (VR) or seamlessly blending real and virtual elements (MR). Advanced tracking: HMDs use advanced tracking technologies (e.g., inside-out tracking, external sensors) to monitor the user's movements and interactions in real-time. Multisensory feedback: HMDs can integrate visual, auditory, and haptic feedback to enhance the sense of presence and realism.	Partial immersion: HMDs offer immersion by providing a wide field of view, but only from a single position, and blocking out the real world. Accessibility: Can be viewed on a mobile devices, or desktop computers with a 360-video player. Fixed installation: optional, may be projected in a dome theatre. Interaction and feedback: None unless combined with other technologies.

(*Continued*)

Table 5.1 (Continued)

	Projection-Based AR (a.k.a. spatial augmented reality)	*Mobile AR*	*Head-Mounted Displays (HMDs) for MR*	*Head-Mounted Displays (HMDs) for VR*	*Cine VR (360 video)*
Interaction and Accessibility	Hands-free interaction: Projection-based AR allows users to interact with digital content without holding or wearing devices, making it more accessible for tasks that require manual dexterity. Natural viewing: Users can view augmented content with their natural field of view, reducing the cognitive load associated with switching attention between a screen and the physical environment.	Device dependency: Requires users to hold devices, which can be limiting in certain contexts.	Device dependency: Requires users to wear devices, which can be limiting in certain contexts.	Device dependency: Requires users to wear devices, which can be limiting in certain contexts. Advanced, high-precision tracking may require external beacons.	Interaction is limited to head movement for looking around the environment, offering a passive experience. Accessibility is high since viewers do not need complex controllers or input devices, making it suitable for users with limited technology resources.

(*Continued*)

Table 5.1 (Continued)

	Projection-Based AR (a.k.a. spatial augmented reality)	*Mobile AR*	*Head-Mounted Displays (HMDs) for MR*	*Head-Mounted Displays (HMDs) for VR*	*Cine VR (360 video)*
Use Case Suitability	Fixed environments: Ideal for environments where tasks are performed at a fixed location, such as assembly lines, classrooms, or stages. Collaborative work: Enables multiple users to view and interact with the same augmented content simultaneously, fostering collaboration.	Portability: Better suited for applications that require portability, such as outdoor gaming or remote training.	Mobility: Better suited for applications that require access to real world and virtual worlds simultaneously.	Mobility: Better suited for applications that require mobility and accurate tracking, such as indoor simulations.	Accessibility: Ideal for scenarios focused on immersive storytelling, virtual tours, and training experiences where an immersive perspective enhances understanding or emotional impact without requiring user interaction.
Technological Integration	Seamless integration: Can seamlessly integrate with existing physical environments, enhancing them without the need for additional devices. Durability and reliability: Projection systems are durable and can operate continuously without the need for charging or battery swaps, making them ideal for industrial applications.	Advanced features: Offers advanced features such as spatial audio, high-resolution displays, providing a richer sensory experience. Durability and reliability: High, since it depends on end user's smartphone.	Advanced features: Offers advanced features such as spatial audio, haptic feedback, and full-immersion displays, providing a very rich sensory experience. Durability and reliability: Moderate, dependent on the specific technology.	Advanced features: Offers advanced features such as spatial audio, haptic feedback, and full-immersion displays, providing the richest sensory experience. Durability and reliability: Moderate, dependent on the specific technology.	Advanced features: Spatial audio with head tracking and HRTF customization is possible. Durability and reliability: Very high due to simplicity of the technology.

elements (MR). HMDs use advanced tracking technologies (e.g., inside-out tracking, external sensors) to monitor the user's movements and interactions in real time. HMDs are usually part of a framework providing multisensory feedback that can integrate visual, auditory, and haptic feedback to enhance the sense of presence and realism.

VR headsets designed for consumer applications are available from multiple vendors and are widely used for immersive gaming experiences, while more specialized devices with advanced capabilities target enterprise applications for training, simulation, and task assistance in fields such as medicine, engineering, and design. VR and MR headsets enable designers to create and manipulate 3D objects in a virtual space while providing high-fidelity haptics, visuals, and sound.

Orchestrating the Technological Substrate of Immersive Experience

Orchestration refers to the process of combining the events and objects in the scenario with technological affordances to create a cohesive experience. This involves synchronizing components, designing participant flow, and plenty of testing and iteration. Given this is a new realm with rapidly evolving technology, plan to

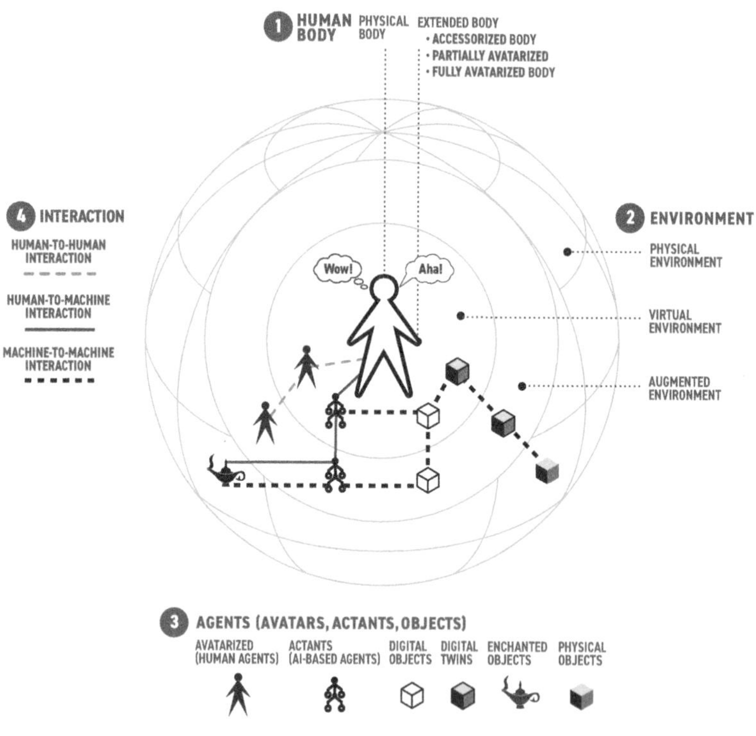

Figure 5.1 Human body in the XRI world framework

Source: Illustration by © 2024 Peter (Zak) Zakrzewski, concept by Peter (Zak) Zakrzewski with David Tamés.

experiment with a range of commercially available and experimental or home-grown tools to determine the best solution for your specific scenario. Orchestrating immersive experiences from the perspective of engineering the XR tech means managing a number of potentially conflicting considerations.

When considering the technology mix, we recommend using the same (1) human body, (2) environment, (3) agents, and (4) interactions framework introduced in the scenario-writing chapter. All of these basic elements need to work together seamlessly to deliver the optimal audience experience. For example, VR experiences rely heavily on synchronizing the participant's avatar movements with real-time inputs and machine feedback. Audience members must be guided through XR consent in a logical progression of interactions that allows them to experience the sense of effortless flow. This may involve designing tutorial stages, setting interaction cues, or planning feedback loops.

The Embodied Human Actor in the XRI World

One of the major strategic decisions XR content makers need to consider is how to physically conceptualize the audience members within the immersive space. To create compelling immersive experiences, we begin with an understanding of the fundamental building blocks available in the new paradigm of immersive media. We start with the human body as the locus of experience, expanding out to include the immersive environment, and agents and objects each offering unique characteristics and opportunities for interaction. The human body can be immersed within the XR experience in purely physical, augmented, or virtual environments. These include direct interaction with tangible objects in the physical environment, enhanced capabilities through augmented environments, and full immersion in virtual environments. Embodiment in immersive media can be realized by extending, accessorizing, partially avatarizing, or fully avatarizing the human body, each providing different levels of interaction and immersion and requiring different engineering approaches.

Extended Body

Technology may be used to extend the body's capabilities, such as through wearables that enhance physical abilities or senses. Exoskeletons, for example, can extend human strength or endurance in a physical environment. Another possibility is allowing participants to perform physically impossible actions in the real world, such as flying or interacting with distant objects.

Accessorized Body

The accessorized body is a body augmented with wearables or external accessories that enhance the user's physical and cognitive abilities, such as AR glasses that overlay digital information onto the physical world or a magic ring that is used as input to the system or a heads-up display that overlays real-time data in the user's field of view while they interact with the physical world.

Partially Avaterized Body

Partial avatarization refers to the representation of specific parts of the body in an augmented or virtual environments. For example, in many VR applications, participants can see a virtual representation of their hands, which they use to interact with objects in the environment using physical controllers. This creates a sense of embodiment as they use their virtual hands to control their movement and interact with physical objects, reinforcing the sense of agency and immersion in the virtual environment. Studies have shown that even partial representations of the body can create a strong sense of presence and immersion for users in virtual reality (Slater, 2009).

Fully Avaterized Body

Full avatarization involves the complete representation of the participant's body as a virtual avatar in the digital environments. Users are able to control a full-bodied avatar that mirrors their physical movements. This form of embodiment can be found in social VR experiences like VRChat, in which users can see their entire virtual body and interact with the virtual world as if they were physically present. However, this does require wearing a set of supported trackers.

Another example is *Lone Echo* (Ready at Dawn, 2017), a VR adventure game set aboard a space station in which players move in zero gravity while controlling their full-body avatar, including arms and hands, as they interact with the environment. Motion tracking enables precise control over the avatar's limbs, enhancing the experience of embodiment. Full avatarization allows for high levels of realism, immersion, and interaction, as participants sense that their entire body is immersed in the virtual environment.

Movement

Accounting for user movement is a vital component of spatial computing. The ability to incorporate physical movement into XR experiences varies significantly among XR devices due to differences in technological capabilities. In mobile AR, digital elements overlay a screen image of the real world, with interactions typically through smartphones, tablets, or AR glasses, limited to the user's physical space and using touch screens or simple gestures. MR blends real and virtual worlds, requiring sophisticated hardware like MR headsets for users to interact with digital objects as part of the real environment.

In VR, users are fully immersed in a computer-generated environment, wearing VR headsets that track head and hand movements for navigation and interaction. Each technology offers unique movement and interaction methods, catering to different use cases and levels of immersion. With a headset, predicting where the user will look at any moment is impossible, so cues must guide their attention to specific objects or characters. Spatial audio and visual indicators are effective in providing such cues.

In VR and MR, head rotation, body movement, and teleportation offer navigation options. Inspiration can be drawn from cinema and architecture, but XR fundamentally differs from traditional media. In MR, users see their physical surroundings through pass-through cameras, complicating the positioning of virtual objects as they must blend with the real world and remain discoverable and interactable.

Cues must guide users toward relevant physical and virtual positions, akin to mise-en-scène in cinema or interior design in architecture, working within existing spatial constraints. MR approaches include overlaying a virtual window on the real world (e.g., Microsoft's HoloLens) or regenerating the real world in virtual space (e.g., Apple Vision Pro). The display field of view (FOV) in XR is typically less than human vision, necessitating design adjustments for limited peripheral vision in AR, often moving critical content with the user while avoiding discomfort from locking content to head movements. Each technology offers unique ways for users to move and interact within their respective environments, catering to different use cases and levels of immersion, as summarized in Table 5.2.

Table 5.2 Comparison of user movement in AR, MR, and VR

Form	*Mobile Augmented Reality (AR)*	*Mixed Reality (MR)*	*Virtual Reality (VR)*
Movement	Users move in the real world and see digital overlays on their devices. Body movements are natural and unrestricted by the technology.	Users can move freely in their physical space, and the MR system tracks these movements to adjust the virtual elements accordingly. This creates a seamless interaction between real and virtual objects.	Users are typically stationary or move within a confined physical space. Movement in the virtual world can be achieved through locomotion, teleportation, walking in place, or using controllers.
Interaction	Limited to tapping on screens or using basic gestures. The digital elements do not dynamically interact with the real world.	Users can manipulate virtual objects using gestures, voice commands, or controllers. The virtual objects respond to real-world physics.	Highly interactive, using hand controllers, gloves, or even full-body tracking systems. Users can pick up, move, and manipulate virtual objects as if they were real.
Devices	Smartphones, tablets, AR glasses	Overlay headsets (e.g., Meta Quest Pro and Apple Vision Pro); pass-through headsets (e.g., Microsoft HoloLens)	VR headsets (e.g., Meta Quest 3, HTC Vive, Varjo, etc.)

Tracking

Tracking technology is a critical component of high-end XR applications that incorporate embodied user interactions in immersive spaces. A variety of devices and technologies are used to determine how accurately the system can track and translate the user's movements into the virtual environment.

Two primary tracking methods are currently used: inside-out and outside-in tracking. Inside-out tracking uses sensors and cameras on the headset, offering easy setup and portability, making it ideal for standalone headsets. On the other hand, outside-in tracking relies on external sensors for higher accuracy and larger play areas, making it suitable for high-end systems and professional applications. Understanding the differences between these tracking methods will help you choose the right system for your needs and will ensure an optimal experience for your participants.

Inside-Out Tracking

Inside-out tracking uses sensors and cameras mounted on the headset itself to track the user's position and movements within the physical space. The headset processes the data from these sensors to determine its location and orientation relative to the environment. The headset is equipped with multiple cameras and sensors that capture the surrounding environment. These cameras typically face outward and scan the room to identify fixed reference points.

Systems typically use SLAM (simultaneous localization and mapping) algorithms to create a map of the environment and track the headset's position within that map. This involves continuously updating the map as the user moves. The camera and sensor data are processed in real time to determine the headset's position and orientation. This allows for accurate tracking of head movements and, in some cases, hand movements if the headset's cameras also track the controllers.

For example, the Meta Quest headsets use inside-out tracking with multiple built-in cameras to track the user's movements. The cameras scan the environment and use SLAM algorithms to provide accurate positional tracking without external sensors. Standalone headsets do not require external sensors or a connection to a desktop computer. This makes them more portable and easier to set up. Inside-out tracking supports room-scale VR experiences, allowing users to move freely within a defined play area without needing external tracking devices.

Outside-In Tracking

Outside-in tracking uses external sensors and cameras placed around the physical space to track the position and movements of the headset and controllers. The external devices capture the user's movements and relay the data to the headset or connected computer for processing. The tracking system relies on external sensors or cameras around the play area to track the headset and controllers by detecting signals or visual markers. In some systems, base stations (a.k.a. beacons) emit infrared signals that are picked up by sensors on the headset and controllers. The

Table 5.3 Comparison table of inside-out and outside-in tracking

Feature	*Inside-Out Tracking*	*Outside-In Tracking*
Setup	Easy to set up, external sensors are not required	Requires installation of external beacons or sensors
Portability	Highly portable, suitable for standalone headsets	Less portable, typically requires a fixed interaction area due to the installation of external devices
Accuracy	Good accuracy but can be affected by occlusion	Excellent accuracy, less affected by occlusion
Cost	Lower cost due to fewer components required	Higher cost due to additional hardware requirements
Flexibility	Limited to the FOV of the of the cameras on the headset	Can cover larger areas with additional sensors
Use Cases	Standalone VR headsets; room-scale VR	High-end headset; room-scale VR systems; professional applications

system triangulates the position of the devices based on the signals received. The data from the external sensors are processed to determine the precise position and orientation of the headset and controllers. This allows for highly accurate tracking of movements within the play area.

These systems also enable the use of additional trackers placed on objects or body parts to track those movements. For example, the Valve Index uses outside-in tracking with two or more base stations (a.k.a. Lighthouse sensors) that emit infrared signals. The sensors on the headset and controllers detect these signals, allowing for precise tracking of movements. Outside-in tracking is commonly used in high-end VR systems that require precise and accurate tracking, such as those used for professional applications and advanced gaming. This tracking method is suitable for larger play areas, as additional sensors can be added to cover a wider space and ensure consistent tracking.

Positional Tracking

Implementing positional tracking ensures that virtual objects and interactions are accurately aligned with the user's physical movements. Tracking is typically accomplished using one of two methods: inside-out tracking using internal sensors or outside-in tracking using external sensors.

Inside-out tracking has been implemented by contemporary VR headsets like the Meta Quest and VIVE XE Elite, which use built-in cameras and sensors to track the user's position and movements within the physical space. This method eliminates the need for external sensors and allows for greater mobility. These sensors capture the user's head and hand movements in real-time, translating them into the virtual environment in order to ensure that virtual objects and interactions are accurately aligned with the user's physical actions. The advantage of inside-out systems is that they provide more flexibility in terms of where they can be used and are cost effective. However, they have issues with position accuracy and occlusion.

On the other hand, outside-in tracking combines external base stations with spatial mapping. Systems like the HTC Vive use external sensors or base stations placed around the room to track the user's movements. These sensors emit infrared signals that are detected by the headset and controllers, providing precise tracking of the user's position and orientation. The system creates a detailed map of the physical play area, ensuring that virtual objects remain accurately positioned relative to the user. The advantage of outside-in systems is that they provide higher position accuracy, suffer less from occlusion issues, and can support larger play areas. The disadvantage is the requirement of installing external base stations in the play area.

The Environment: Planning the Immersive Spatial Experience

The environment is the context in which the immersive experience occurs. It can take place in one or more of three levels: virtual, augmented, and physical. No matter what technological architecture you choose to work with (e.g., mobile AR, MR, VR, etc.), you must always consider and design activities in all three layers of the environment simultaneously. The physical environment refers to the real-world surroundings, while the augmented environment overlays digital information onto the physical world, enhancing perception and interaction. The virtual environment fully immerses users in a computer-generated world, engaging multiple senses to create a convincing feeling of mediated presence.

Virtual environments are completely synthetic spaces in which all elements are computer-generated. The canonical example is fully immersive VR worlds in which participants interact primarily within a virtual world, and most of their sensorium is fully mediated by the computer. Augmented environments enhanced physical spaces with digital information. AR experiences, such as using a smartphone to view virtual objects overlaid on the real world, fall into this category. The physical environments blend the physical world with the virtual world to expand the work or entertainment opportunities for the audience. In some cases, physical environments may be enriched with digital content through AR or MR technologies. It's not unusual for participants to pick up physical objects with a digital twin in the virtual world (see Digital Twin below).

Spatial Mapping and Calibration

To set up a room-scale VR environment, users must define a clear play area free of obstacles. The recommended minimum space for consumer systems is typically 6.5 ft × 6.5 ft (2 m × 2 m) to allow for comfortable movement. The VR system is calibrated to match the virtual environment with the physical space. This involves mapping the boundaries of the play area and ensuring that virtual objects are correctly aligned with real-world coordinates.

To prevent users from colliding with walls or obstacles, room-scale VR systems often include a chaperone system. This feature provides visual or haptic warnings when users approach the boundaries of the play area, guiding them back to the center. Some VR applications use teleportation to help users navigate larger virtual environments without hitting physical boundaries. Users can point to a location within the virtual world and instantly "teleport" there, allowing for seamless exploration.

Agents and Objects

Various entities in XR, such as other human agents, their avatars, AI-based agents (actants), digital twins, and enchanted objects, play roles in creating realistic and dynamic immersive interactions. Avatars represent users in the virtual environment, while AI agents simulate realistic behaviors and provide assistance. Digital twins are virtual counterparts of physical entities. Enchanted objects are physical objects augmented with digital capabilities that allow users to engage in what is sometimes referred to as tangible user interactions (TUI) or natural user interactions (NUI), which do not require the users to extend their bodies with cognitive or sensory enhancements.

XR environments need to be capable of accommodating a great diversity of objects and agents from physically present human agents to avatarized human agents who are digital representations of participants. AI-based characters can also play a significant role in immersive experiences such as games and educational apps. These can take the form of avatars in virtual meeting spaces. Digital objects are virtual items that exist in virtual or augmented environments. They could be, for example, models of body parts or medical instruments that provide realistic, hands-on learning experiences for medical students to practice surgical techniques. These objects allow the students to interact with complex systems in a safe, controlled environment.

Digital twins are digital replicas of physical objects used for simulation, monitoring, or control. For example, a digital twin of a robot arm can be used in a virtual environment to mimic the actions of the real arm at a separate location. These are physical objects enhanced with digital capabilities. They're described by David Rose as "enchanted objects," physical items enhanced with digital capabilities (Rose, 2014). These objects that appear in the real world may have embedded sensors or displays that allow them to interact with participants in novel ways.

Finally, participants need to be accommodated in interacting with physical objects such as tools or furniture that are tangibly present within the physical environment.

Engineering the Immersive Interactions

Interactions define how participants engage with the environment, objects, and other agents within the scenario. They include the range of human-to-human, human-to-machine, and machine-to-machine interactions. Human-to-human interactions encompass direct interactions between human participants, either in the physical world or augmented or virtual worlds mediated by technology. For example, participants can communicate through avatars in a virtual space (e.g., VRChat). Participants provide input to machines, such as through gestures, voice commands, or touch screens. In VR, this might involve using motion controllers to manipulate virtual objects.

In machine-to-human interactions, the computer provides feedback to the participant, which is usually visual, auditory, or haptic. For example, the display in a headset often provides visual feedback in response to participant movements, and controllers incorporate haptic feedback to enhance interactions.

Machine-to-machine interactions are automated interactions between machines that do not require direct human involvement. This could involve IoT devices communicating with each other to adjust the environment based on participant preferences or AI agents engaging in the scenario under programmatic control.

Determining Technological Affordances

Technological affordances refer to the potential actions that can be performed with a technology within a given context. Determining what affordances are required to implement your scenario involves understanding the technology's capabilities and how they can be used to support or enhance the scenario. This process includes identifying necessary technologies, assessing interaction modes, and describing the digital assets you will integrate into the experience.

Identifying Necessary Technologies

Based on the scenario breakdown, identifying necessary technologies requires an analysis of both the environment and the user interactions within it. For scenarios involving full avatar representation, technologies such as VR headsets are essential for immersive visual experiences, enabling users to enter a virtual space that mimics real-world depth and presence. Motion-tracking devices, including gloves and body sensors, enhance realism by capturing the user's movements in real time, allowing avatars to reflect accurate, natural motion.

To bring these avatars to life, 3D modeling and animation software are crucial, as they facilitate the creation of detailed avatars and the environment around them. These tools also allow for personalization and detailed rendering, ensuring that each avatar and element within the space meets the visual and functional needs of the scenario. In scenarios in which multiple users interact within the same virtual environment, network and cloud infrastructure becomes necessary to support real-time communication and data synchronization.

Cloud computing services can handle the processing load required for realistic graphics rendering and data sharing across different devices, while networking technologies ensure low latency and reliable connectivity. AI and machine learning algorithms may be deployed to provide dynamic, responsive environments that adapt to user behavior, making the experience more immersive and interactive. This selection of technologies lays the groundwork for building a cohesive virtual environment that aligns with the scenario's goals and technical requirements.

Assessing Interaction Modes

The types of interactions within a scenario determine the selection of input and output devices, tailoring them to user needs and engagement goals. For human-to-machine interactions, touch screens are commonly used when users need to directly manipulate objects or navigate interfaces, while gesture recognition can be beneficial in hands-free environments or immersive VR scenarios. Gesture-recognition technology translates physical movements into digital commands, allowing for intuitive, noninvasive control in virtual spaces.

For multimodal interactions, voice recognition systems can be incorporated, especially in scenarios where natural language processing can facilitate smoother, hands-free navigation or command execution. On the output side, feedback mechanisms are equally crucial in creating a sense of presence and realism. Haptic devices, such as gloves or wearable feedback bands, provide tactile sensations that mimic real-world touch, which is especially valuable in training simulations or collaborative environments requiring physical interaction.

Spatial audio is an important component to consider, using directional sound cues to enhance immersion and spatial awareness within the virtual environment. Visual output can be further enhanced by high-resolution displays or VR headsets that deliver a broad field of view and realistic rendering, which deepens the sense of immersion and ensures users remain fully engaged with their tasks or interactions within the scenario.

Integrating Digital Content

Digital content integration, such as adding 3D models, animations, and interactive elements, requires careful attention to detail and alignment with the scenario's goals. For instance, if the scenario involves creating a realistic digital twin of a physical space, high-resolution 3D scans or detailed models must be used to capture the exact proportions, textures, and spatial layout of the environment. This often involves using photogrammetry or LiDAR scanning to ensure that the digital reproduction is accurate and interactive, enabling users to explore and manipulate objects in the virtual space as they would in reality.

Interactive elements, such as clickable objects, NPCs, or dynamic animations, can be added to enrich the experience and provide users with meaningful engagements within the environment. The process of integrating digital content also requires ensuring compatibility across different devices and platforms to facilitate a seamless experience. When avatars are present, the underlying animation and rendering technologies must support smooth, real-time interaction across VR headsets, desktops, and mobile devices, ensuring that the user experience remains consistent regardless of the hardware used.

The content you create or select will need to be optimized for performance, balancing visual fidelity with system requirements to prevent lag or loss of detail in high-activity scenarios. This approach to content integration assures that the virtual environment remains visually engaging and responsive to participant inputs, ultimately enhancing the realism and functionality of the XR experience.

Multimodal Immersive Composition: Aligning the XR Technology With the Eight Sensory Channels

Human senses play a vital role in human+machine immersion, enabling two-way communication and collaboration between humans and computers. As an immersive media designer, it is important to develop your understanding of how our senses function and contribute to immersion and communication. We cover the topic of multimodal mediation throughout the book.

The five externally focused senses of sight, hearing, touch, smell, and taste enable us to interact dynamically with the physical world. The internally focused sense of proprioception helps us with the awareness of the position, movement, and orientation of the body. The vestibular sense helps us with balance and motion while performing physical activity. Proprioception is sometimes referred to as our "sixth sense." We also have an eighth sense, interoception, which provides feedback on our internal bodily sensations. This section will discuss all of these in the light of immersive mediation, emphasizing aspects of each that relate to the design of immersive content design.

As humans, we connect to the physical world and the inner world of our physical bodies through our eight sensory channels that provide us with the modalities through which we interact with the world. All of the senses connect to the nervous system and the brain to work together to provide us with a cohesive perception of the environment and the internal state of the body, contributing to an overall sense of being embodied, being alive, and having agency.

As described in Section II, each sensory channel begins with specialized receptors that detect specific types of stimuli from the environment or within the body. For instance, photoreceptors in the retina respond to light, mechanoreceptors in the skin detect pressure and vibration, and chemoreceptors in the taste buds and olfactory epithelium sense chemical molecules responsible for taste and smell. Once these receptors detect a stimulus, the information is encoded into electrical signals that are transmitted along sensory neurons. These neurons carry the sensory data to the spinal cord, where initial processing and reflexive actions may occur. For example, in the case of pain or sudden touch, the spinal cord can initiate a reflex action, such as pulling a hand away from a hot surface, even before we consciously process the sensation.

Our sensory channels can be broadly categorized as outward-facing (exteroceptive) and inward-facing (interoceptive). Exteroceptive channels gather information from the external environment and help us interact with the world around us. These include vision, hearing, smell, taste, touch, and proprioception. While proprioception is often considered inward-facing, it is related to our awareness of our body's position and movement in space, providing information about limb positioning.

Interoceptive channels monitor the internal state of the body and include interoception and the vestibular system. Interoception communicates the internal physiological condition of our body, such as hunger, thirst, and visceral sensations (e.g., feeling full after dinner or the feeling at the pit of our stomach when something does not feel right, the source of the phrase "trust your gut"). The vestibular system is responsible for sensing the orientation and movement of the head relative to gravity, helping to maintain balance and equilibrium.

From the spinal cord, sensory information travels via specific pathways to the brain, particularly to the thalamus. The thalamus acts as a relay station, filtering and directing the sensory signals to the appropriate regions of the cerebral cortex for further processing. For instance, visual information is sent to the occipital lobe, auditory information is sent to the temporal lobe, and somatosensory data is sent to the parietal lobe.

In the cerebral cortex, higher-order processing occurs, integrating the raw sensory data into coherent perceptions. For example, the primary visual cortex (V1) in the occipital lobe processes basic visual elements like edges and colors, while higher visual areas interpret these into recognizable objects and scenes. Similarly, the somatosensory cortex in the parietal lobe creates a detailed map of the body's sensations, allowing us to perceive the texture, temperature, and shape of objects we touch.

Beyond these primary sensory areas, further integration and interpretation occur in association areas of the cortex. These regions combine inputs from multiple sensory channels, enabling complex perceptions and actions. For instance, when holding a cup of coffee, you don't just see the cup; you also feel its warmth, smell the coffee, and perhaps even anticipate its taste. This multisensory integration allows for a rich and unified experience of the environment.

The ability of our nervous system to process and integrate sensory information is not just limited to external stimuli. Interoception, for example, involves organ receptors like the heart and stomach that send information about internal bodily states to the brain. This allows for the perception of hunger, thirst, or the need to breathe, which are critical for maintaining homeostasis. These sensory channels and their associated neural pathways are crucial for survival and adaptation. They enable the brain to create a continuous and dynamic representation of both the external world and the body's internal state, which guides decision-making, emotional responses, and motor actions.

When any part of this sensory processing system is disrupted, it can lead to sensory processing disorders, affecting perception and behavior. For example, in conditions like neuropathy, the nerves that transmit sensory information may be damaged, leading to altered or diminished sensations. Similarly, in autism spectrum disorder, sensory processing differences can result in heightened or reduced sensitivity to sensory inputs, affecting how individuals interact with their environment.

Our understanding of the complexity of these sensory channels, from receptors to cortical processing, comes from fields like neuroscience, psychology, and medicine. The multimodal nature of XR makes this knowledge essential to practical applications in designing both the XR technology and immersive experiences, given that interaction with human sensory processing is intimate and integral to the medium.

This section sets the stage for a detailed exploration of the eight sensory channels, emphasizing their importance and the potential applications and outlining the interaction provided by currently available technology architectures. Traditionally, immersive media applications have only stimulated one or two senses, often vision and hearing. Research has demonstrated that adding more sensory channels to an experience can improve the participant's sense of immersion, engagement, task performance, and even cognitive functioning (Apostolou & Liarokapis, 2022).

Mediating the Visual Sensory Channel

The visual sensory channel, which involves the detection of light and visual information, is the cornerstone of not only our sensing ability but most mediation from photography, to movies, TV, personal and mobile computing, and XR. Visual stimuli within the visible spectrum of light, coming from our surroundings, arc processed by a complex system of interconnecting neurons, beginning with the optic nerve in the eye and ending in the visual processing center in the forebrain called the visual cortex.

Vision is the dominant sense in many species. It is estimated that 30% to 55% of the brain's processing is devoted to visual perception, as compared with 8% for touch and 3% for hearing (Sheth & Young, 2016). Like in the physical world, visual perception provides the primary means through which users perceive and interact with virtual environments. Much of the effectiveness of immersive experience relies on the quality and fidelity of visual information, which includes depth perception, motion, spatial orientation, and object recognition.

Visual information is crucial for depth perception and spatial orientation within immersive experiences. Techniques such as stereoscopic displays, which present slightly different images to each eye, enhance the perception of depth and three-dimensional space, making virtual environments more realistic (Mihelj et al., 2014, pp. 97–101).

High-quality, real-time graphics are essential for orchestrating immersion. Real-time rendering ensures visual updates occur without noticeable lag, maintaining the illusion of a coherent and interactive virtual world (Loftin et al., 2005). Although the visual sense is dominant, it is important to appreciate that humans process inputs simultaneously using multiple sensory channels. Thus, the visual channel often works with other sensory channels, such as auditory and haptic feedback, to create a multisensory experience.

XR systems are able to integrate visual and proprioceptive information over time, with increased latent information, parallax, depth, and other factors playing a role in participant performance in tasks. Proper synchronization between visual and other sensory inputs is crucial to prevent sensory dissonance and reduce discomfort, such as motion sickness (Kim & Lee, 2022).

Visual Human-to-Computer Inputs

Computer "vision" is a subfield of artificial intelligence (AI) that enables computers to interpret and understand visual data from the world, such as images and videos, using techniques that mimic the functions of human vision. Cameras and other sensors (e.g., LiDAR) collect a stream of data, and algorithms process the data to detect objects, recognize objects, and reconstruct scenes, among other tasks. Computer "vision" is fundamental to XR, providing the means for machines to respond dynamically to participants and physical objects in the real world.

Despite its anthropomorphic name, computer "vision" is quite different from human vision. Humans visually perceive the world not as a series of static images but as a flowing assemblage of shapes, colors, textures, motion, depth, and environmental context. For reasons of performance and reliability, engineers block out as much information as possible to focus computer "vision" on specific visual features without considering anything else.

For instance, computer vision devices are frequently optimized to work within specific restricted wavelength ranges to maximize their performance. Humans perceive the visible spectrum of light wavelengths from 380 nm, which looks violet to us, to around 780 nm, which appears to us as red. We refer to the invisible wavelengths shorter than 380 nm as the UV range and to the light wavelengths longer than 780 nm as infrared wavelengths (PSC, 2024). Unlike human vision, computer vision devices do not have to work within the visible spectrum. They can also be designed to work within specific wavelength ranges to maximize their performance and usefulness.

Computer vision technology is used in XR media to detect and track objects within the physical environment, allowing immersive systems to integrate these objects into the virtual space. This can include recognizing and tracking everyday objects or even the position of the user's head, hands, and other body parts for more natural interactions.

Gesture and Movement Recognition

Gesture and movement recognition using computer vision enhances immersive experiences by enabling intuitive and natural interactions. For example, *PianoVision* (ZarApps, 2023) uses hand tracking to align a virtual piano to a MIDI-enabled keyboard in the physical world to help you learn songs by displaying notes falling in sequence toward the keys. It achieves this through hand tracking and real-time responses to what you play through the connection to the keyboard. *PianoVision* is one of the best immersive education apps we have used, and it points to new possibilities for natural user interfaces (NUI) stemming from hand tracking.

Full-Body Tracking

Full-body tracking allows the participant to control their avatar's movements with their entire body, including walking, crouching, jumping, and dancing, to create a strong association between the participant and their avatar. The VIVE Ultimate

Tracker is a small device worn on the body parts you want tracked that incorporates two wide-FOV cameras and computer vision for real-time spatial recognition, keeping virtual and physical movements synchronized. Full-body tracking is popular with social VR platforms like VRChat, which typically support tracking of feet, knees, hips, chest, elbows, and shoulders.

While the Meta Quest primarily uses internal cameras for hand tracking and movement recognition, it can be combined with third-party solutions that enhance the Meta Quest platform with full-body tracking. SlimeVR can be used with Quest headsets to provide additional tracking points such as feet, legs, and torso for complete avatar control, ideal for virtual dancing and fitness routines. The KAT Walk C2 treadmill enables users to walk, run, and crouch within a confined space, translating these movements into the virtual world. It can be used in virtual worlds like VRChat and games like *No Man's Sky*, where users explore vast virtual environments.

Arm Movements

Current computer vision–based tracking systems can detect broad arm movements to perform actions like throwing, pushing, or lifting of virtual objects. This is often combined with hand gestures to create more dynamic interactions. Arm movement is typically implemented using the same technology used for full-body tracking. Standard cameras can be used to detect arm movements and gestures by analyzing color and shape data.

Computer vision algorithms use image-recognition techniques to identify key points on the user's body, such as the arms, hands, and joints. These key points are used to track broad movements like throwing or pushing. Some devices enhance their computer vision capabilities with depth sensors, providing time-of-flight data to calculate depth, which is critical for determining an object's position in 3D space. Depth sensing helps in tracking the distance of objects in virtual or real environments, allowing for accurate response to such gestures as pushing or pulling.

Leaning, Tilting, and Crouching

Leaning, tilting, and crouching the body in different directions can be creatively incorporated into XR content to allow users to dodge obstacles, peek around corners, or fly in a certain direction in MR and VR applications. Incorporating these body movements can be particularly useful in simulation contexts. Body leaning, tilting, and crouching is typically implemented using the same technology used for full-body tracking.

One of the best examples of this XR user interface (UI) can be found in *Beat Saber* (Beat Games, 2018). We suggest spending time in this game because it offers a creative application of body tracking that allows participants to have a strong sense of agency and embodiment as they move their bodies to avoid virtual obstacles.

Computer Vision and AI

The present state and the future advancements in computer vision are only possible because of developments in AI. VOXReality is a European Commission–funded initiative that integrates AI models with XR technologies to create immersive, multi-modal XR experiences by combining natural language processing (NLP) and computer vision models to enable more intuitive and natural interactions between humans and XR systems. The system has been used to enhance virtual meetings by providing real-time language translation and interactive visual effects. VOXReality is pushing the boundaries of how humans interact with digital environments by leveraging AI to create more immersive and context-aware XR experiences, providing a glimpse into the future of intelligence augmentation (IA), the combination of AI and XR (Maniatis et al., 2023).

Imagine a worker is wearing an AR headset while navigating a complex factory floor, seeking guidance on operating a machine by speaking to the system. VOX-Reality's personal assistant can interpret the user's request and identify the machine the user is referring to while it scans the environment, recognizing and identifying the machine in question through image recognition. The response might include a 3D overlay that appears on the machine, highlighting important parts of the equipment, along with step-by-step visual and verbal instructions projected in the user's field of view.

Visual Computer-to-Human Outputs

VR headsets use separate small high-resolution monitors for each eye to deliver a stereoscopic image, enhancing depth perception and making the virtual environment appear three dimensional and realistic based on three primary factors: resolution, parallax, and occlusion. If an image is low resolution, the user will be able to notice individual pixels, commonly described as the "screen door" effect.

The visual system integrates parallax and occlusion to create a coherent perception of depth, allowing us to navigate and interact with our three-dimensional environment efficiently. Parallax refers to the perceived displacement of an object when viewed from different positions. In stereoscopic vision, two types of parallax are relevant. Binocular parallax (disparity) occurs because each eye responds to light from a slightly different position, creating a discrepancy between the left and right eyes. This disparity provides the primary cues that make the world appear three-dimensional. Motion parallax results from the relative movement of objects at different distances from the observer. Closer objects appear to move faster than those that are farther away. This provides depth information that is complementary to binocular cues.

Occlusion occurs when one object blocks another, providing cues regarding their relative positions in space. Partial occlusion occurs when an object partially blocks another object. The visual system interprets occlusion to construct the spatial arrangement of objects, even when only one eye can see the occluded surface (half occlusions). This enables segmenting the visual world into coherent objects positioned in space.

Motion parallax describes surfaces appearing and disappearing as the observer moves. This is a form of dynamic occlusion that provides additional depth cues.

VR Head-Mounted Displays

Unlike computer vision devices, displays meant for human visual input are optimized for the properties of the human eye. High-resolution displays are critical for providing clear and detailed visual information. Head-mounted displays (HMDs) with high pixel density and low latency ensure that users experience a seamless and immersive virtual environment. Several platforms are available, ranging from devices intended for consumer applications (HTC Vive) to high-end headsets designed for enterprise applications (Varjo). Current HMDs are a far cry from Ivan Sutherland's Sword of Damocles that required a room-size setup, but they are still rather bulky and somewhat obtrusive devices. Mental and physical fatigue associated with using HMDs must be factored into the design of any VR experience.

MR Head-Mounted Displays

High-resolution displays that mix real-world and synthesized images are critical for providing clear and detailed visual information for MR applications. Head-mounted displays (HMDs) with high pixel density and low latency ensure that users experience a seamless and immersive blend of virtual and physical environments. In contrast to VR headsets, MR headsets blend the physical and digital worlds without completely blocking out the user's surroundings.

There are two primary implementations of this approach: pass-through and see-through. Pass-through headsets (e.g., Meta Quest and Apple Vision Pro) allow the user to see both the real and the virtual worlds while wearing the headset. Pass-through is implemented using cameras on the headset that capture the external environment and display it on the visual display inside the device. This enables the placement of virtual elements over the represented real-world visuals generated by the headset's cameras. Some devices implement monochrome pass-through, as seen in early VR headsets, showing the real-world view only in black and white. Newer headsets display a full-color view of the real world with higher fidelity.

The alternative to pass-through is optical see-through technology (e.g., Microsoft HoloLens). While pass-through devices use cameras to capture and digitally display the real world on the screen, optical see-through technology lets users see the real world through transparent visors, with virtual elements optically overlaid onto the user's view. While proponents suggest that see-through provides a more natural view of the real world, with less image distortion or latency, the computational complexity of integrating virtual objects with the real-world view has proven to be too complicated.

The limited field of view (FOV) of see-through headsets is more of a usability than a technological feasibility limitation. The above reasons are, at least in part, why Microsoft discontinued development of its HoloLens see-through headset, which faced stiff competition from such headsets implementing pass-through as Meta Quest and Apple Vision Pro.

Digital Overlays (Mobile AR)

With the growing popularity of mobile AR, manufacturers have made significant advancements in smartphone and tablet display technologies. High-resolution displays with high pixel density and low latency are now standard on smartphones and tablets, providing a widely deployed platform ready for mobile AR experiences. Current devices support smooth interactions, crisp visuals, responsive touch interactions, and spatial audio. Apple's newest iPhones feature high-dynamic-range (HDR) displays with high pixel density.

Along with the hardware, Apple's ARKit development framework delivers precise motion tracking, scene understanding, and integration of digital objects into the real-world view on the screen. Android devices have seen similar improvements in display technology and hardware, supported by Google's ARCore development environment. XR development platforms like Unity facilitate development that can be deployed to both iPhone and Android devices, allowing XR content makers to deploy their apps to the vast majority of smartphones and tablets in use today.

Digital Overlays (Projection AR)

Projection AR is a form of augmented reality that uses projectors to overlay digital information directly onto physical surfaces, objects, or environments. Unlike mobile AR, which typically involves viewing augmented content on a screen of a smart device, Projection AR allows users to engage with augmented content projected onto real-world surfaces without the need for intermediary devices like smartphones or AR glasses.

High-resolution projectors are used to display images, animations, or information directly onto physical surfaces. The surfaces onto which the images are projected are mapped and analyzed using sensors or cameras to ensure that the projections align correctly with the physical features of the surface. In more advanced systems, Projection AR can be combined with sensors like depth cameras or touch sensors to allow users to interact with the projected content. For instance, touching a projected button on a table could trigger an action, or moving an object could change the projected image on it. Projection AR can adapt to the characteristics of different surfaces, such as color, texture, and shape, to optimize the visibility and clarity of the projected content.

Visual Design Considerations

Display resolution, pixel density, field of view, refresh rate and motion clarity, user comfort and eyestrain, and other factors of the technology related to sensory system alignments are critical considerations for choosing the displays for immersive experiences. Regardless of whether your team's project involves virtual reality (VR), mixed reality (MR), or augmented reality (AR), these technology choices will directly impact the quality of the user experience in the dominant visual channel. Let us discuss each one of these in detail.

Display Resolution

The resolution of a display affects the clarity, realism, and immersion of the virtual environment as well as user comfort. Higher resolution requires more graphical processing power, which can strain the hardware and affect performance, leading to lower frame rates and potential lag.

Several factors relate to display resolution. The nature of your project's visuals—whether it's a highly detailed simulation or a simple low-fi gaming experience—will dictate the resolution requirements. Content that relies heavily on detailed visuals, such as architectural simulations or medical training, requires higher resolutions to be effective. Conversely, less graphically intensive applications can function well at lower resolutions. Designing with higher resolutions in mind can help future-proof VR/MR devices, ensuring they remain competitive and capable of running more advanced applications as they become available.

Pixel Density (Pixels per Inch—PPI)

High pixel density ensures that the individual pixels on the display are less noticeable, reducing the "screen door effect" (where the gaps between pixels become visible). For VR and MR, displays need to have a high PPI to ensure sharp and clear images, especially when the display is close to the user's eyes. A higher PPI leads to finer detail and a more realistic visual experience. With AR applications, contemporary smartphones and tablets offer high pixel density, making them imperceptible to the eye from a close distance. You can count on graphics to be sharp and clear on any smartphone or tablet that supports AR functionality.

Field of View (FOV)

For headset-based applications, the field of view is the extent of the observable world seen at any given moment. A wider FOV enhances immersion by allowing the user to see more of the virtual environment. As FOV increases, the resolution must also increase to maintain visual clarity across the entire view. If the resolution is not sufficiently high, a wide FOV can lead to pixelation and a reduction in image quality.

Screen Door Effect

For headset-based applications, the screen door effect occurs when the gaps between pixels become visible, creating a grid-like appearance. To minimize this effect, headset designers aim for high resolution with small pixel sizes. Advanced display technologies like OLED with subpixel rendering can help reduce the screen door effect but lead to more expensive hardware.

Rendering Resolution

The rendering resolution is the resolution at which the content is generated by the computer on the screen before being displayed. It should match or exceed the display resolution to avoid blurring and loss of detail. Upscaling lower-resolution content to match a higher-resolution display can lead to artifacts and reduced image quality.

Refresh Rate and Motion Clarity

The refresh rate, measured in Hz, is how often the display updates with new images per second. A high refresh rate is required for smooth motion and reducing motion blur, which is especially important for headset-based experiences in which head movement is frequent. Higher resolutions require more processing power to maintain high refresh rates.

Balancing resolution with refresh rate is necessary to ensure both visual clarity and smooth motion. A common goal is to achieve at least 90 Hz for VR to reduce motion sickness and provide a comfortable experience. Some headsets implement foveated rendering, which uses eye tracking so that only the center of the field of view is rendered at full resolution, which can help optimize hardware performance.

User Comfort and Eyestrain

Low-resolution displays can cause eyestrain, headaches, and discomfort during prolonged use as the eyes struggle to focus on blurry or pixelated images. Ensuring, whenever possible, the use of high-resolution devices will go a long way toward helping to minimize these issues by providing clear and sharp visuals, reducing the strain on the user's eyes. Additionally, the display's brightness and contrast should be adjusted to comfortable levels.

Dynamic Lighting and Shadow Rendering

Depending on the technology platform you choose to work with, the realistic lighting and shadow effects that enhance the visual realism of immersive environments will vary. Some applications will require visual cues to help users perceive depth and spatial relationships more accurately, contributing to more convincing virtual experiences.

Sound, Tactile, and Visual Synchronization

Immersive experiences rely on close integration between all of the mediated channels. Combining visual stimuli with audio enhances the immersive experience by providing spatial audio cues that align with the visual environment. This synchronization enhances realism and improves task performance demonstrating the value of applying visual-audio and visual-haptic multimodal feedback for improving task performance (Burke et al., 2006; Kim et al., 2019).

Numerous studies have shown that adding an additional modality to visual feedback improves the user experience and task performance overall. Both the combined visual-auditory feedback and the visual-tactile feedback provide advantages in reducing reaction times and improving performance and increasing the sensation of immersion (Burke et al., 2006). Combining haptic stimuli with visual content enhances the immersive experience by providing tactile cues that align with the visual environment.

As with visual and auditory synchronization, this alignment enhances multimodal realism and improves task performance. Numerous studies have shown that adding an additional modality to visual feedback improves the overall performance of tasks in the virtual and augmented environments (Kim et al., 2019). Both visual-auditory feedback and visual-tactile feedback provide advantages in reducing reaction times and improving performance and increasing the sensation of immersion (Burke et al., 2006).

Mediating the Auditory Sensory Channel

Sound plays a significant role in shaping our experience and interactions within real and virtual worlds. The sense of hearing involves the detection and processing of sound waves. The auditory channel is connected to human emotion and thus enhances engagement. By providing spatialized sound, auditory cues contribute to the user's sense of immersion, improve spatial awareness, and can significantly affect emotional responses and interactions within the virtual environment.

Auditory cues are often delivered through spatial audio techniques, which simulate how sound behaves in a three-dimensional space. This allows users to perceive the direction and distance of sounds, enhancing their immersion and presence in the virtual environment. Background music, sound effects, and ambient sounds can evoke specific emotions and set the atmosphere, making the virtual environment feel more alive and engaging. The congruence between auditory and visual stimuli is essential for a cohesive experience. When auditory cues are aligned with visual events, they reinforce each other, leading to a more convincing and immersive experience. Conversely, incongruence between these cues can disrupt immersion and lead to motion sickness (Kim & Lee, 2022).

Auditory Human-to-Computer Inputs

Speech Recognition

Multimodality of XR experiences can be enhanced by enabling audience members to interact with immersive environments and the objects in them through voice commands. The auditory sensory input method allows for a natural hands-free interaction that can be particularly useful in scenarios where traditional input devices such as touch screens or joysticks might be cumbersome. A speech-recognition feature is currently available on most of the commercially available XR platforms and is relatively straightforward to implement for teams whose members have some technical background.

Sound Detection

XR systems can incorporate sound-recognition algorithms to detect environmental sounds or user-generated noises, triggering specific responses or actions within the virtual environment. While speech recognition is the most accessible sound trigger to implement with commercially available equipment out of the box, with additional software, XR systems can respond to customized other sound triggers if your scenario requires it. The Apple Vision Pro, for instance, can distinguish between claps and snaps, allowing your audience to trigger actions based on a variety of specific nonspeech sounds detected by the device.

Auditory Computer-to-Human Outputs

Spatial Audio

Spatial audio simulates how sound interacts with the human ear from different directions and distances and thus enhances the participant's ability to locate sounds and objects in a virtual environment. Static, dynamic, and environmental sound localization cues that are accurately reconstructed with perceptually unnoticeable latency contribute to more convincing experiences for participants (Zotkin et al., 2004).

Ambisonics is a spatial audio technology widely used by immersive media creators to capture and reproduce a three-dimensional sound field. Unlike stereo or surround-sound audio formats, which are channel based and limited to specific speaker configurations, Ambisonics represents a three-dimensional sound field including above and below the listener, providing a complete 360-degree audio experience that is not restricted to any device, channel, or speaker format. Ambisonics sound mixes can be encoded for a wide variety of contexts including headphones (binaural), headsets (binaural with spatial tracking) and physical spaces with multispeaker setups.

In addition, mono, stereo, and surround materials can be incorporated and positioned in an Ambisonics sound mix. The Ambisonics format is scalable, and the spatial resolution and accuracy of Ambisonics can be extended with the addition of more audio channels. Widely used sound editing and mixing applications like ProTools and Reaper support Ambisonics. Development platforms like Unity and Unreal are able to play Ambisonics audio materials. Several immersive platforms support Ambisonics, including YouTube 360 and Facebook 360 for delivery of 360° video content.

Dynamic Soundscapes

In the context of XR, the auditory dimension significantly enhances the overall sense of immersion and realism. By generating sounds that respond to user actions and interactions in a dynamic fashion, your XR system can create a more engaging experience that better aligns with the visual and spatial components of the environment. Auditory feedback may be expressionistic or mimic real-world physical and environmental cues, helping users to feel more connected to the activities they perform in XR.

It is common in the cinematic arts that the sound of footsteps changes depending on the surface a character is walking on, or the intensity of background noise is adjusted based on the proximity to certain objects or events. These conventions may also be applied in XR and are essential to consider when the participant is the center of the XR experience. Roundware, for example, provides a framework for supporting location-based mobile audio. And the *Pokémon GO* mobile AR experience is enhanced by dynamic audio elements. The sound of nearby Pokémon changes based on your proximity to the creature, with cries or sounds becoming louder and more distinct as you get closer. Environmental sounds, too, adapt to your location, with the soundtrack changing depending on whether you are in a park or an urban area.

Auditory Feedback

Sound increases the sense of presence in XR experiences because it contributes to a multisensory experience that more closely resembles real-life interactions. Studies have demonstrated that auditory feedback, when synchronized with user inputs, enhances our ability to navigate and perform tasks efficiently within the virtual space, improving overall task performance and interaction (Skalski & Whitbred, 2010).

Auditory feedback reduces cognitive load by providing extra sensory information that supports the participant's situational awareness. When auditory cues reinforce interactions, users are less dependent on visual feedback, which can reduce sensory overload, particularly in complex scenarios. By hearing a sound that confirms they have used a tool (e.g., the lever being pulled) or completing a task (e.g., notification chime), the participant can focus on the most salient aspects of the experience without always having to check visual feedback, enhancing immersion.

At this time, 3D spatial audio is available with all of the commercial immersive platforms. Therefore, you can count on the ability to provide auditory feedback to enhance augmented, mixed, and virtual experiences when the user interacts with digital objects in the virtual environment or enchanted objects in their physical environment. In immersive games like *Half-Life: Alyx*, auditory feedback contributes to the sense of presence in the storyworld. The game provides spatial audio cues like distant footsteps and radio chatter to help the audience navigate and react to their virtual surroundings.

Speech Synthesis

Speech synthesis, or text-to-speech (TTS) technology, transforms written text into spoken language. When incorporated into XR experiences, it helps you design more natural, intuitive interactions between humans and machines. TTS is widely used to support multimodal interactions for more natural interactions. Speech synthesis supports the development of natural user interfaces (NUIs), where users interact using natural behaviors like speaking, gestures, or touch.

Speech can be used to deliver verbal responses, commands, or feedback to the participant and is particularly valuable in situations where hands-free interaction is essential, such as machinery operation or medical training, providing clarity without requiring the participant to take their eyes off their task. Speech synthesis also improves accessibility for users with visual impairments or reading difficulties.

The advantages of TTS technology are illustrated by Microsoft's Remote Assist, which enables people to solve problems in real time. Technicians or workers in the field use a HoloLens to collaborate hands-free with experts who provide remote guidance. TTS is integrated into the system to enhance communication. When experts type instructions or messages, the TTS system reads them aloud through the HoloLens or AR device, allowing the field worker to hear the guidance without needing to look at a screen or stop their task. This feature ensures that the user can stay focused on their work while still receiving critical instructions in real time.

Participants on existing iOS and Android devices can share live video feeds and capture instructions using 2D annotations, not limiting the app to the use of headsets. This approach to communication in complex environments, such as industrial repair, maintenance operations, healthcare, or anywhere visual attention and hands-free interaction are essential to the task at hand.

Personalized Spatial Audio

Spatial audio improves immersive experiences in XR by simulating how sound behaves in the real world. With spatial audio, users are more likely to feel as if they are situated within a virtual world.

One approach to enhancing spatial audio experiences is by tailoring sound reproduction to each participant's unique head-related transfer functions (HRTFs). HRTFs describe how sound waves are modified by the shape of a person's torso, head, and pinnae (outer ear) before reaching the eardrums. As sound travels toward the listener, it gets scattered and filtered by these anatomical structures, causing changes in the intensity, timing, and frequency spectrum of the sound at each ear. This phenomenon allows the brain to determine the direction and distance of sounds, giving us a sense of the spatial origin of sounds in our environment.

Because every person's body is slightly different, HRTFs are unique for each individual. As a result, using generalized or averaged HRTFs in spatial audio systems can lead to less precise sound localization. A sound that should seem to be coming from behind might be perceived as coming from a different direction, breaking the immersive illusion. By personalizing spatial audio using each participant's unique HRTFs, your XR systems can simulate more accurately how sounds are perceived. Customized HRTFs make it easier for listeners to identify where sounds are coming from and how far away they are.

Measuring individual HRTFs requires specialized equipment, including anechoic chambers and microphones placed inside the user's ears. This process can be time-consuming, costly, and impractical for consumer devices. Therefore, most XR systems use standardized or averaged HRTFs, which work adequately for most users but do not fully take advantage of the potential realism of personalized spatial audio.

Given the value of improving the sonic experience in XR, researchers are currently exploring machine learning algorithms to estimate personalized HRTFs based on easily obtainable data, such as a 3D scan of the user's ears or facial structure. This will allow user-generate customized HRTFs without complex measurements. As more consumers use headsets for immersive experience, manufacturers may find value in integrated HRTF measurement tools directly into headsets. We predict that in the not-too-distant future, XR headsets and AR glasses will use built-in sensors to measure a user's ear shape and estimate their HRTFs on the fly.

The intimate connection between sound and emotion is too important to pass up. We have seen a parallel to this. When the movie industry found itself in competition with television in the 1950s, screens got bigger. Later, in the 1980s, theaters added high-fidelity surround sound to continue their lead over home viewing of movies available on videocassette.

Auditory Design Considerations

Soundscapes and Emotion

Sound plays a pivotal role in shaping human emotions. The auditory sensory channel is closely linked to the brain's limbic system, which is responsible for processing emotions. The sound–emotion connection has been extensively explored by filmmakers, who use powerful cinematic scores and evocative sound effects to increase the dramatic effect of the visual footage. We can only imagine how much more powerful of an effect sound can have when woven into an immersive experience.

Sound design should be emphasized whenever a design team's goal is to trigger a strong emotional response. The emotional effect of sound can increase the drama or create a calming effect. Immersive media is increasingly being used in therapeutic settings, where auditory stimuli can be designed to soothe or stimulate, depending on the therapeutic goals. For example, calming nature sounds might be used in a VR environment designed to reduce anxiety (Andreano et al., 2009).

VR App is a mindfulness and meditation experience designed to promote relaxation, emotional well-being, and mental clarity using VR technology. It combines guided meditation, calming visuals, relaxing music, and soothing environments to create a personalized and engaging therapeutic experience. The auditory component of *VR App* has been crafted to enhance relaxation. Binaural beats and calming soundscapes are used in the background to help users achieve deep relaxation or focus. The combination of spatial audio and dynamic visual environments creates a holistic sensory experience.

Music

Music is one of the most direct ways to influence the audience members' emotional state. Filmmakers since the silent film era have used live and recorded soundtracks to evoke nostalgia, sudden loud noises to induce a sense of fear, or a fast-paced, upbeat soundtrack to energize the action on the screen and to create a sense of tension. Consider the use of music to set the tone and influence how users feel as they navigate through your XR experience with an added effect of spatial dimension.

Auditory Cues

Sound has a strong link to memory, which can be leveraged in immersive experiences to evoke specific emotions. Subtle sound effects, like the rustling of leaves or distant thunder, can create an atmosphere that influences the user's emotional state. These cues, when strategically placed, can enhance the realism of the environment, making the experience more convincing and emotionally engaging.

In training simulations, spatial audio can help trainees better understand the environment and respond to auditory cues, such as alarms or communication signals, that are critical in real-world scenarios. This is particularly important in aviation or emergency response, where sound cues are crucial for decision-making (Bohil et al., 2011). Auditory cues can be used to create a sense of tension or urgency, guide players through the environment, and enhance the realism of interactions. By accurately positioning sounds in a three-dimensional space, designers can make users feel as though they are truly "inside" the environment.

Synthesized Speech

Despite the usability and natural user interface potential of speech synthesis, there are challenges achieving authenticity and emotional expressiveness in synthetic voices. Current research is focused on improving the quality and naturalness of synthesized speech. We expect it will soon become indistinguishable from human speech thanks to advances in artificial intelligence and machine learning, but it's not quite there yet as this book goes to press.

Mediating the Haptic Sensory Channel

The sense of touch, or the haptic sensory channel, plays a vital role in enhancing immersion. While both tactile haptic and kinesthetic haptic inputs can affect immersive experiences, it is important to clarify the difference between tactile haptic and kinesthetic haptic technologies. Haptic feedback uses haptic devices and vibration motors to simulate touch-related experiences, while tactile feedback focuses on tangible, physical touch experiences.

Haptic feedback, which includes tactile and force sensations, allows users to "feel" virtual objects and environments, creating a more realistic and engaging experience. This feedback can significantly affect user perception, interaction, and overall sense of presence within a virtual environment. Tactile feedback involves the sensation of texture, vibration, temperature, and pressure.

In VR, this can simulate the feeling of different surfaces or the impact of interactions with virtual objects. For example, a user might feel the roughness of a virtual stone wall or the vibration of a virtual tool in their hand. This type of feedback is essential for creating realistic interactions within the immersive environment. Force feedback provides resistance or force sensations, enabling users to feel the weight, inertia, or resistance of virtual objects. This is particularly important in simulations requiring precision and realism, such as surgical training or virtual prototyping. The sensation of pushing against a virtual object or the resistance encountered when lifting a heavy virtual item enhances the realism of the experience.

Kinesthetic feedback involves the sensation of movement and position. This feedback is crucial for tasks that require the user to sense their body's position and movement within the VR environment, such as in sports simulations or training exercises. This type of feedback can be provided through wearable devices that track and respond to the user's physical movements.

Haptic Human-to-Computer Inputs

Haptic Gloves, Controllers, and Motion-Tracking Sensors

These devices capture the participant's hand movements and provide feedback through vibrations or resistance. Examples include gloves or controllers that detect gestures and apply force feedback based on virtual interactions. Sensors can track human body movements and translate them into the virtual environment. For instance, a glove might track finger movements and provide corresponding tactile feedback when interacting with virtual objects. Sensors may include infrared cameras reading light beams from dedication beacons. Haptic gloves are wearable devices that are able to track the movement of different parts of the hand and detect when the fingers are touching a surface or gripping an object.

Haptic Computer-to-Human Outputs

Haptic Gloves and Wearables

Haptic gloves are wearable devices that provide detailed tactile feedback to the user's hands and fingers. They can simulate the sensation of touching, grasping, and manipulating virtual objects by applying varying levels of force and vibration to different parts of the hand using a variety of technologies. Vibrotactile devices use vibrations to simulate the sensation of touch. These devices can be integrated into gloves or other wearables, allowing users to feel textures, impacts, or other surface details of virtual objects.

Advanced haptic devices can simulate temperature changes, adding another layer of realism. For example, an immersive experience might include a haptic glove that warms up when the user interacts with a virtual fire or cools down when touching virtual ice.

Force Feedback Devices

Haptic technology refers to the modality that simulates the feeling of touch when interacting in a virtual environment. Haptic feedback technologies have major implications for XR interactions, since the UI might not likely involve the use of familiar physical input devices like a keyboard, mouse, joystick, or touchscreen. For instance, smartphone keyboards vibrate gently to let us know that they recognize our input.

XR haptic technology uses a combination of vibrations and motions to confirm user inputs such as lifting an object in VR. They let the users know if they have placed their hands in the right position, performed the right gesture, and applied the necessary amount of pressure. Used appropriately, force feedback devices can enhance the sense of agency and presence by providing force feedback by applying resistance to the user's movements, simulating the physical properties of virtual objects, such as weight, elasticity, or hardness. There is currently a lot of research and development effort focused on haptic technologies.

Ultrasonic mid-air haptic technology allows users to experience tactile sensations without physical contact with any device. Ultrasonic waves create the sensation of touch in mid-air, enabling interactions with virtual objects that feel tangible despite being intangible (Fuchs et al., 2011). For instance, an MIT-based research team has developed a chip-free flexible, elastic electronic skin that uses a collection of sensors, vibrating magnets, and wireless transmitters that can be attached to any part of the body and turned into a haptic-based input system for XR. The electronic skin requires only 350 microseconds to convey its haptic signal when operating on Wi-Fi. That is actually a faster response rate than an average human reaction of 550 microseconds (Chu, 2022).

Researchers at Meta are working on their own haptic glove for use in multiplayer gaming scenarios and other MR applications. Meta Glove, according to XR Today, is densely packed with tactile sensors that can convey such tactile inputs as pressure, weight, and other feedback factors (XR Today Team, 2022).

Ekto VR is working on addressing a longstanding problem of "moving" around virtual environments where users traverse long distances without really moving at all. Ekto VR's haptic-enabled boots use a set of motorized wheels to mimic movement, eliminating the problem of providing users with the equivalent physical space. The company is also working on an omnidirectional treadmill, inspired by *Ready Player One*, that would make walking in virtual spaces even more seamless (XR Today Team, 2022).

By leveraging haptic technologies that provide tactile, force, and thermal feedback, your team can create more engaging and interactive virtual environments. In medical training applications, haptic feedback allows trainees to feel the resistance of tissues, the pressure required for incisions, and the texture of organs, providing a realistic training environment. Designers can use haptic feedback to interact with virtual models, feeling the texture, weight, and flexibility of materials before they are physically produced. This application is crucial in industries like automotive and aerospace engineering.

Haptic feedback has been used for motor rehabilitation, where patients can perform exercises in a virtual environment that provides real-time tactile feedback, enhancing the effectiveness of the therapy (Feng et al., 2022). Consider the use of haptic feedback to enhance immersion by allowing participants to feel the impact of interaction with objects and to feel the texture of virtual surfaces, making the experience more engaging.

Mediating the Olfactory Sensory Channel

The olfactory sensory channel, which involves the sense of smell, plays an increasingly important role in enhancing immersion and realism in immersive media. While sight and sound have traditionally dominated VR experiences, incorporating olfactory stimuli adds a new dimension that can significantly increase the user's sense of presence, emotional engagement, and memory recall. Adding olfactory stimuli to experiences can greatly enhance the sense of presence, making the virtual experience feel more real and immersive. For example, the smell of burning wood in the simulation of a campfire can make the scene more convincing and engaging.

Smells are closely linked to emotions and memory. Incorporating olfactory cues in VR can evoke specific emotional responses or memories, enhancing the overall impact of the experience. Adding olfactory cues can increase the realism of virtual environments, leading to more natural user behaviors. For example, the smell of smoke in a fire simulation can prompt more realistic evacuation behaviors, making the training more effective. Olfactory simulation has been effectively used in therapy protocols treating posttraumatic stress disorder (PTSD) that use specific smells to trigger traumatic memories in a controlled environment.

Olfactory Human-to-Computer Inputs

Biometric Sensors

Users' responses to olfactory simulation can be tracked using a range of biometric sensors that track physiological changes occurring within the nervous system in response to smells. These include changes in heart rate or skin conductance. The gathered data can be used to adjust the experience dynamically based on the user's reactions, making the experience more personalized and responsive.

Olfactory Computer-to-Human Outputs

Olfactory Delivery Devices

Modern advancements in the fields of biotechnology and nanotechnology have enabled the development of digital scent technology that can be applied to medicine, fashion, and entertainment. Digital scent technology, also known as scentography devices, can affect users on neurological, emotional, and psychological levels. Research in olfactory simulation has been ongoing since at least the 1950s.

Today, the field of scentography is actively working on developing new and more advanced devices that can be commercialized for a variety of immersive experiences. Olfactory displays can release specific odors at precise times to align with visual and auditory cues, enhancing the overall sensory experience. Some systems can deliver scents directly to the user's nose without diffusing them into the environment. This can be achieved using techniques like air cannons or vortex rings, which ensure that scents are perceived only by the participant and at the intended time.

The current examples of olfactory simulation systems include the Vaqso VR Dev kit, which was introduced in 2018 by Vaqso VR. OVR Technology introduced a dedicated XR Architecture of Scent Platform in 2020. The kit includes the ION Scent Device, which can be attached to AR and VR headsets, a software suite with a plug-in that allows content developers to add scent triggers and geometries to XR objects, and the company's in-house developed ION Scentware. This unique combination of hardware, software, and scentware platform can deliver precise scent molecules based on the user's interactions with the virtual environment to deepen the sense of immersion and mediated presence for XR-based training and wellness applications.

In 2021, Northwestern University researchers Daniel Dombeck and Brad Radvansky developed a system that can control scents as a function of the user's movement in XR space. Of note to content makers is the 2023 Darren Aronofsky film *Postcard From Earth*, which mixes high-resolution imagery, sound, and scents specifically created to be shown at the Las Vegas Sphere.

Olfactory Design Considerations

Olfactory cues that can provide additional multimodal information about the environment are still new to XR experiences; however, currently, there are a variety of application-specific and experimental output devices available.

Several studies have examined whether the introduction of smells enhances users' experience, including their sense of presence. Results demonstrate that the addition of odors increased participants' sense of spatial presence in the virtual environment compared with no odor. Engaging your participant using the olfactory channel provides an opportunity to create a more immersive experience to increase the subjective experience of presence.

Olfaction has demonstrated potential for use in virtual training environments and virtual-reality exposure therapy (Archer et al., 2022; Baus et al., 2022; Egan et al., 2017). The timing and intensity of scents can be precisely controlled to match the virtual environment's context. For example, the scent of food can be stronger near a virtual kitchen and fade as the user moves away from it. In therapeutic settings, olfactory cues can be used to trigger memories or emotions that are critical for the therapy's effectiveness. For instance, VR therapy for PTSD might incorporate the smell of a specific environment to help patients confront and process traumatic memories (Aiken & Berry, 2015).

For training, especially for scenarios involving safety or emergency response, the use of olfactory stimuli can make the simulation more realistic and effective. For instance, the smell of hazardous chemicals in a VR industrial safety training scenario can help trainees better understand the dangers and respond appropriately.

Mediating the Gustatory Sensory Channel

Another less explored but now actively researched immersive mode involves the gustatory sensory channel, which has to do with the sense of taste. Gustation can play a unique albeit still largely experimental role in immersive media. While more commonly explored sensory channels include sight, sound, and touch, integrating taste into experiences presents new design opportunities. The perception of flavor is a multisensory experience that integrates gustatory (taste) and olfactory (smell) stimuli, often influenced by visual cues.

Leveraging this multisensory integration can create more convincing experiences. The sense of taste is closely tied to emotions and memory. Incorporating gustatory feedback in your design can evoke strong emotional responses, making the experience more memorable and impactful. For instance, tasting a particular flavor in a VR experience could evoke nostalgia or even influence behavior, such as increasing appetite in a virtual dining experience. Experimental devices have been implemented with which the taste of a virtual food item can be enhanced by the combination of appropriate visual and olfactory stimuli, leading users to perceive a more realistic flavor (Halabi & Saleh, 2021).

Gustatory Human-to-Computer Inputs

Taste Sensors and Input Devices

Though not commercially available, several experimental devices exist that can detect and simulate basic tastes (sweet, sour, salty, bitter, umami). While experimental, these devices could be used in your prototype to determine if users interacting with virtual foods and receiving real-time gustatory feedback enhances the experience you are designing.

Biometric Sensors

These sensors can track physiological responses related to taste, such as salivation or facial expressions, to adjust the VR experience dynamically. For example, detecting a user's preference for sweet flavors could lead the VR system to emphasize sweet notes in subsequent interactions.

Gustatory Computer-to-Human Outputs

Gustatory Displays

These are experimental devices designed to deliver taste sensations to the user. One approach involves electric or thermal stimulation of the tongue to create the perception of different tastes without the need for actual food. Such devices could be integrated into VR headsets or accessories to simulate the taste of virtual foods (Marey & Alsabbagh, 2023).

Olfactory Integration

Since taste is closely linked to smell, VR systems can use olfactory displays to emit scents that complement the visual and taste stimuli. For example, smelling freshly baked bread in a VR bakery scenario can enhance the taste perception of virtual bread.

Gustatory Design Considerations

Though still in its early stages of exploration, the gustatory sensory channel for immersive media design presents significant potential. Integrating taste with other sensory channels allows you to create more holistic and engaging experiences that tap into the powerful connection between taste, emotion, and memory.

As technology advances, including gustatory feedback could become a significant component in a wide range of applications, from entertainment and education to therapy and marketing. Systems can augment flavor perception by altering the visual or auditory context in which food is consumed. For instance, changing the color of a virtual beverage might alter the perceived taste, making it seem sweeter or more refreshing (Apostolou & Liarokapis, 2022).

In a virtual restaurant setting, users could taste and smell virtual dishes, allowing for a unique dining experience that could be used for entertainment, culinary education, or even dieting support. For example, a VR experience could simulate a gourmet meal, with each course offering different gustatory and olfactory feedback. Gustatory devices could be used in therapy, particularly for patients with eating disorders or those undergoing rehabilitation. Simulating the taste and smell of food could help patients reacquaint themselves with eating experiences in a controlled environment.

Consider the potential of using a gustatory display to simulate new flavors or food products before they are physically produced. This could be particularly useful in taste testing and product development, allowing companies to gather user feedback in a virtual environment.

Mediating the Proprioceptive Sensory Channel

The proprioceptive sensory channel, which involves the sense of body position and movement through receptors in muscles and joints, plays a critical role in the design of immersive experiences by enhancing the user's sense of immersion, embodiment, and interaction with the virtual or augmented environment. Proprioceptive feedback helps users feel as if they are truly within the virtual space.

This sense of embodiment is crucial for immersive experiences in which users interact with virtual objects and navigate through virtual environments. When the proprioceptive system is accurately stimulated, users can experience realistic movements and interactions, increasing the overall immersion (Zakharov et al., 2021). Accurate proprioceptive feedback aids in spatial orientation and navigation within virtual environments. This is particularly important for tasks that require precise movements, such as reaching or navigating complex environments. The integration of proprioceptive cues with visual information enhances the user's ability to judge distances and coordinate movements (Valori et al., 2020).

Proprioceptive Human-to-Computer Inputs

Motion-Capture Systems

Motion-capture systems are already widely adopted for generating insights in sports performance, ergonomic assessment, workplace safety, biomechanical research, and rehabilitation and injury prevention. In content creation, they are used by such companies as Marvel, Netflix, and Ubisoft in game design, animation, cinematic VFX, and XR production to capture and track body movements in real time using sensors placed on the body. The data collected from these sensors provide the XR system precise information about the user's limb positions and movements. These can be used to create realistic avatars and more dynamic interactions within the XR environments.

Inertial Measurement Units (IMUs)

Inertial measurement units (IMUs) are often integrated into VR controllers or suits to detect movements and orientations of the user's limbs. This data is essential for providing accurate proprioceptive feedback and ensuring that virtual actions correspond to real-world movements.

Proprioceptive Computer-to-Human Outputs

Haptic Feedback

Devices such as gloves, suits, or controllers with haptic feedback provide tactile sensations that mimic real-world touch and movement. This feedback can simulate the feeling of interacting with virtual objects, enhancing the user's sense of presence and engagement. More advanced XR setups may include robotic exoskeletons or manipulanda that provide physical resistance and support, mimicking the sensation of weight and force. These interfaces are particularly useful in rehabilitation and training applications, where realistic proprioceptive feedback is crucial for effective therapy.

Proprioceptive Design Considerations

XR applications are currently being used to aid in the rehabilitation of motor functions, particularly for stroke patients. Proprioceptive feedback can help patients relearn movements by providing real-time feedback and immersive exercises that mimic real-world tasks. In gaming, proprioceptive feedback enhances gameplay by making movements more intuitive and realistic. Games that involve physical activities, such as sports or combat, benefit greatly from accurate proprioceptive feedback, allowing players to perform actions that closely mimic real-life movements.

Training for tasks such as piloting or surgery often incorporates proprioceptive feedback to improve the realism and effectiveness of the training. This feedback helps trainees develop the necessary motor skills and spatial awareness in a controlled, repeatable environment. When teaching anatomy and physiology, making use of proprioceptive feedback has been demonstrated to help students understand the spatial relationships and movements of different body parts, enhancing their learning experience.

Mediating the Vestibular Sensory Channel

The vestibular sensory channel, responsible for the sense of balance and spatial orientation, plays a crucial role in immersive media design by influencing how users perceive motion and maintain equilibrium. This system, located in the inner ear, detects head movements and provides feedback to the brain to help coordinate balance and eye movements.

The vestibular system contributes significantly to the sensation of immersion. When vestibular inputs are congruent with visual stimuli, the sense of being in the virtual environment is enhanced, leading to a more immersive experience (Virre, 1996). Mismatches between visual and vestibular inputs can lead to simulator sickness, characterized by nausea, dizziness, and disorientation. Reducing this sensory mismatch is critical for enhancing user comfort and usability in MR and VR environments (Weech et al., 2018).

When designing high-immersion experiences, it is critical to ensure vestibular inputs are congruent with visual stimuli. This enhances realism by reducing sensory mismatch and, more importantly, will reduce the likelihood of motion sickness. This integration is critical for applications ranging from entertainment to training and rehabilitation.

Vestibular Human-to-Computer Input

Inertial Measurement Units

Vestibular input can be created using inertial measurement units (IMUs) which are electronic devices that measure and report data on a body's specific force, angular rate, and the orientation of the body by using a combination of accelerometers, gyroscopes, and magnetometers. IMUs can be used as vestibular system feedback by detecting the user's head and body movements and orientations in real time, providing a crucial vestibular channel dimension to both the user and the XR system. The obtained data can help the XR system synchronize visual stimuli with the user's physical movements to reduce sensory mismatch.

Vestibular Computer-to-Human Output

Aligning the Vestibular, Visual, Haptic Feedback, and Audio Cues

High-fidelity visual environments that accurately reflect the user's head and body movements help align visual and vestibular inputs, reducing sensory conflict. The ability for the XR system to align the visual, auditory, and haptic feedback with the vestibular feedback will greatly enhance the sense of presence. Providing tactile sensations through wearables or controllers can help users feel more grounded and reduce the disparity between expected and actual sensory feedback, improving balance and spatial awareness. Spatial audio can further enhance the sense of orientation and movement, providing an additional layer of sensory input that complements visual and vestibular feedback.

Galvanic Vestibular Stimulation

The vestibular system has two main groups of receptors. One is composed of three semicircular canals, the other of two otolith organs: the utricle and the saccule. The vestibular system nerves in the ear responsible for maintaining balance can be stimulated with the aid of galvanic vestibular stimulation (GVS) devices. GVS involves applying small electrical currents to the vestibular system to simulate the sensation of motion. This technique can be used to provide vestibular inputs in sync with visual motion, thereby enhancing the realism of VR experiences (Groth et al., 2022).

Vestibular Design Considerations

The sense of balance has a powerful effect on our overall well-being. Vestibular system stimulation capable XR applications are already being used in rehabilitation to help patients recover from balance disorders. Virtual environments can simulate conditions that challenge the vestibular system, aiding patients in reorganizing and compensating vestibular functions (Matsumura & Murofushi, 2021). Such immersive applications as flight simulators and similar gaming and training experiences use IMUs to track head movements and provide realistic visual feedback, creating a more immersive and training-effective environment. This reduces the risk of simulator sickness and improves pilot training outcomes (Kennedy & Lilienthal, 1995).

Mediating the Interoceptive Sensory Channel

Interoception, the sense of internal body states, including heartbeat, breathing, and visceral sensations like hunger, thirst, etc., can be conscious or subconscious and includes the processes by which we sense, interpret, integrate, and regulate signals within ourselves.

Interoception plays a crucial role in the design of immersive media experiences. Interoception is essential for creating a more immersive experience by enhancing the user's sense of embodiment. Leveraging the human interoceptive sensory channel can enhance the user's sense of immersion and embodiment. Integrating human-to-computer inputs such as breathing and heart rate monitors and providing computer-to-human outputs like visual, auditory, and haptic feedback are key to designing interoceptive-rich experiences.

Interoceptive Human-to-Computer Inputs

There is no straightforward way to leverage interoception as an interaction modality, in contrast to the ubiquity of visual, auditory, and haptic interfaces. We can offer you some starting points if you wish to tap into a participant's internal bodily sensations. Immersive systems can track physiological states by incorporating biofeedback sensors (e.g., heart rate monitors and galvanic skin response sensors). This data can alter the scenario in real time, creating adaptive experiences. For example, if you are working on a meditation application, you could adjust the soundscape and visuals based on the participant's heart rate to provide a more responsive and personalized experience. A growing number of biofeedback sensors are commercially available to provide human to machine inputs of user biometric data.

Heart Rate Monitors

XR technology takes advantage of knowledge gained in medical research regarding the role of biomarkers such as body temperature used for detecting fever or blood pressure used to determine the risk of stroke. Combining biomarker sensing with virtual reality (VR) has proven to be highly effective in identifying mild cognitive impairment (MCI) by capturing behavioral data associated with hard-to-detect deficits in instrumental activities of daily living such as challenges in using a food-ordering kiosk.

Biomarkers obtained from magnetic resonance imaging (MRI) have demonstrated their efficacy in quantifying observable structural brain changes that can aid in early detection of cognitive malfunctioning (Park et al., 2024). Heart rate variability (HRV) is one of the key noninvasive biomarkers used in cardiovascular health and mental health research. HRV measures the fluctuations and variations in time intervals between successive heartbeats or R-R intervals (RRi). Heart rate monitors such as the Polar H10 chest strap can be integrated with Bluetooth or ANT+ connectivity applications to monitor heart rate and heart rate variability (HRV) to assess stress levels, relaxation, and overall cardiovascular responses during immersive experiences.

Electroencephalography (EEG)

Electroencephalography (EEG) sensors measure brainwave signals produced when parts of the brain send messages to each other. They have been used in immersive meditation and mental health applications to provide feedback on focus and relaxation levels.

One of the most creative uses of EEG sensors we've come across is *The MOMENT,* an interactive film developed by Richard Ramchurn while a doctoral student at the University of Nottingham. Viewers wear inexpensive NeuroSky MindWave headsets to pick up their EEG signals, and small drops in the viewer's focus will alter the signals. In response, the film adapts the editing, music, and animation based on the viewer's level of attention. *The MOMENT* provides a glimpse of the future and how brain–computer interfaces might transform storytelling and audience engagement (Metz, 2018).

Galvanic Skin Response (GSR)

Galvanic skin response (GSR) is a measure of the change of electrical properties of skin. Under emotional arousal and stress, our body temperature increases. As we sweat, skin conductance increases. Such electrodermal activity (EDA) can be computed by applying a small current to measure the resistance of skin between two placed electrodes (Can et al., 2019). GSR sensors evaluate the user's emotional states and stress levels by measuring skin conductance. GSR sensors can be integrated into immersive systems to adjust system responses based on the user's biomarker data.

Electromyography (EMG)

Electromyography (EMG) sensors detect muscle activity. They can track hand and arm movements, enabling more natural interaction within an immersive experience. For example, the Delsys Trigno is a high-precision EMG sensor that can measure muscle activity in various body parts during exercise or rehabilitation. It is used for physical therapy, sports training, and other scenarios in which muscle activity needs tracking and responding to.

Breathing Monitors

Respiration monitors which measure breathing rate can be integrated into XR applications to monitor and respond to the user's emotional states, as indicated by breathing patterns, to adjust the system feedback accordingly. For instance, virtual reality exposure therapy (VRET) assesses and monitors a patient's anxiety levels by capturing their respiratory rate, among other critical biomarkers. The corresponding respiratory biosignals are captured by external hardware and recorded within the VR experience to correlate specific anxiety triggers to respiratory rates (Mevlevioğlu et al., 2021).

Eye-Tracking Sensors

Highly accurate eye-tracking devices are currently integrated into several headsets, such as Apple's Vision Pro, to bring XR experiences closer to the goal of attaining natural user interfaces (NUI). Eye-tracking technology aids immersive user interfaces by monitoring where the participants direct their gaze, improving system performance and graphics rendering through foveated rendering. Devices equipped with eye tracking can also detect pupil dilation, providing valuable insights into users' attention and focus that can further support the sense of mediated presence.

Temperature Sensors

Temperature sensors are available to monitor skin temperature (e.g., GENTHERM's FlexSense with wireless connectivity) and can be integrated into wearable devices or clothing. These are often used in health monitoring and wellness applications and can be used to infer stress levels when used together with other data sources such as heart rate and respiration rate.

Stress can cause skin temperature to drop slightly, caused by vasoconstriction (narrowing of blood vessels) associated with our fight-or-flight response. A sudden decrease in skin temperature along with an elevated heart and respiration rate provides a fairly reliable signal of heightened stress or anxiety. For example, Cho et al. (2017) used a combination of temperature and respiration sensors to detect stress levels in participants given an arithmetic subtraction task while in VR, and they found that a decrease in skin temperature and an increase in respiration rate were consistently associated with higher stress levels, demonstrating the effectiveness of this combination to detect stress within immersive environments.

Interoceptive Computer-to-Human Outputs

Visual and Auditory Feedback

As mentioned before, the interoceptive system is responsible for addressing the self-awareness of the internal physiological and physical states of our body. The interoceptive system provides such critical signals from the body as the sensation of pain, temperature level, muscular and visceral sensations such as tension, vasomotor activity, and the perceived need for air. The system is associated with autonomic motor control, which is different from mechanoreception, which sensed in the skin, or proprioception, which is sensed in the muscles and joints of the body.

The primary interoceptive activity involves providing the basis for subjective feelings of our physical and emotional awareness occurring in the right anterior insula. XR systems can visually represent interoceptive states, such as showing an avatar's chest rising and falling with the user's breath or visualizing the heartbeat as a pulsing light. Auditory feedback, such as sounds synchronized with the user's heartbeat or breathing, can further enhance this immersive experience.

Haptic Feedback

Advanced haptic devices can provide tactile sensations that mimic internal bodily states. For instance, subtle vibrations on the chest area can simulate the sensation of breathing, while rhythmic pulses can mimic the feeling of a heartbeat, providing a richer interoceptive experience.

Thermal Feedback

Temperature changes can simulate various internal states, such as the sensation of warmth during physical exertion or cooling during relaxation exercises. This type of feedback can be delivered through wearable devices that adjust temperature in response to the user's physiological signals.

Interoceptive Design Considerations

Integrating interoceptive cues, such as breathing patterns, into an immersive experience can significantly enhance immersion. For example, a study demonstrated that synchronizing a virtual avatar's breathing with the user's breathing rate can lead to adjustments in the user's own respiration, thereby creating a deeper connection to the virtual body (Czub & Kowal, 2019).

Fitness programs can use real-time interoceptive feedback to adjust workout intensity. For example, they might increase virtual obstacles or pace in response to the user's heart rate to ensure optimal workout intensity or use breathing feedback to guide cool-down periods. Incorporating interoceptive feedback in games can create more engaging and realistic experiences. For example, a horror-themed experience could adjusts environmental stressors based on the user's heart rate to maintain a constant level of tension and immersion.

VR environments designed for therapy can use interoceptive inputs to tailor experiences for stress reduction, pain management, and emotional regulation. Traditional mindful breathing, for instance, focuses on internal sensations. Immersive media can enhance this experience through visual-auditory synchronization, providing an immersive distraction from pain based on the user's breathing and heart rate and thereby modifying pain-related brain activity to optimize relaxation (Hu et al., 2021).

When considering the use of interoceptive data, a multimodal approach will usually yield the best results. For example, El Ali et al. (2023) observed that including audio feedback increased interoceptive awareness, suggesting that auditory cues can enhance the participant's awareness of their internal states in virtual reality. They used the inexpensive Polar H10 heart rate monitor, opening new opportunities for innovative human–machine interaction. There is a nuanced interplay between external modalities and interoception, which has implications for the practical application of biofeedback in immersive experiences.

Context Scenario Breakdown

In media production, breaking down a script involves a process of a detailed, systematic analysis of the inscribed narrative to identify the dramatic building blocks necessary to tell the story. Screenplay analysis reduces all production elements into organized categories and lists that will form the foundation of such management tools as a production board, schedule, and budget. During the scenario breakdown process, the production team analyzes the script to identify and code the entire assemblage of dramatic elements, including the cast, extras, props, wardrobe, make-up, hairstyles, stunt artists, special effects and other technical equipment, film score, sound effects, etc. All categories are typically highlighted and color-coded within a shooting script. The highlights are used by production management to organize the shooting schedule and by individual departments to prepare for each production day. We have applied this standard filmmaking practice to the production of XR applications.

Follow these steps to perform a scenario breakdown that will provide the information needed to specify the technology package and design the technical architecture required to realize the context scenario. This is where you translate your magical scenario into the reality of procedures, actions, content, inputs, and outputs. Performing a scenario breakdown involves defining the context, identifying the needed content, interactions, and outputs, and creating a detailed technical architecture to support the context scenario. It is important to revisit the deliverables you produced during prior steps and revise them as needed. What are you aiming to achieve? What is your primary use case? Clearly defining your objectives will help guide the technology assessment process and prototype creation. By following these steps, you can ensure a comprehensive and effective design for immersive media applications.

Step 1. Read and Review the Context Scenario Thoroughly

Start by reading the entire context scenario multiple times. The first read should help you understand the overall flow of the experience. Subsequent readings should focus on identifying specific details that must be addressed during production. This process is similar to the script breakdown process widely used in film production, in which the script is highlighted with different colors to help identify the key elements. We suggest a similar approach for crafting scenario breakdowns. Begin by thoroughly reading the script or design document for the XR experience. Focus on understanding the narrative and interactive elements and how the real and virtual worlds will coexist and interact.

Step 2. Identify Key Elements

During the first or the subsequent read through the context scenario, start thinking critically about the key elements such as hardware, software, and digital assets that will be needed to enact the proposed experience. Various headsets like Quest, HoloLens, or Vision Pro will offer different levels and qualities of immersion that will affect the user experience. Note all of the digital assets included in the scenario. They will all need to be either acquired or created by the production team. Do the agents require AI capabilities? That will also have to be factored into the production schedule and budget. Go through the entire scenario, critically assessing the technologies and production elements that will bring it to life. This process needs to involve the entire production team. Your programmers need to look for any code they will need to create. Visual designers will look for environments, characters, and objects needed, etc. The list you create will become the foundation of your breakdown document.

Step 3. Create a Breakdown Document

A breakdown document is where you list all the highlighted elements scene by scene. This document (often implemented as a spreadsheet) is typically divided into categories including human agents, actors, settings, digital objects, digital twins, enchanted objects, physical objects, human inputs, system outputs, etc. For each scene, list every element that has been highlighted in the context scenario. This document serves as a reference for all members of your team to understand specific needs for each beat through each scene (or level) of the experience.

Step 4. Detail the Elements

This step involves breaking down each component of the scenario into detailed, actionable elements to guide the prototyping process. Before diving into the specifics of each element, it is essential to engage in a collaborative discussion with all project stakeholders. This includes designers, developers, artists, engineers, the project manager, and any other relevant team members. The goal is to ensure that everyone has a clear understanding of the objectives, the desired experience, the technical requirements, and most important of all, the hypothesis being tested. When you involve all stakeholders in the discussion, you can identify potential challenges early. This collaborative approach allows the team to brainstorm solutions or alternative approaches before building prototypes. Break down each element in detail.

A. Interactive elements. Detail the interactive components that users will engage with, such as buttons, levers, menus, or objects within the virtual environment. Specify how users will interact with these elements—whether through hand gestures, voice commands, or controller inputs—and outline the expected behavior of each interaction. For example, describe what happens when a user presses a button: Does it trigger an animation, open a new menu, or activate a specific event in the scenario?
B. Environments. Provide a detailed description of the physical, augmented, and virtual environments, including layout, aesthetics, and the role each part of each environment plays in the scenario. This includes defining the textures, lighting, and visual effects that will be used to create the desired atmosphere. For instance, if the scenario takes place in a forest, describe the density of the trees, the quality of light filtering through the canopy, and the sounds of wildlife that contribute to the immersive experience.
C. Digital objects. Outline the design and functionality of the digital objects. Detail how these elements will be integrated into the environment. Specify the placement, size, and interaction methods for essential elements. However, don't focus on fidelity at the prototyping phase. Most important is how these objects will appear in the scenario and how they will respond to user actions.
D. Movement. Explain how participants will navigate through the virtual, augmented, and physical environments. Describe the methods of movement, such as teleportation in VR, walking in physical space with augmented pathways, etc. Detail any special mechanics, like flying or swimming, and how these will be implemented. Consider accessibility options to ensure that all users can navigate comfortably.
E. Sound design. Detail the audio elements, including background music, sound effects, and spatial audio cues. Explain how sound will be used to enhance immersion, guide users, or signal important events. Describe the placement of sound sources within the environment and how they will interact with the user's position and movement.
F. Haptic feedback. If applicable, provide details on how haptic feedback will be incorporated to enhance the sensory experience. Specify the types of feedback

(vibrations, forces, textures) and when they will be triggered during the scenario. For example, haptic feedback could simulate the feeling of walking on different surfaces, the impact of objects, or wind blowing on the participant's face.

G. Experimental interactions and interfaces. If applicable, provide details on how the olfactory and gustatory channels will be incorporated to enhance the sensory experience and when they will be triggered during the scenario. For example, the olfactory sense could be used to create a sense of place, e.g., the scent of the trees in a pine forest.
H. Immersion and flow. Discuss what contributes to the sensation of immersion, focusing on how the scenario will feel from the participant's perspective. Consider the flow of interactions, the pacing, and how participants will be guided through the experience. You can maximize engagement by providing clear options, intuitive interactions, and meaningful feedback.

Step 5. Create a Production Schedule and Budget

After completing the script breakdown and technology assessment and making decisions on the technology frameworks you will be working with, you will be ready to prepare the production schedule and budget. This will cover all of the digital assets, software modules, and hardware components required to build your first prototype. The production schedule should outline when and how each element will be developed, integrated, and tested. The schedule should also account for iterations based on user feedback during testing phases, with the scale and scope driven by the specific details of your project.

Step 6. Review and Revise

As the project progresses, the scenario breakdown and associated documents will need to be revised based on budget constraints, technological feasibility, resource availability, or other logistical factors. Continuous communication among key stakeholders ensures that the scenario breakdown remains accurate and actionable and that everyone has a clear understanding of what is required for each scene (or level) of the experience. Testing new technologies and workflows is crucial to catching potential issues early, giving the production team time to find solutions. As the project progresses, continuously review and update the breakdown to reflect any changes in the design or technical requirements. XR experiences often require iterative development, so flexibility and ongoing communication with the team are key.

Summary

In this chapter, we tackled the important subject of engineering the immersive experience. We highlighted the fundamental link between immersion and presence, which can only be achieved through a close alignment between the assembled technology and the participants' embodied way of experiencing the world. With that in mind, we covered a number of currently available XR technologies capable of mediating the eight sensory channels.

The field of sensory mediation technologies is shifting and advancing so rapidly that we strongly recommend paying close attention to new developments from research labs, existing XR leaders such as Meta, Apple, and Microsoft, and new commercial start-ups offering breakthrough technologies.

Finally, we outlined the scenario breakdown process to help the production team bring the vision presented in the context scenario to life in all of its potential immersive glory. Breaking down scenarios brings the team one step closer to building and testing the proof-of-concept prototypes, which is the topic of our next chapter.

References

Aiken, M. P., & Berry, M. J. (2015). Posttraumatic stress disorder: Possibilities for olfaction and virtual reality exposure therapy. *Virtual Reality*, 19(2), 95–109. https://doi.org/10.1007/s10055-015-0260-x

Andreano, J., Liang, K., Kong, L., Hubbard, D., Wiederhold, B. K., & Wiederhold, M. D. (2009). Auditory cues increase the hippocampal response to unimodal virtual reality. *CyberPsychology & Behavior*, 12(3), 309–313. https://doi.org/10.1089/cpb.2009.0104

Apostolou, K., & Liarokapis, F. (2022). A systematic review: The role of multisensory feedback in virtual reality. *2022 IEEE 2nd international conference on intelligent reality (ICIR)*, 39–42. https://doi.org/10.1109/ICIR55739.2022.00024

Archer, N. S., Bluff, A., Eddy, A., Nikhil, C. K., Hazell, N., Frank, D., & Johnston, A. (2022). Odour enhances the sense of presence in a virtual reality environment. *PLoS ONE*, 17(3), e0265039. https://doi.org/10.1371/journal.pone.0265039

Arthur, W. B. (2009). *The nature of technology: What it is and how it evolves*. Free Press.

Baus, O., Bouchard, S., Nolet, K., & Berthiaume, M. (2022). In a dirty virtual room: Exposure to an unpleasant odor increases the senses of presence, reality, and realism. *Cogent Psychology*, 9(1), 2115690. https://doi.org/10.1080/23311908.2022.2115690

Bohil, C. J., Alicea, B., & Biocca, F. A. (2011). Virtual reality in neuroscience research and therapy. *Nature Reviews Neuroscience*, 12(12), 752–762. https://doi.org/10.1038/nrn3122

Burke, J. L., Prewett, M. S., Gray, A. A., Yang, L., Stilson, F. R. B., Coovert, M. D., Elliot, L. R., & Redden, E. (2006). Comparing the effects of visual-auditory and visual-tactile feedback on user performance: A meta-analysis. *Proceedings of the 8th international conference on multimodal interfaces*, 108–117. https://doi.org/10.1145/1180995.1181017

Can, Y. S., Arnrich, B., & Ersoy, C. (2019). Stress detection in daily life scenarios using smart phones and wearable sensors: A survey. *Journal of Biomedical Informatics*, 92, 103139. https://doi.org/10.1016/j.jbi.2019.103139

Cho, D., Ham, J., Oh, J., Park, J., Kim, S., Lee, N.-K., & Lee, B. (2017). Detection of stress levels from biosignals measured in virtual reality environments using a kernel-based extreme learning machine. *Sensors*, 17(10), 2435. https://doi.org/10.3390/s17102435

Chu, J. (2022, August 18). Engineers fabricate a chip-free, wireless electronic "skin." *MIT News Office*. https://news.mit.edu/2022/sensor-electronic-chipless-0818

Clarke, A. C. (1999). *Profiles of the future: An inquiry into the limits of the possible* (Millennium ed.). Gollancz.

Czub, M., & Kowal, M. (2019). Respiration entrainment in virtual reality by using a breathing avatar. *Cyberpsychology, Behavior and Social Networking*, 22(7), 494–499.

Deco, G., & Roland, P. (2010). The role of multi-area interactions for the computation of apparent motion. *NeuroImage*, 51(3), 1018–1026. https://doi.org/10.1016/j.neuroimage.2010.03.032

Egan, D., Keighrey, C., Barrett, J., Qiao, Y., Brennan, S., Timmerer, C., & Murray, N. (2017). Subjective evaluation of an olfaction enhanced immersive virtual reality environment. *Proceedings of the 2nd international workshop on multimedia alternate realities*, 15–18. https://doi.org/10.1145/3132361.3132363

El Ali, A., Ney, R., Van Berlo, Z. M. C., & Cesar, P. (2023). Is that my heartbeat? Measuring and understanding modality-dependent cardiac interoception in virtual reality. *IEEE Transactions on Visualization and Computer Graphics*, 29(11), 4805–4815. https://doi.org/10.1109/TVCG.2023.3320228

Feng, G., Zhang, J., Chai, G., Li, M., Zuo, G., & Yang, L. (2022). An effective training strategy for upper-limb rehabilitation robots based on visual-haptic feedback using potential field. *2022 12th international conference on CYBER technology in automation, control, and intelligent systems (CYBER)*, 678–681. https://doi.org/10.1109/CYBER55403.2022.9907062

Fuchs, P., Moreau, G., & Guitton, P. (Eds.). (2011). Theoretical and pragmatic approach to virtual reality. In *Virtual reality: Concepts and technologies* (pp. 33–68). CRC Press. https://doi.org/10.1201/b11612-8

Groth, C., Tauscher, J.-P., Heesen, N., Hattenbach, M., Castillo, S., & Magnor, M. (2022). Omnidirectional galvanic vestibular stimulation in virtual reality. *IEEE Transactions on Visualization and Computer Graphics*, 28(5), 2234–2244. https://doi.org/10.1109/TVCG.2022.3150506

Halabi, O., & Saleh, M. (2021). Augmented reality flavor: Cross-modal mapping across gustation, olfaction, and vision. *Multimedia Tools and Applications*, 80(30), 36423–36441. https://doi.org/10.1007/s11042-021-11321-0

Hu, X.-S., Beard, K., Sherbel, M. C., Nascimento, T. D., Petty, S., Pantzlaff, E., Schwitzer, D., Kaciroti, N., Maslowski, E., Ashman, L. M., Feinberg, S. E., & DaSilva, A. F. (2021). Brain mechanisms of virtual reality breathing versus traditional mindful breathing in pain modulation: Observational functional near-infrared spectroscopy study. *Journal of Medical Internet Research*, 23(10), e27298. https://doi.org/10.2196/27298

Kennedy, R. S., & Lilienthal, M. G. (1995). Implications of balance disturbances following exposure to virtual reality systems. *Proceedings Virtual Reality Annual International Symposium*, 95, 35–39. https://doi.org/10.1109/VRAIS.1995.512477

Kim, H., & Lee, I.-K. (2022). Studying the effects of congruence of auditory and visual stimuli on virtual reality experiences. *IEEE Transactions on Visualization and Computer Graphics*, 28(5), 2080–2090. https://doi.org/10.1109/TVCG.2022.3150514

Kim, S., Izdebski, K., & König, P. (2019). The effectiveness of multimodal sensory feedback on VR users' behavior in an L-collision problem. In J. Trojanowska, O. Ciszak, J. M. Machado & I. Pavlenko (Eds.), *Advances in manufacturing II* (pp. 381–389). Springer International Publishing. https://doi.org/10.1007/978-3-030-18715-6_32

Loftin, R. B., Chen, J. X., & Rosenblum, L. (2005). Visualization using virtual reality. In *Visualization handbook* (pp. 479–489). Elsevier. https://doi.org/10.1016/B978-012387582-2/50026-5

Lombard, M., & Ditton, T. (2007). At the heart of it all: The concept of presence. *Journal of Computer-Mediated Communication*, 3(2). https://doi.org/10.1111/j.1083-6101.1997.tb00072.x

Maniatis, A., Bourou, S., Anastasakis, Z., & Psychogios, K. (2023). VOXReality: Immersive XR experiences combining language and vision AI models. *14th international conference on applied human factors and ergonomics (AHFE 2023)*. https://doi.org/10.54941/ahfe1002938

Marey, A., & Alsabbagh, N. (2023). Simulating human senses to improve thermal comfort. *Architecture and Planning Journal (APJ)*, 28(3). https://doi.org/10.54729/2789-8547.1236

Matsumura, M., & Murofushi, T. (2021). Vestibular rehabilitation after vestibulopathy focusing on the application of virtual reality. *Journal of Otorhinolaryngology, Hearing and Balance Medicine*, 2(2), 5. https://doi.org/10.3390/ohbm2020005

Metz, R. (2018, May 25). Now playing: A movie you control with your mind. *MIT Technology Review*. www.technologyreview.com/2018/05/25/142725/now-playing-a-movie-you-control-with-your-mind/

Mevlevioğlu, D., Murphy, D., & Tabirca, S. (2021). Visual respiratory feedback in virtual reality exposure therapy: A pilot study. *ACM international conference on interactive media experiences*, 1–6. https://doi.org/10.1145/3452918.3458799

Mihelj, M., Novak, D., & Beguš, S. (2014). *Virtual reality technology and applications* (Vol. 68). Springer Netherlands. https://doi.org/10.1007/978-94-007-6910-6

Park, B., Kim, Y., Park, J., Choi, H., Kim, S.-E., Ryu, H., & Seo, K. (2024). Integrating biomarkers from virtual reality and magnetic resonance imaging for the early detection of mild cognitive impairment using a multimodal learning approach: Validation study. *Journal of Medical Internet Research*, 26, e54538. https://doi.org/10.2196/54538

PSC. (2024, February 21). The art of seeing: How computer vision and human vision differ. *PSC*. https://plasticservicecompany.com/en/articles/computer-vision-vs-human-vision/

Rose, D. (2014). *Enchanted objects: Design, human desire, and the Internet of things*. Scribner.

Shaer, O. (2009). Tangible user interfaces: Past, present, and future directions. *Foundations and Trends in Human–Computer Interaction*, 3(1–2), 1–137. https://doi.org/10.1561/1100000026

Sheth, B. R., & Young, R. (2016). Two visual pathways in primates based on sampling of space: Exploitation and exploration of visual information. *Frontiers in Integrative Neuroscience*, 10. https://doi.org/10.3389/fnint.2016.00037

Skalski, P., & Whitbred, R. (2010). Image versus sound: A comparison of formal feature effects on presence and video game enjoyment. *PsychNology Journal*, 8(1).

Slater, M. (2009). Place illusion and plausibility can lead to realistic behaviour in immersive virtual environments. *Philosophical Transactions of the Royal Society B: Biological Sciences*, 364(1535), 3549–3557. https://doi.org/10.1098/rstb.2009.0138

Valori, I., McKenna-Plumley, P. E., Bayramova, R., Zandonella Callegher, C., Altoè, G., & Farroni, T. (2020). Proprioceptive accuracy in immersive virtual reality: A developmental perspective. *PLoS ONE*, 15(1), e0222253. https://doi.org/10.1371/journal.pone.0222253

Virre, E. (1996). Virtual reality and the vestibular apparatus. *IEEE Engineering in Medicine and Biology Magazine*, 15(2), 41–43, 69. https://doi.org/10.1109/51.486717

Weech, S., Moon, J., & Troje, N. F. (2018). Influence of bone-conducted vibration on simulator sickness in virtual reality. *PLoS ONE*, 13(3), e0194137. https://doi.org/10.1371/journal.pone.0194137

XR Today Team. (2022, February 9). The best haptic VR devices and innovations for 2022. *XR Today*. www.xrtoday.com/virtual-reality/the-best-haptic-vr-devices-and-innovations-for-2022/

Zakharov, A. V., Kolsanov, A. V., Khivintseva, E. V., Fedorovich Pyatin, V. F., & Yashkov, A. V. (2021). Proprioception in immersive virtual reality. In J. A. Vega & J. Cobo (Eds.), *Proprioception*. IntechOpen. https://doi.org/10.5772/intechopen.96316

Zotkin, D. N., Duraiswami, R., & Davis, L. S. (2004). Rendering localized spatial audio in a virtual auditory space. *IEEE Transactions on Multimedia*, 6(4), 553–564. https://doi.org/10.1109/TMM.2004.827516

6 Building the Proof of Concept

The Road From Ideas to Experiences

The ultimate advocate for digital product building, Steve Jobs, pointed out the "tremendous amount of craftsmanship" involved in going from "a great idea" to "a great product" (Sen, 2012). Building an immersive product or experience is a multistep iterative process that requires the identification of clear target outputs and appropriate design methods at each stage. Building an XR experience for a few users utilizing a few headsets is a lot easier to manage than building and maintaining an infrastructure for hundreds or thousands of users, with potentially different hardware devices, software, and licenses to manage. The important thing is to start somewhere, and not just anywhere but at the right place from which scaling will be doable.

The place to start the building and scaling process is with a proof of concept (PoC). A proof of concept is an important middle stage of the design process that builds the bridge between the concept and execution stages and allows the team to assess the proposed solution's technical feasibility and practical doability. PoCs are early stress tests conducted to falsify the teams' hypothesis about the idea's desirability and feasibility and the proposed technology. PoCs focus on a particular aspect of the experience under development.

As we mentioned in Chapter 3, there are four main hypotheses involved in the creation of immersive content.

1. There is an audience hypothesis that answers the questions of who is the audience and what will resonate with them.
2. There is the value hypothesis which answers the question regarding the audience needs and resonance. Will the proposed experience perform a job-2B-done in the life of your audience?
3. The usability/feasibility hypothesis helps you to assess how the system will perform; how immersive it will be.
4. The growth hypothesis will help you to start assessing how well your XR media franchise can scale. All of these hypotheses can be tested with proof-of-concept prototypes.

DOI: 10.4324/9781003442462-11

Build to Fail Early and Succeed Sooner

Polish-born mathematician and humanist Jacob Bronowski poetically suggested that "The hand is the cutting edge of the mind" to argue that "The world can only be grasped by action, not by contemplation" (Ratcliffe, 2017). IDEO's famous dictum, "Fail early to succeed sooner," points to the important interplay between the mind and the hand—the thinking to build and building to think cycles in the design process.

There is no doubt that building products requires thinking and planning, but ideation, so necessary to experience design, is located solely in the theoretical world, while the audience you seek to enchant lives in the world of embodied experiences. As immersive experience designers, we must transfer our constructs into things we can actually work with, test, and learn from. Developing operational definitions entails transferring the constructs created in our minds into things our audiences can actually work with.

> *The world can only be grasped by action, not by contemplation ... The hand is the cutting edge of the mind.*
>
> JACOB BRONOWSKI

The creative force behind such market blockbusters as the iPod and the Nest line of smart home products, Tony Fadell, has defined key characteristics of product development in his book *Build: An unorthodox Guide to Making Things Worth Making* (Fadell, 2022). Fadell's big four include:

1. Visionary Thinking, referred to in innovation as abductive thinking. Visionary thinking is all about developing a value hypothesis that can stand the test of time to scale upward and outward. It involves envisioning groundbreaking experiences that meet and anticipate consumer needs and desires.
2. Attention to Detail means paying close attention to every aspect of a product, from its functionality to its aesthetics, ensuring that it supports a compelling user experience. This can only be achieved by recruiting a dedicated team and
3. Setting High Standards and expectations for everyone involved. Setting high standards in the real world is very different from perfectionism, which only exists in the realm of fantasy and is, in fact, the enemy of creativity.
4. Resilience and Adaptability mean that the team is agile enough to follow the iterative design process of ideation, prototyping, and hypothesis testing to adjust and even pivot to navigate obstacles while focusing on achieving long-term goals.

Mixing building-to-think and thinking-to-build phases is a fundamental part of media design. Ideation is the part of the process that allows us to think in order to build. Creative imagination is the first step toward visualizing possibilities and speculating on the consequences of our ideas. We sketch and describe our ideas to capture and make them visible to ourselves and others.

Prototyping, on the other hand, is an act of getting our ideas out of our heads and into the hands of the prospective audiences. Prototyping is a way to get fast and inexpensive feedback from the users, which will allow us to iterate, evolve, and improve the products we build. Prototyping is a way to build things in order to think with our hands. Prototyping makes our ideas come to life so we can learn from them quickly in order to iterate and evolve them.

What a Proof of Concept Is and Why You Need One

A proof of concept is an initial demonstration your team will produce of some aspect of the immersive experience to confirm that the conceptual design work will translate well into the final product. PoCs are used by digital product creators to validate the feasibility of new product or content ideas. PoCs are standalone projects that validate various aspects of your product idea.

The main goal of creating a PoC is to test your assumptions and to verify that your immersive experience concept can be executed from the engineering and production point of view. PoC can take the form of a working part of the final product, developed with specific technology, to prove that the hardware and the software can deliver particular XR content. Even though PoC is a type of an early prototype, full-fledged prototypes, which we will discuss in Chapter 8, are more refined and elaborate simulations of the final feature or a product that give the media makers a clear view of its real-world operation.

PoCs and prototypes are used at different stages of the development process. A PoC process offers the media makers a practical verification of whether the immersive application can be built with the given resources, while prototyping involves building a working model of the application with limited functionalities. Proofs of concept can be thought of as preprototypes, as they are used to help the team to decide whether particular functionalities of the application are feasible to implement. Prototypes, by comparison, focus on testing the full functionality of the included features. PoCs are typically created by the team to be used internally. Their main benefit lies in minimizing the risk of technical deficiencies arising during the development process. A PoC is one of the most efficient and accurate methods of testing key hypotheses and validating fundamental assumptions about the desirability and feasibility of your XR app idea.

Building a proof of concept is one of the best ways to validate your creative intuitions and test your assumptions, especially in areas that require the XR product to satisfy a high degree of contextual complexity or expert knowledge, such as healthcare or manufacturing.

A PoC in medical VR could take the form of a clinical trial that is used to gain insight into the actual product use in a clinical setting. A negative result from such a trial can save a lot of time and development costs while providing clear guidance on how to correct any faulty assumptions. Positive results, on the other hand, prove that the team has the right understanding of the problem and access to the appropriate technology and that the application is feasible to build. PoCs can also help the team to secure funding for further product development.

The Proof-of-Concept Development Process

With a wide range of proof-of-concept categories, the development and testing process can vary quite substantially depending on the team's learning objective. The common steps in the proof-of-concept development process include:

1. Identifying the hypothesis to be tested. A hypothesis is an assumption or an idea that is proposed for the sake of argument so that it can be tested. In science, it refers to a precise, testable statement of what the researchers predict will be the study's outcome. A hypothesis usually involves proposing a relationship between the independent variable (what the researchers change) and the dependent variable (what the research measures). Decide what unknown (variable) dimension of the project you want to test. Do you have an existing hypothesis on how to approach it? A value hypothesis example might include a belief that an immersive AR-guided adventure through the Colosseum in Rome will resonate with the prospective audience interested in travel, history, and exploration. A usability/feasibility hypothesis might include the ability of the team to create the entire package of the hardware and software components to support such an AR experience effectively.
2. Design and build a prototype that enables you to conduct the PoC experiment. This does not necessarily mean building a fully functional piece of software or hardware. There are many ways to test early ideas, including role-playing, bodystorming, and paper prototyping. The important measure here is based on the scientific concept of validity, which essentially refers to measuring the right thing. For instance, getting audience feedback after asking them to walk around the Colosseum with Bluetooth headphones and an iPad Pro in the middle of the summer might elicit responses indicative of general fatigue and dehydration rather than the fatigue caused by the AR application. Building prototypes requires resources and expertise. Make sure that your core team can handle the building process. If not, you can hire an outside agency to facilitate prototyping.
3. Once the prototype is ready, your team can conduct the experiment to either confirm or falsify the chosen hypothesis. Make sure that at this stage, you clearly define the success criteria for your PoC. What will prove or disprove your hypothesis? Success criteria can be experiential or outcome focused. An audience segment engaging with the VR Colosseum tour experience and offering to promote it to their friends might be one such success criterion for value hypothesis testing. Cobbling together basic software and hardware and having test subjects successfully navigate a portion of the experience using it might be success criteria for usability/feasibility hypothesis testing. Clearly defining success criteria will

help the entire team and project funders understand if the project is on the right track based on the metrics for a successful PoC. Experiments provide evidence that gives the media makers real-world feedback on their key assumptions. It is critical to document the feedback by collecting data from the experiment in a way that can be organized and processed in the next step.

4. This step involves understanding the experiment results and gaining insights from them. Insights can be organized around the project management concept of "lessons learned," which refers to the knowledge your team gains during the execution of your PoC experiments. Some people erroneously associate identifying lessons learned with failure. This view misses the point of failing often to succeed sooner. Small calculated failures provide the cheapest way to avoid large, costly mistakes.
5. Finally, comparing the results to the stated success criteria allows the team to confirm or revise the original hypothesis to determine the next actionable steps needed to be taken toward the goal.

Leveraging Foundations

Bernard Francois, a principal at Preview Labs specializing in rapid prototyping, emphasizes the importance of leveraging the foundational user interaction frameworks provided by the operating system or platform you are working with when

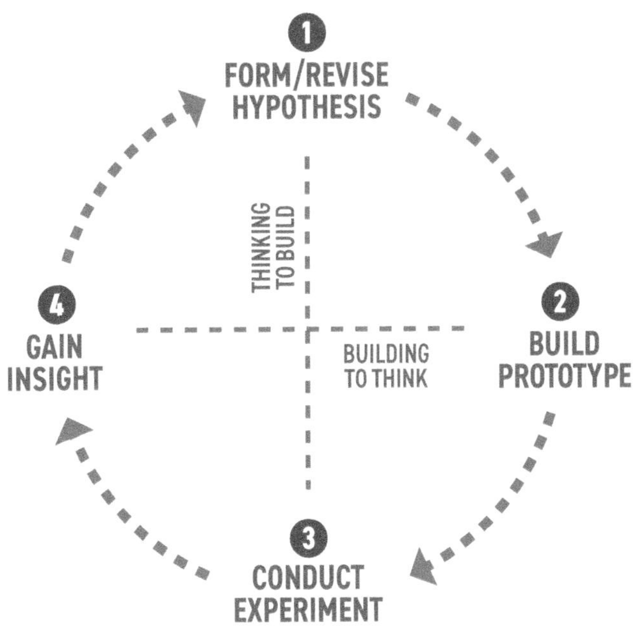

Figure 6.1 Hypothesis-testing feedback loop

Source: Concept and illustration by © 2024 Peter (Zak) Zakrzewski, based on Eric Ries', Build-Measure-Learn Feedback Loop.

designing immersive experiences. During the early stages of PoC prototyping, designers and developers should focus on their unique core value propositions rather than reinventing basic functionalities already handled well by the operating system or platform. This approach allows for innovation where it truly matters—within the app-specific experiences—while ensuring that foundational elements are robust and engaging (personal communication, August 13, 2024).

A practical application of this principle was demonstrated in a project that Preview Labs developed in collaboration with Dr. Andrew Hashikawa at the Department of Emergency Medicine at the University of Michigan. The goal was to create a VR prototype to raise awareness about the dangers of distracted driving without inadvertently encouraging users to "practice" multitasking behind the wheel. This could lead to a misinterpretation of the simulation as a training tool. The team developed three distinct prototypes, each focusing on simulating the effect of distractions in a controlled, nondriving environment, drawing clear parallels to the experience of driving while distracted.

In the Distracted Sorter prototype, users were tasked with sorting gems by color and shape, both with and without distractions, such as flashing lights. The goal was to highlight how even simple tasks can become challenging while distracted, mirroring divided attention while driving.

In the Distracted Cook prototype, the player had to slice food and complete orders within a time limit while managing additional distractions like a smoking oven. This prototype emphasized the increased difficulty of task management under stress and distraction.

The Distracted Navigator prototype (ultimately selected for further development) placed the user in control of a spaceship, navigating through space while managing distractions inside the vessel. It characterized the challenges of maintaining focus on multiple tasks simultaneously, similar to distracted driving.

Andrew Hashikawa and his team used these prototypes to determine the most effective approach for their project objectives. By selecting the Distracted Navigator prototype and refining it based on user feedback, they created a more refined experience that introduced even more distractions to emphasize the risks of divided attention.

The method of iterative prototyping successfully achieved their educational and awareness goals. By focusing on what made their scenario unique and leveraging established technologies for essential functions, Andrew Hashikawa and his team were able to focusing their innovation on the specific core experience being developed in order to create the most impactful experience (Cootmans, 2021).

The Proof-of-Concept Design Document

The proof-of-concept design document is a structured plan that outlines the objectives, features, and development process for creating a PoC prototype. It serves as a blueprint for the critical PoC prototyping step, helping to ensure that the entire team stays focused on the key goals and requirements and providing a reference point for evaluating the success of your work. This document typically includes the following elements:

1. Objectives articulate the purpose of the prototype and the essential ideas you aim to test.
2. "Hypothesis to be tested" statement articulates the hypothesis or multiple hypotheses you plan to test. Several different hypotheses may be tested with one prototype. A hypothesis can be as basic as asking: Will the microphone data be good enough with the software package X we got from Y to detect if the participant is comfortable or not?
3. Key features section identifies the core features that must be developed and tested in the prototype. These are the essential elements that define the functionality of the prototype.
4. Requirements section offers a detailed description of the technical and design requirements necessary to build the prototype. This may include hardware, software, and specific technologies that will be used.
5. Scope and limitations define what the PoC prototype will and will not include. This helps prevent feature creep by setting boundaries for the project.
6. A development plan is a step-by-step plan for developing the PoC prototype, including timelines, milestones, and iteration cycles.
7. Assumptions and risks section document any assumptions made during the planning phase and potential risks that could impact the development or testing of the prototype.
8. Evaluation criteria states the metrics or criteria that will be used to evaluate the PoC prototype's success.

Keep in mind that the PoC design document is a living, breathing document that needs to evolve as the project evolves and as various team members are added in the development process.

Types of Proof-of-Concept Prototypes: What Are You Trying to Learn?

PoC prototypes answer similar questions to other forms of prototypes. Is this idea worth developing further? Even though proofs of concept require investments in time, effort, and capital, media makers, start-up founders, and project leaders in institutional settings have discovered that building a PoC can help to significantly reduce the uncertainty involved in new product development by validating the market resonance, product usability, and feasibility of the immersive experience quickly and at a relatively low cost. PoCs can start as conceptual guidelines such as the Apple Knowledge Navigator marketing video we mentioned in Chapter 4. They can also go as far as building functional hardware and software environments indistinguishable from the final product.

PoCs are particularly important for teams entering unknown markets or introducing new types of XR experiences. They can help teams validate a project idea, gauge resonance, and get valuable feedback from the prospective audience. The goal of PoCs is to obtain maximum feedback at a minimum cost. Build your PoC with a small team. Above all, be clear about what you are trying to learn. Some goals include visualizing the basic features of the experience, validating the technology, and collecting feedback from your intended audience.

As discussed, there are three major categories of hypotheses PoC can help to validate in digital and immersive product design. The first of these is the set of assumptions dealing with the following:

1. The value hypotheses that address the desirability and resonance of the proposed XR experience. Testing assumptions within this set means answering such fundamental questions as: Does the product perform a job-to-be-done in the life of the audience?
2. The usability/feasibility hypotheses center around the technological feasibility, ease of use, and the level of engagement the proposed system provides to the audience. Within this category, you will ask questions such as: How will the system support the users in reaching their goals? How compelling will the experience be? Do we have the technological resources to support the desired experience? The last set of assumptions concerns the spreadability of the immersive experience.
3. The growth hypothesis questions test your assumptions about the scaling potential of the proposed digital product. Addressing The growth hypothesis assumptions means answering the following questions: How will the word about our digital product spread? How big can the franchise grow? The *Mario* series, one of the most successful game franchises ever, has grown across six games to gross nearly $900 million dollars worldwide. This, of course, is dwarfed by the *Star Wars* media franchise, estimated to be worth over $65 billion. Resonant themes can lead to sustained growth over many geographical markets and long periods of time if the effort is guided by accurate value, usability, and growth hypotheses.

There are at least six types of prototypes you can build to test the three types of hypotheses your team might generate about the project.

1. Physical prototypes can test the functioning of XR system hardware such as the *Birdly* flight simulator developed by the Swiss firm Somniacs, which provides an audience with a unique VR experience of flying around the historic German city of Ulm. To create this unique XR media content, the design team had to build and test a human-sized simulator unit that allows the participants to direct the path of their flight with their stretched arms, akin to the flapping of a bird's wings (Demodern, 2017).
2. Physical prototypes can also test the form of the objects in the environment such as the lunar or Mars mission objects used by NASA to train astronauts. Even the headsets themselves can be tested using physical prototypes. Industrial designers behind both Apple's Vision Pro and Meta Quest have put a lot

of effort into making these HUDs as ergonomic, aesthetically appealing, and easy to use as possible.

3. Environment prototypes are extensively used in spatial computing to model and test the virtual environments the users will navigate. For the Birdly's Ulm environment, the town of 2,000 buildings and 6,000 various other objects like trees and animals was constructed virtual brick by brick by the Berliner Stiftung Interactive Media Foundation and Hamburg digital agency Demodern, whose team had previously been involved in creating the immersive worlds of the well-known computer game *World of Warcraft*. *Birdly's* participants can fly around the historical Ulm of the 1890s to pass by such places as the highest (in its day) church tower in the world, or the unassuming number 20 Bahnhofstrasse, where the author of the relativity theory Albert Einstein was born in 1879 (Demodern, 2017).
4. Person-to-person prototypes have been extensively used to prototype services and other difficult-to-simulate interactive experiences. This type of approach can be used by the designers as a way of giving the coding team more specific directions for key coded behaviors needed for the experience.
5. Person-to-technology prototypes test the usability of both the software and the hardware.
6. Technology-to-technology prototypes test the engineering feasibility and compatibility of software and hardware sourced from different suppliers.

Any form of prototyping involves creating an incomplete version of a product, service, or experience to explore ideas, test hypotheses, examine functionality, assess usability, and gather feedback from representative users. Functional prototypes operate as a bridge between your project vision and value proposition and real-world engagement with end users. They approximate the functionality of your project, demonstrating the project's potential even if performance is not perfect or the feature set is incomplete.

Observing people using the prototype helps your team to identify any usability issues, learn about user expectations, and assess task efficiency. This information will help your team make informed design, scope, and functionality decisions before committing to full-scale project development.

Functional prototypes are not intended to be final or production-ready deliverables. Instead, they are part of your iterative design process, providing an object to think with and a tool for validating the hypotheses you set out to test. A functional prototype should be quick and inexpensive to produce and thus limited to the most representative or critical functionality. Once the functional prototype has served its purpose, your team can move closer to building the final product.

Creativity in PoC Testing

Your team's creativity is the only limit on how you test your hypotheses. There are many types of prototypes we can build depending on what we are trying to learn and the resources at our disposal. Some physical prototypes allow the development team to test the functional aspect of the design. Other physical prototypes allow us to test the form and the aesthetics. Person-to-person prototypes are quick ways of testing out interactions. Person-to-technology prototypes let us test the usability. Technology-to-technology prototypes test the engineering feasibility of the proposed tech stack. Environment prototypes are a great way to test any design experience involving physical spaces.

One of our favorite prototyping categories is known as the "Wizard of Oz" tests. It allows us to build virtual prototypes that test the desirability of themes, concepts, and ideas. For instance, a poster or thumbnail can be used to find out how many people find the game or show concept to be resonant. One of the well-known examples of the Wizard of Oz tests was the Dropbox demo in which Drew Houston showed how easy the app was to use. The video, which was posted on YouTube, attracted thousands of beta testers showing the founders and their funders how resonant the need to replace flash drives with cloud storage tools really was.

Low-Fi PoCs

There are many ways to prototype various aspects of the immersive system, from low-fidelity paper prototypes to hacked and coded functional ones. Prototypes can be categorized based on their purpose of testing the value hypothesis or the usability hypothesis. Typically, value-hypothesis prototypes come early in the process, and once the concept is validated, usability becomes the focus. Low-fi PoCs are simplified, early-stage models used to quickly visualize and test core ideas for immersive media experiences without the time and resource investment required by high-fidelity prototypes. These prototypes are invaluable for rapidly iterating on ideas and gathering early user feedback, as they focus on basic interactions, structure, and flow rather than intricate graphics or detailed environments.

Typically constructed using sketches, storyboards, physical mock-ups, or basic digital tools, low-fi prototypes allow your team to map out the participant's journey and key functionalities in a way that can be easily adjusted based on testing and feedback.

In immersive media, low-fi prototyping is particularly useful for capturing spatial relationships, guiding navigation, and experimenting with core interactive elements, giving stakeholders and participants a tangible sense of the experience's layout and feel without extensive development. By emphasizing quick iteration and low costs, these prototypes support a flexible, creative design process, enabling designers to explore diverse concepts, refine interactions, and troubleshoot potential issues early in the development pipeline, all before investing in the complexities of a fully detailed build that requires more extensive development resources and financial commitments.

Mid-Fi PoCs

Mid-fi prototypes represent a more extended mode of immersive media prototyping, where the foundational ideas are fleshed out with a moderate level of detail and interaction, offering a clearer representation of the experience than low-fi prototypes. Unlike low-fi prototypes, which focus primarily on spatial organization and flow, mid-fi prototypes incorporate visual and interactive elements to provide a realistic sense of the experience's core features and aesthetics, though still without full polish. This level of prototyping often involves a combination of both classic and digital tools to simulate interactions and environment visuals in a controlled way, allowing for user testing that is closer to the final immersive experience.

Tools like Photoshop and After Effects can be used to create layered visuals and animations that reflect how elements will appear and interact in the final experience. Simple 3D models made in Blender can simulate spatial relationships and movement, giving participants a rough but effective sense of scale and placement within a virtual environment. Video walkthroughs, created with video editing tools (e.g., Adobe Premiere Pro, DaVinci Resolve), screen recordings, and presentation software (e.g., Keynote, Google Slides), can provide mock-ups that can be used to guide participants through key interactions, demonstrating essential functionalities and sequence with little or no code.

Mid-fi prototypes strike a balance between simplicity and realism, allowing designers to explore participant interactions in a contextually relevant environment. They can help identify critical usability issues and fine-tune interactions without the full commitment of high-fi development. This approach enables more advanced feedback and is ideal for testing narrative flow, transitions, and user engagement in immersive media experiences. Mid-fi prototypes are particularly advantageous in immersive projects because they will communicate the mood, visual tone, and interactivity, helping both stakeholders and participants engaged in testing to envision how the final product might look and feel.

By combining rapid, flexible prototyping techniques with a moderate level of fidelity, these prototypes will enable your team to iterate effectively on detailed aspects, bridging the gap between conceptual low-fi ideas and the highly realistic, interactive demands of a high-fi prototype.

High-Fi PoCs

High-fidelity (high-fi) prototypes represent the most advanced stage of prototyping for immersive media, delivering a close-to-final version of one or more specific aspects of the experience with almost fully realized visuals, interactive elements, and spatial details that closely simulate the intended experience. These prototypes are built using powerful tools like game engines, such as Unity or Unreal Engine, or specialized immersive prototyping platforms like Adobe Aero, Hoverlay, and ShapesXR, allowing designers to achieve a functional and realistic model of the final product.

Unlike low-fi and mid-fi prototypes, which focus on basic structure and flow, high-fi prototypes incorporate close-to-finished graphics, responsive interactions, and realistic spatial design to provide participants and stakeholders with an experience that feels nearly production-ready. High-fi prototyping offers the ability to test specific functionalities, verify technical performance, and refine detailed design elements with high accuracy, making it an essential step before committing to full-scale development.

The Unity and Unreal Engine game engines are popular choices for high-fi prototyping due to their capabilities in rendering realistic 3D graphics, simulating physics, and enabling complex interactive behaviors. For instance, in VR game development, a prototype built with Unity can simulate user movement and object interaction within a lifelike environment, offering a level of immersion that allows designers to assess comfort, usability, engagement, and presence and collect data in real time during the experience.

Game engines have been used for augmented reality (AR) prototyping. For example, a retail company might create an AR shopping experience to visualize product placement in a customer's home, testing interactions and transitions in a controlled but realistic scenario. For AR-focused prototypes, platforms like Adobe Aero and Hoverlay support the design of interactive AR scenes without extensive coding. Both Hoverlay and Adobe Aero allow designers to bring 2D and 3D assets into an AR environment, where participants can interact with elements as they would in the final product. This makes it ideal for prototyping experiential marketing campaigns or virtual product demonstrations or training scenarios that can be tested in a real-world setting.

ShapesXR is especially useful for VR and spatial design; it provides a collaborative space where teams can build and iterate on VR prototypes working in VR, exploring spatial arrangements, object scaling, and user interaction within a virtual setting. Once a prototype has been constructed in ShapesXR, the project can be exported as a Unity Project for further refinement, taking advantage of the many affordances game engines provide in terms of programming, user data collection tools, etc.

By creating prototypes with high-fidelity tools, you can gain insights into how the immersive experience will function under more realistic conditions, which can suggest both technical constraints and user experience enhancements. High-fi prototypes are essential for final-stage user testing, providing valuable feedback that guides final adjustments in design, performance, and interactivity. Through the use of high-fi tools, creators can thoroughly test and perfect their immersive media experiences, ensuring that the final product meets creative, technical, usability, and participant experience expectations.

Methods for Testing XR Experiences

The choice of methods for evaluating XR experiences varies based on your goals for the evaluation. Your team may want to test specific hypotheses or compare new interaction techniques against existing ones, focusing on effectiveness or efficiency metrics. Developers might prioritize hardware performance. Marketers might prioritize measurement of customer satisfaction. High-level evaluation criteria often include usability, usefulness, immersion, and ultimately presence.

Traditional usability assesses how effectively users interact with a product or experience, and these methods have been expanded to address the unique characteristics of XR. It's important for your team to decide which hypotheses are the most important to measure at this phase of the project. The choice of evaluation methods will depend on the design phase and available artifacts. System walkthroughs can be employed at any stage, allowing you to review processes and features conceptually, while experience testing requires a functional prototype for empirical analysis.

Gathering feedback is crucial throughout the evaluation process. While not all issues identified will demand immediate resolution, reflecting on and prioritizing them based on severity and impact can effectively guide your iterations of proof-of-concept and functional prototypes. If you select the evaluation methods that are right for your project, all of the stakeholders will gain valuable insights into the overall effectiveness and engagement of the experience.

Wizard of Oz PoCs

"Wizard of Oz" testing is a usability testing technique in which users interact with what they believe is an automated system, but in reality, a human controls some or all of the system's responses. This method is particularly useful in the early stages of development when fully implementing certain features would be overly time-consuming or too costly. It allows your team to test hypotheses about user interactions and behaviors before investing heavily in full-scale prototype development. This type of testing is cost-effective since it allows you to test and validate your hypothesis without needing to develop the full system up front.

The method is adaptable and can be used to explore different interaction models or quickly test new ideas based on user feedback. By focusing on real user behavior and feedback, you can ensure that the full prototype implementation will align with actual user expectations and needs. This method facilitates direct observation of how users interact with your scenario. Pay attention to how intuitive they find the interaction process and whether they struggle with any particular actions. Gather qualitative feedback to learn how natural the interactions felt, whether any actions were confusing, and if they had any expectations that weren't met.

Evaluating PoCs With Crowdfunding

Product concepts can be evaluated with crowdfunding. Finding out if a segment of an audience is prepared to pay for your creation is more valuable than asking people for an opinion. When Palmer Luckey developed a prototype for a VR headset in 2011, despite his technical skills, he needed more financial resources and industry connections to bring the headset to market. In August 2012, Luckey launched a Kickstarter campaign to fund the development of the Oculus Rift. The campaign aimed to raise $250,000 but raised over $2.4 million from more than 9,500 backers (Oculus, 2016).

The campaign's overwhelming success demonstrated a significant demand for VR technology and validated Luckey's vision. The campaign promised backers early access to development kits of the Rift, allowing them to create and test VR content before the commercial release. The success of the Kickstarter campaign garnered significant attention from the tech community and major investors and was instrumental in proving the viability of consumer VR technology. The campaign attracted the interest of Facebook, which acquired Oculus VR for $2 billion in 2014 and sent a powerful market signal that immersive media was ready for prime time.

This Kickstarter campaign is widely cited as a defining moment in both immersive media and crowdfunding history. The benefit of crowdfunding PoCs is that it validates user interest and raises precious funds required to build prototypes with feedback from early adopters who connect with your concept. Kickstarter, Indiegogo, and GoFundMe are rife with product concepts in search of funding from early adopters who will become evangelists for your creation.

Concierge-Service PoCs

The concierge-service test is a proof-of-concept technique used to validate the usability and functionality of a project idea before committing to full-scale development. This testing method involves simulating the experience using real people performing the tasks that will eventually be handled by hardware and custom software in the working prototype. The primary goal of a concierge-service test is to gather feedback to assess the usability of the project (or a specific aspect of the project) without building the working system.

In some variations of the concierge-service test, the participants will be unaware that humans are driving the experience behind the scenes, leading them to believe they are interacting with a functional system. This testing method makes it possible to observe user behavior, collect feedback, and adjust the scenario before investing in costly software development for a working prototype. Because humans operate the system, your team can quickly iterate on different aspects of the user experience, allowing for rapid iteration and identifying essential features and requirements before working prototype development begins.

Thinking-Aloud PoCs

Thinking-aloud proofs of concept allow design teams to obtain usability feedback with cheap, rapid prototypes. In thinking-aloud tests, designers ask the participants to use the system while continuously verbalizing their thoughts, feelings, and impressions (thinking out loud). The test requires skillful facilitation that prompts users to keep talking as they engage with the system.

According to Jakob Nielsen (2012), running a basic thinking-aloud usability study consists of a working prototype and recruiting representative users, giving them strategically chosen tasks to perform while letting them do the talking. Thinking-aloud, like crowdfunding tests, are much more valid than surveys and focus groups because they test the audience's response to the selected variables and features of the actual concept. Nielsen argues that the method's most significant advantage is that it offers the design team a window into the honest thoughts and feelings of the participants, including their misconceptions, which can turn into actionable redesign directions.

Nielsen suggests that even the most "arrogant designers," "hard-boiled developers," and "tight-fisted" business folks can see the light when faced with direct exposure to what test participants think about their work (Nielsen, 2012). Nielsen goes so far as to suggest that "Thinking aloud may be the most valuable usability engineering method" (Nielsen, 1993, p. 195). The method is inexpensive to implement, and special equipment is not needed. It is also very robust compared to quantitative (statistical) usability studies because even less-than-average facilitators can still get reasonably good results by listening to the users and recording the findings. The method is particularly suited for agile projects because it is very flexible. It can be applied at any stage in the development lifecycle, from low-fi paper prototypes to fully coded systems.

Borrowed-Experience PoC Tests

Borrowed-experience tests leverage the heavy lifting performed for a previous product or service (or the efforts of another company) to craft a differentiated value proposition aimed at different audiences. For example, Niantic utilized the AR and mapping technologies they initially developed for their game *Ingress* to create *Pokémon GO*. By applying these technologies to the Pokémon franchise, they produced a unique gaming experience that appealed to fans of the Pokémon franchise and new users who were curious about AR technology. *Pokémon GO* became a global phenomenon, demonstrating how the lessons learned in implementing systems and technologies for other products or services can be leveraged to create new experiences.

3D Storyboards and Wireframing for XR

Storyboards are series of drawings used in film, animation, TV programs, advertisements, and digital game production to depict how the story of a film will unfold. Wireframes are visual guides that represent the structure and layout of a web page or an application in UX/UI design. While storyboards show the discrete scenes of a film or a video unfolding over time, wireframes help to organize the graphical elements and layouts of multiple pages or screens based on a product concept or creative idea to provide early visualizations of potential user interfaces.

Both of the tools facilitate quick iterations and informed design decisions by focusing on the participant's point of view of the experience or user interaction flows. Combining both of these methods makes a lot of sense for immersive content, which borrows heavily from both film and digital app production. XRXD wireframes can be valuable for explaining concepts and gathering feedback from within the design and development team and from external stakeholders and funders. They can play a critical role in decision-making during both early and advanced stages of product design and development. They can be reviewed by product teams, stakeholders, and representative users. Wireframes help refine features, information architecture, user flows, and other product elements.

Immersive storyboards can help to account for the participants' ability to move freely within the virtual space, offering media designers a fluid process of brainstorming and planning the participants' flow through the immersive experience. Depending on the purpose, wireframes can either be static approximations or support basic interactions, flows, and navigation using clickable elements. Interactive wireframes can be generated and iterated using simple presentation tools such as Google Slides or Apple Keynote that are easy to share and modify.

Figure 6.2 shows the template Peter has adapted from traditional storyboarding and interactive wireframing to the wireframing of spatial computing and XR content. The template can help the design team to set the scene by following the user, environment, objects, interactions framework introduced in the previous chapters to conceptualize the participant's journey through the immersive experience in space and in time.

The XRXD wireframing process involves planning the virtual space with maximum degrees of freedom (DoF), which refers to the number of ways an object can move in the virtual space. Granting users the freedom to move around a virtual space is closely linked to the system's ability to track the movements of the participant using the AR, MR, or VR headset.

A notable example of immersive wireframing was used by the team behind the *Titanic VR* (Immersive VR Education, 2018), an interactive experience in which participants witness the 1912 sinking of the RMS *Titanic* and explore its wreck. The developers, led by animator David McDermott, utilized 3D storyboarding and "immersive wire-framing" techniques. They began by creating 3D models in Blender and using the Grease Pencil tool to draw directly within the 3D scene, which allowed for planning camera movements, user perspectives, and interactions that traditional 2D storyboards couldn't effectively capture. This method enabled the mapping of user navigation and interactions with elements like the submersible or key events such as the lifeboat sinking.

The 3D storyboards then guided immersive wireframing sessions, in which scenes were prototyped in virtual reality to refine spatial relationships, user interactions, and visual storytelling. Quick character sketches within the 3D space helped map out scenes, ensuring the virtual environment's accuracy and engagement. The team further utilized *Blender's* headset viewing mode to visualize key scenes in VR, allowing for early testing and refinement of the immersive experience (Engage VR, 2017).

Colorscripts

Color has a powerful impact on human emotions. It is a vital element of all multisensory content. Colorscripts are a unique method for outlining the emotional flow of mediated experiences, developed by the creative team at the digital animation studio Pixar. Unlike storyboards, which show the specific characters, objects, and actions within various scenes, colorscripts allow the design team to view the entire experience as a continuous flow of emotions the story attempts to portray and elicit from the audience.

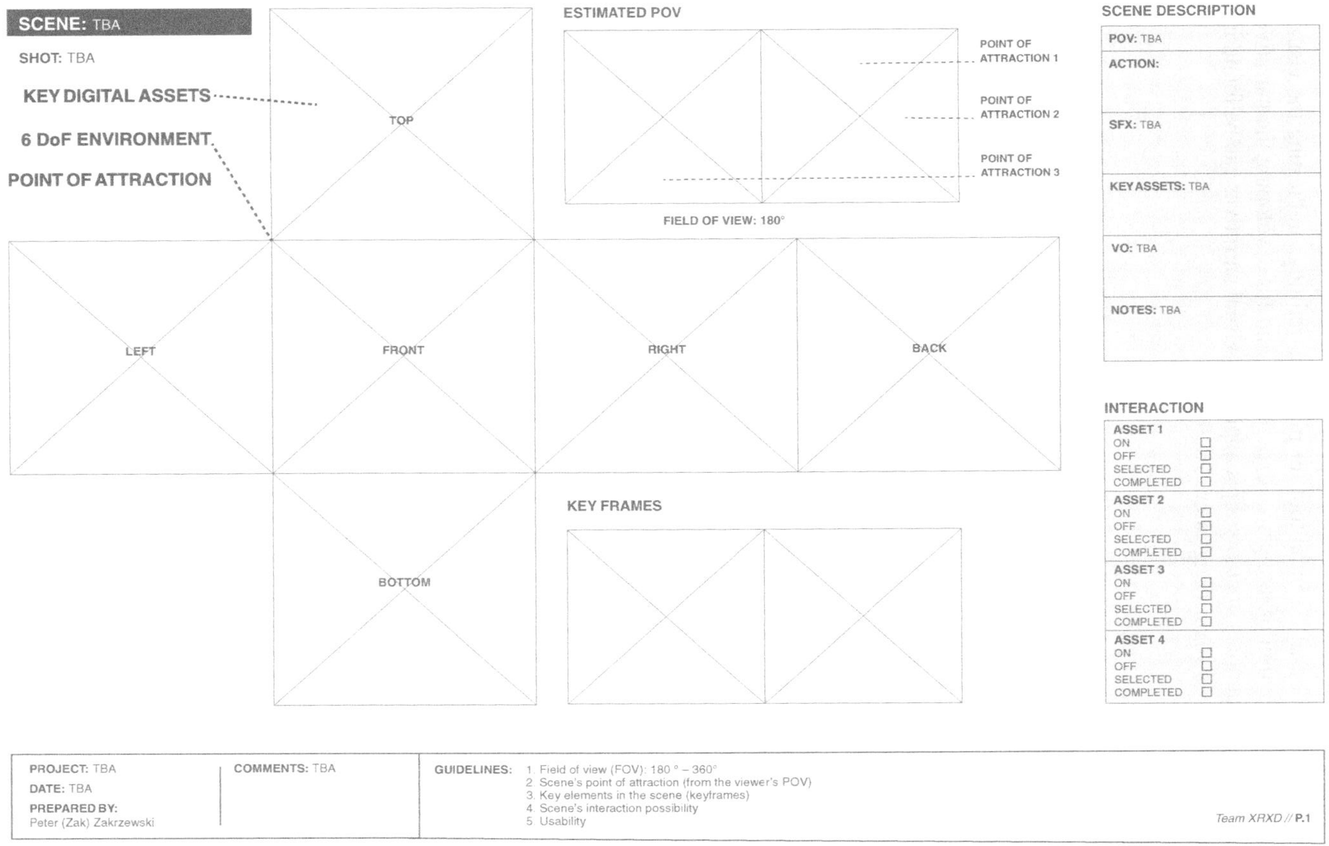

Figure 6.2 The XRXD wireframing template

While storyboards can often be created using shapes and grayscale, colorscripts map the emotional arc of the media production's story through light and color. Colorscripts can be an essential tool for planning, refining, and testing the visual and emotional rhythm of immersive experiences and for developing the narrative arc. Excellent examples of colorscripts applied to visual narratives can be found in the book *The Art of Pixar* (Eggleston, 2020).

Testing Functional Fidelity

Prototyping and simulation-based testing methods involve creating simulated XR environments and user experiences before deploying them in real-world settings. The goal is to assess functional fidelity to ensure that the prototype can approximate real experiences effectively. Jan Hendrik Plümer and Markus Tatzgern have proposed a framework that enables designers to systematically validate XR prototypes by establishing performance equivalence between real-world experiences and XR simulations. This approach emphasizes the importance of achieving visual, haptic, audio, interaction, and functional fidelity while accounting for the simulation overhead introduced by XR hardware (e.g., wearing head-mounted displays).

By emphasizing validation, the framework can help you identify and control for confounding factors, ensuring that XR simulations are ecologically valid and translatable to real-world applications. This may be relevant to your project if you are performing early-stage quantitative evaluations of XR experiences that require a high level of similarity between real-world experiences and the XR simulation of the final experience (Plümer & Tatzgern, 2023).

System Walkthroughs

Using a system walkthrough to test XR experiences involves simulating user interactions in a structured manner to evaluate user interfaces, usability, and functionality. The goal is to identify potential usability issues. Start by identifying key tasks users are expected to perform in the XR environment. Develop scenarios that include these tasks, focusing on critical interactions such as navigation, object manipulation, or responding to system feedback. Include five key groups in the walkthrough: the design team, subject matter experts (particularly important for training and simulation experiences), expected users, developers, and XR usability experts. Each group provides a different perspective on the XR experience. Users engage with the interface, developers observe technical performance, and experts can assess usability and experience-specific concerns that users or designers may not pick up on or be able to articulate.

Divide the walkthrough into stages reflecting different interactions. (1) Focus on specific user goals and how effectively they are achieved (task oriented). (2) Assess the ease of moving through the scenario and locating relevant information or elements (exploration and navigation). And (3) evaluate the system's feedback and how intuitively users respond to it (responsiveness). During the walkthrough, employ a checklist to assess different design elements and interaction points.

Have a facilitator or expert observe participants as they perform tasks, documenting where difficulties occur. This step is vital for highlighting specific usability problems like unclear navigation paths or confusing feedback. Once the walkthrough is complete, analyze the issues that came up and prioritize them based on impact and frequency. Discussions should involve all stakeholders to make informed decisions on necessary changes.

Projection AR

For interactive experiences, you will most likely combine 3D models with a game engine like Unity or Unreal Engine to create your PoC or functional prototype. TouchDesigner provides another option for prototyping projection mapping scenarios with a node-based visual programming environment. You can import Adobe Substance 3D models and textures into TouchDesigner or create models and textures within TouchDesigner. The Kantan Mapper tool in TouchDesigner supports projection mapping and allows you to draw geometric shapes over the surfaces you want to project onto. These shapes will help you map digital content accurately onto the physical world. With TouchDesigner, you can easily add real-time interactivity using depth cameras or other sensors to add interactive elements to your projection. For example, people moving in space can influence what gets projected, for instance, animations that respond to sound, motion, or other environmental factors in real time.

TouchDesigner is a node-based tool, meaning that all processes (materials, geometry, logic, effects) are built using nodes that you can visually connect. This can be more intuitive for visual artists, designers, or those who prefer to see everything laid out graphically in contrast to a game engine like Unity, which primarily uses C# for programming. While it offers a rich graphical interface, much of the logic and interactivity requires coding, making the learning curve steeper if you're unfamiliar with programming. Unity's workflow is more traditional and requires more setup when compared to the node-based approach of TouchDesigner. It's better suited for game developers or those with a background in coding and software development. Unity does not have native tools specifically for projection mapping. You'll need to manually set up virtual cameras and projection surfaces and rely on third-party plugins like MadMapper or custom shaders/scripts for accurate mapping.

While TouchDesigner is probably best for most live installations, performances, interactive exhibits, and installations, TouchDesigner isn't designed for complex physics, game mechanics, or large 3D environments, which is the primary strength of game engines. Therefore, for performance-heavy, real-time visuals on high-performance machines, Unity can scale to handle more complex scenes across different platforms. At the end of the day, TouchDesigner's community is artist centric, while Unity offers a broader range of plugins and resources for more technically oriented designers. Each tool has its strengths, and the choice depends on the specifics of your Projection AR project and your expertise.

Choose TouchDesigner if your Projection AR project involves real-time visualizations or interactive installations; quick prototyping of visuals or integration with physical sensors; dynamic content that requires frequent adjustments in a live environment; and an emphasis on creative visual effects, projection mapping, or sensor-based interaction.

Choose a game engine like Unity or Unreal Engine if your AR project involves more complex, game-like interactions with 3D environments; high-quality graphics, lighting, and physics; building applications for a broader range of devices (AR/VR platforms). These engines are good choices if there is a need for more detailed scripting and extensive custom logic.

Heuristic Evaluation

Alistair Sutcliffe and Brian Gault (2004) have extended Nielsen's usability heuristics in light of the unique interaction styles and requirements of XR experiences. They introduced 12 heuristics that address both usability and the immersive quality of presence. Usability heuristics used for traditional WIMP interfaces are insufficient for XR based on the differences in user expectations and interaction methods.

Sutcliffe and Gault also propose combining these heuristics with a technology audit to assess the technological capabilities and limitations of the XR system. Given the complexity of XR experiences that combine navigation issues, visuals, sound, and multimodal feedback, a tailored evaluation method is essential. By incorporating principles of natural engagement, realistic feedback, and navigation support, this method provides a comprehensive framework to assess your XR experience. The heuristics they set forth include:

1. Natural engagement. Interactions should approach the participant's expectation of interaction in the real world in the context of their activity as closely as possible. The interpretation of this heuristic depends on the user's sense of presence and engagement.
2. Compatibility with the participant's task and domain. The experience and behavior of objects should correspond as closely as possible to the participant's expectation of real-world objects, their behavior, and affordances for task action.
3. Natural expression of action. The representation of the self/presence in the experience should allow the participant to act and explore in a natural manner. However, this quality will be limited by the available technology. For example, when haptic feedback is absent, the natural expression of many gestures is degraded.
4. Close coordination of action and representation. The representation of the self/presence and behavior should be faithful to the participant's actions. For example, the response time between user movement and update of the display (this applies across the range of mobile AR, MR, and VR) should be less than 200 ms to avoid motion sickness.

5. Realistic feedback. The effect of a participant's actions on virtual and enchanted objects should be immediately visible and conform to the laws of physics and the participant's perceptual expectations.
6. Faithful viewpoints. The visual representation of the virtual world and digital objects or the mix of physical world and digital objects should correspond to normal perception, and viewpoint changes by head movement or movement of a smartphone held in the participant's hand should be rendered by the display without delay.
7. Navigation and orientation support. The participant should always be able to find where they are in the virtual, augmented, or physical world and return to known, preset positions. Magical actions such as moving through surfaces or flying may help, but these have to be judged in a trade-off with naturalness and consistent with the rules you have articulated for the virtual or augmented world.
8. Clear entry and exit points. The means of entering and exiting from the XR experience world should be clear to the participant.
9. Consistent departures. Design compromises, if required, should be consistent and clearly marked—for example, cross-modal substitution (use of one sensory modality to substitute or replace the functions of another) and power actions for navigation.
10. Support for learning. Active objects should be cued and, if necessary, explain themselves to promote learning how objects in the virtual or augmented world function.
11. Clear turn-taking. If the system takes the initiative in the scenario, this should be signaled clearly, and the conventions for turn-taking should be consistent.
12. Sense of mediated presence. The user's perception of engagement and being in a "real" world should be as natural as possible.

Every project will require a bespoke heuristic evaluation to effectively identify usability issues. XR projects require expanded usability practices and context-specific heuristics in order to orchestrate effective experience assessments (Sutcliffe & Gault, 2004, p. 3).

User Testing With Gaze Tracking and Physiological Signals

XR technology has enabled content makers to not only offer their audiences an expanded way to interact with digital content but also new ways to collect user data from eye tracking and sensing of physiological signals. Once your team reaches the phase of having a working system, you can use software packages that work with development frameworks like Unity and Unreal Engine to collect, analyze, and share user behavior data in 6DoF XR experiences.

Among these packages is the PLUME system (Javerliat et al., 2024), an open-source toolbox that captures detailed XR interaction data, such as object movements, user inputs, gaze data, and physiological signals like heart rate. The captured data can be analyzed and visualized, allowing your team to visualize data like user trajectories, heat maps for gaze and positional data, and interaction highlights. This enables a deeper understanding of user behavior and interaction patterns within the 3D environment. For example, the Viewer can show the participant's gaze patterns over dynamic objects, highlighting areas of focus and interactions during the session. Data can be exported to common formats such as CSV or XDF for further processing with statistical tools. You can use tools like this to validate your interaction designs as well as conducting detailed cognitive and behavioral studies.

Surveys and Interviews

When testing your XR proof of concept, surveys and interviews, when conducted with a structured approach, can help capture insights on usability, engagement, and presence. Start with a pre-experience survey to gauge the participant's familiarity with XR technology, expectations, and specific accessibility needs. This must be tailored to your specific questions and the aspects of the project you are testing. Following the PoC test, a postexperience survey can be used to assess the level of immersion, impact, and engagement of the immersive content.

We suggest using both quantitative and qualitative questions to balance objective and subjective feedback. Likert scale questions can help you measure usability, engagement, and comfort, while semantic differential scales are useful for capturing the emotional spectrum, such as how "present" or "detached" users felt. Open-ended questions allow participants to share reflections on what they found particularly engaging, confusing, or frustrating.

For interviews, a semistructured format works well, as it provides a foundation to ensure important areas—like usability, emotional response, and engagement—are covered while allowing room for participants to express unanticipated insights. Guided reflection can be especially effective, where users recount their experience step by step, sharing thoughts and feelings as they recall specific interactions in the XR environment.

Another effective method is scenario-based questioning, where users are presented with hypothetical situations that could arise within the XR environment. This approach helps assess how users might respond to new challenges or changes in real time, providing valuable insights into the adaptability and flow of the experience.

In both surveys and interviews, we suggest you focus on usability, engagement, and presence questions. Consider using cognitive walkthroughs, in which participants explain their thought process while navigating the environment, to highlight confusing or frustrating points. Since XR experiences can vary widely in complexity, it's vital to gather data on how immersive and engaging participants found the experience and whether they experienced physical discomfort or ease of use with the controls. Questions should also probe the learning curve to determine if users could accomplish tasks without prior XR experience.

To avoid bias, keep questions neutral and use accessible language, particularly for those unfamiliar with XR terms. Make sure to include questions about comfort and physical impact, such as potential VR-induced motion sickness, as these are important indicators of overall accessibility and usability.

For analysis, quantitative responses from Likert or differential scales will help identify trends in reactions and ease of use, while qualitative analysis of open-ended responses and interview transcripts can reveal common themes like immersion and usability challenges. This approach to surveys and interviews will give you a comprehensive view of the user experience, guiding you toward improvements in the XR environment's usability, accessibility, engagement, and presence.

For evaluating presence and immersion in XR experiences, several standardized questionnaires are widely used to assess how "present" participants feel in the virtual space and the depth of their engagement. Following is an overview of several widely used tools, each of which provides a unique perspective on measuring presence and immersion, allowing you to gather insights into how participants perceive and interact with your XR experience. Using these questionnaires can help identify areas for improvement before designing the system prototype.

The Slater-Usoh-Steed (SUS) Presence Questionnaire is one of the most common tools for measuring participants' sense of "being there" within a virtual environment. This questionnaire includes items like, "To what extent did you feel you were inside the environment rather than just observing it?" and "How often did you think of the virtual environment as a place you visited rather than as images you saw?" Such questions aim to capture participants' subjective sense of immersion.

Another widely used tool is the ITC-Sense of Presence Inventory (ITC-SOPI), which provides a broader look at presence by evaluating dimensions like spatial presence, engagement, ecological validity, and adverse effects, such as motion sickness. Sample questions include, "I had a sense of being in the scenes displayed" and "I felt as though I was physically present in the environment." This inventory is beneficial for identifying various aspects of the immersive experience.

Similarly, the Presence Questionnaire (PQ) developed by Witmer and Singer measures presence based on factors like control, sensory cues, and the consistency of the environment. Questions like "How much did the visual aspects of the environment involve you?" and "To what extent did your experiences in the virtual environment seem consistent with your real-world experiences?" help assess how realistic and engaging participants find the XR environment.

The Immersive Experience Questionnaire (IEQ) is another important tool, especially relevant for measuring the depth of cognitive and emotional involvement and is commonly used in gaming and other highly interactive XR environments. This questionnaire includes questions such as "To what extent did you feel as though you were focused on the experience?" and "Did you feel like time passed quickly during the experience?" making it suitable for assessing how absorbed participants feel.

In addition, the Game Experience Questionnaire (GEQ) Core Module, initially created for gaming experiences but applicable to XR, looks at factors like immersion, flow, and both positive and negative effects. Questions here include "I was completely absorbed in the experience" and "I forgot everything around me," which are valuable for understanding the extent of participant engagement in the XR environment.

Proof-of-Concept Testing Considerations

Testing user experience and usability goes beyond the scope of traditional usability testing methods. The unique characteristics of immersive media introduce challenges that must be addressed through conventional techniques and innovative approaches tailored to your project. The ultimate goal is presence, the participant's subjective perception of "being in" the experience. Testing usually involves subjective measures, such as the Presence Questionnaire (Usoh et al., 2000) or the Immersive Tendencies Questionnaire (Rózsa et al., 2022). In some cases, you may want to complement subjective data with objective measures, such as tracking users' physiological responses, e.g., heart rate, skin conductance, EEG, etc., to gauge immersion in the experience.

Safety

Safety is critical, particularly when users physically move in a real-world space while interacting with a virtual environment. Ensuring user safety requires consideration of environmental awareness, physical strain, and fatigue. In VR, users can lose awareness of their real-world surroundings, leading to potential accidents or injuries. Testing should include real-world conditions, such as room-scale setups, to evaluate how well users can maintain their spatial awareness.

Additionally, it is important to implement and test boundary systems, e.g., VR "guardians," that will alert users when they are approaching physical limits of the play space. Immersive media often involves prolonged use of HMDs, controllers, and other hardware, which can lead to physical strain or fatigue. Testing should assess how long participants can comfortably engage with the experience under test and identify physical discomforts, such as neck strain or eye fatigue, that could have a limit to the duration of use.

Input Modality Flexibility

Unlike traditional interfaces, immersive media supports a variety of input modalities, such as hand gestures, voice commands, gaze detection, and motion controllers. Each input modality, as well as multisensory combinations (e.g., gesture and speech commands), presents unique advantages and challenges that must be carefully evaluated. Gesture-based inputs must be assessed for accuracy, ease of use, and the potential for user fatigue. This is particularly important if participants are interacting with virtual objects through hand gestures, as prolonged use could lead to fatigue.

In a study focusing on gesture-based controls for VR, researchers found that user fatigue was a significant issue, requiring the design of more ergonomic gestures to improve user experience (Roosink et al., 2016). Voice commands should be tested for recognition accuracy across different contexts and speaker accents, as these can significantly impact the effectiveness of voice recognition.

Comfort and Ergonomics

Comfort is a significant factor with XR, especially for prolonged use of head-mounted devices or holding a mobile device in your hand for long periods. The physical comfort related to the use of the equipment and ergonomic design considerations are critical. The design and fit of hardware, such as headsets and controllers, contribute to comfort. Testing should involve a diverse group of users to assess the comfort of the hardware across different head sizes, body types, and usage durations. Adjustments in weight distribution, padding, and strapping mechanisms should be evaluated to minimize discomfort.

The spatial arrangement of interactive elements within the virtual environment should be tested for ergonomic efficiency. Elements should be placed within easy reach and at natural angles to reduce strain. Testing should also consider the physical space users require to interact comfortably without causing strain or fatigue.

Cognitive Load

Immersive media experiences often involve complex interactions and multitasking, which can increase cognitive load. Managing this load is crucial to ensuring usability and a positive user experience. Testing should evaluate how much mental effort is required to interact with the system. High cognitive load can lead to errors, frustration, and decreased engagement. Techniques such as task analysis, usability walkthroughs, and think-aloud protocols can be used to identify points of high cognitive load. Based on testing results, interactions should be streamlined to reduce cognitive load. This may involve simplifying navigation, minimizing unnecessary steps in tasks, or providing clear and consistent feedback to guide the user.

Real-Time Data Capture

Conducting tests using immersive media prototyping tools or game development engines provides your team with the opportunity to observe and evaluate user behavior in real time, capturing more authentic responses to the experience to understand better how users interact with the system's virtual and physical aspects. Immersive media projects often require extended use to fully understand their impact on users. Insights from users will evolve as you observe more users, as you can see emerging issues related to sense of presence, task completion, comfort, safety, and cognitive load.

Outlining PoC Project Scope

Once you have done your information gathering, it is time to present your understanding of the project. Depending on the timeline, I would present some images interpreted from the important points of the project representing effects, scenes, or potential assets. Longer scope might require representations as complicated as wireframes, storyboards, animatics, concept sketches, and/or maquettes. As you assemble your first presentation, it is a worthwhile step to map out your notes on assets required by each scene, noting the options for creating or acquiring those assets through different methods and the style and time trade-offs between them.

Managing Expectations

Whenever you show something, be it to a project lead or to a contract programmer, etc., be very explicit about the stage it is at, why you are showing it off, and what you want feedback on. For example, we find that the ways that you transition from scene to scene as well as how you transition assets and effects into and out of view are often things that get lost in the shuffle. Make sure to ask a lot of questions about whether these nitty gritty details are hitting the mark when you present your work. It's up to you to ask the direct questions. This is particularly important at the prototyping phase, when some assets will only be placeholders.

Scheduling

Another aspect of managing expectations is to be generous about the amount of time you think it will take to finish your task. If it's something you've done before, give it your best general guess. If you know the topic of the thing you're being asked to do, but you haven't gotten to really do it yet, at least double the duration estimate! If you don't have any experience, triple or maybe even quadruple your time estimate. Also, in these cases, seek knowledge on anything you haven't tried before. Subcontracting is not a dirty word if you find someone capable of halving your stress.

Successful project management hinges on a proactive approach, fluid communication, thorough preparation, and the use of effective project management tools. To achieve consistent results, it's essential to understand your team's strengths and weaknesses, allowing you to strategically delegate tasks that play to their abilities while addressing any skill gaps. A clearly defined project scope set from the outset is critical for avoiding scope creep and budget overruns, and involving your team in scheduling ensures collective accountability. Breaking down the PoC prototype into manageable tasks with realistic deadlines will keep your team motivated and prevent burnout.

In our experience, every team needs to designate a project manager, someone on the team who coordinates and tracks the tasks that need to be completed. Utilizing collaboration tools to centralize communication, share documents, and reduce the chaos of endless email chains makes a huge difference, but it is not enough. A human with a solid overview of the project is essential. Project management is a discipline unto itself, so to stay on topic, we'll keep this brief and leave you with this point to consider.

We recommend trusting team members to execute their responsibilities without the need for micromanaging. Collaborative projects benefit from consistent, clear communication that fosters an environment where team members feel comfortable raising questions and concerns in a timely manner. Open communication allows everyone to be proactive in identifying and addressing red flags before they derail the project. Likewise, celebrating milestones to acknowledge contributions boosts morale. We also recommend putting contingency plans in place to adapt swiftly to unexpected changes or challenges. By prioritizing these practices, your team will deliver the proof-of-concept prototype on time, within budget, and aligned with your project vision. Flexibility is important at every stage of development, as is compromise.

Managing the PoC Process

Creating and managing all of these aspects sounds like a huge task—and it is—but technologists and 3D artists working in video games, simulation, and immersive media have been practicing character and asset creation, and to them, this is all doable. The important thing for anyone without the full knowledge or skills is that you fill in your technologist or creative developer or 3D artist on the full details of the puzzle piece they are going to tackle well in advance. Things always need edits, things always need to change, and therefore, most professionals account for an amount of revisions, but being able to start on a piece of a project that has some concrete foundations is important to finishing the project.

This is often the stage in a project when you start to bring on people with specialized skills that expand the capabilities of your team. You will start working with creative developers, programmers, and a whole range of people working in disciplines that may be new to you. Trust the professionals, and give them everything they need to work. When they give you updates, try to focus on the stage they are at, or if it looks perfect then, ask how they feel about it and next steps. Trust their judgment if you trust the work they are presenting.

However, still make sure to be honest. Ensure that their work is actually reaching your goals, with the understanding that it isn't complete. Even with an expert, if you say nothing, nothing will change, and if you change too much, the project will end up confused. Finish your project. The next one will be better.

The Case for High-Fidelity XR PoC Prototypes

You should consider moving to high-fidelity prototyping more quickly with an XR project compared to more traditional screen-based experiences due to the embodiment and immersion participants' experience with XR. Creating a sense of presence and agency within a virtual or augmented prototype is essential for accurate user feedback and interaction validation. Unlike traditional screen-based design, XR engages the participant's entire body and sensorium to interact within the virtual, augmented, and physical worlds, creating an embodied experience that engages proprioception (sense of body position) and kinesthetic awareness (sense of motion).

As discussed throughout this book, embodied cognition—the theory that our cognitive processes are influenced by our interaction with the environment—is essential for designing experiences that feel natural and intuitive in XR. High-fidelity prototypes enable you to observe how participants physically navigate, react to, and engage with touch points in a manner that paper or screen-based testing simply cannot replicate, making it a critical distinction between XR experience design and traditional human–computer interaction (HCI) approaches.

Prototyping within an embodied context ensures that user feedback reflects the nuances of the final experience. When testing XR prototypes, having participants immersed in the environment allows your team to observe not only their interactions with objects and agents but also subtle, subconscious responses like body orientation, reach patterns, and gaze direction—all of which contribute to the intuitiveness and comfort of an experience. Observing how users naturally reach, navigate, or even hesitate within the XR space provides insights into spatial usability and accessibility, which are difficult to capture in traditional usability testing setups.

Creating a Plan for Your Proof of Concept

Begin the PoC process by creating a scene-by-scene breakdown (we covered the Context Scenario Breakdown process in Chapter 5) for each scene to be tested. Give a short description of what should happen in this scene. Some aspects of the scene breakdown should include: (1) a floor plan of each scene, which will simplify laying out all of the props and the placeholders that you need for the first draft. (2) A timeline of what happens and when. No one working on your project should be questioning when certain props appear or in what order to trigger sounds, for example. (3) An asset list that includes existing assets and those that need to be created. For example, a scene involving a storybook introduction will be described as:

Scene Description

- A book descends and opens, accompanied by voice over and dramatic music. Purple magic dust swirls out of the opened book. A page rips itself out and floats in front of the viewer then burns up. The book slams shut and falls through the floor. A character falls onto the stage, tells a monologue, then fades out to black.

Assets Description

- Book to be opened with a page to be flipped. Page needs animation of ripping out and sending a wave through before straightening. Purple magic dust makes 2–3 loops before vanishing. Narrator voiceover (VO) to be recorded. Narrator (linked to the character model) to be recorded.

A reference folio will contain the full functionality and desired look of each asset. When assembling your reference folio, make sure to take care answering these questions:

- What is important about the reference?
- What do you like about the object, poses, effect, or place?
- What changes would you make in comparison to the reference?

Working With 3D Components

Shaders

Shaders can be nodes or code that tend to utilize images that represent a template that defines how things are drawn. An example might be rendering a model that fades out from the bottom up. There can also be shaders that place lots of little pictures like particles or grass and shaders that distort the almost-final screen image. The latter two are advanced use cases, and your technologist should be able to guide you as to what is and isn't available.

Materials

Materials are a unique instance of a shader, assigned to renderers to use on meshes, with a custom version of the shader's variables. If you have a scene that fades in from top to bottom, containing a prop and a character, you would have one material for the prop and one for the character. Both of the materials would be using the fadeBottomToTop shader but have the right textures for each model. You could theoretically stagger the start of the fade for different materials. The shader doesn't need to be remade, but you just need to set the two numbers on the materials differently.

Code

Crafting code will constitute one of the most important parts of your PoC project. Designers without a core development team can opt for setting up low-code solutions, but having some coding ability on your team is a huge benefit. Your developer should have some level of familiarity with a pipeline or workflow to control the experience. Because you can do so much with code, there is a danger of trying to push too many changes too quickly near the end. Keep in mind that code breaks all the time. Dependable code only comes from existing and being used for as long as possible.

Make sure you describe functionality early and try to brainstorm alternatives in case something doesn't work. Things will probably break near the end of the project, so having contingencies in the form of backup options and the parts you are willing to cut in case of emergency will ensure you will always know what to do.

Meshes

Meshes are geometries, points in space, connected by triangles. The simplest are triangles or squares and are commonly used to display complex objects like smoke or fire. They may represent objects far enough away that you wouldn't be able to tell an image from a 3D model. Meshes can be as complex as the most intricate museum piece and can animate like a Pixar character, though those take time. A billboard is an example of a mode of mesh that always faces the camera. It can play a flipbook or an animation sequence as well as still images.

Many software programs allow you to render out simulations to be used as billboards. To acquire meshes, you can either buy them on various asset stores, use photogrammetry to make models of real objects via photos or video, or you can have meshes modeled or sculpted by an artist. The process for creating complex models takes more time the more details are required. Any sculpt that needs to be painted will need to have UVs made; otherwise, one of several UV-less options available in certain 3D packages will have to be used. If a model is to be animated, then a rig is needed, and with that, skin weights need to be painted. Animation itself takes an amount of time related to the length of animation.

Describing Special Effects

Group similar effects when describing any effects you want in your project with a name and a description, then mark in the times and/or places throughout the project when each type of effect will be used. With any luck, after you do this audit, some of the effects can be combined. For example, an implosion and an explosion could be one effect, with a simple input for the speed and particle direction. Or in another example, a curve system could be used for both creating magic wisps and steering a car model down a winding path.

Combining and reusing effects and functionality will help reduce the effort required to bring your work to life. It is largely up to the developer to determine what shared functionality can be taken advantage of. At the end of the day, every developer will have different workflows and processes that will make some things harder or easier. Let the developer you're vetting give you an idea of how difficult different parts of your project can be for them. That way, you can plan for multiple fallback plans for tricky features or effects.

Vague descriptions, while cool sounding, can result in many interpretations. The process of giving a description and seeing what comes out is fascinating, but it takes time and ends up being a gamble over whether you end up liking the result. A bit of exploration can be okay if done in a limited context, but you should ultimately understand what you want and why you want it that way. Be descriptive about what happens and when with each effect. How it starts, what conditions trigger changes, and how it ends. You don't need to know how to do it, but like all other things, there shouldn't be many questions about what you want. It is someone else's responsibility to figure out how to make it work. However, make sure you are fair with your effects. When it comes to special effects, less can be definitively more.

Here is a brief description of a carbonated bubbles effect created for a VR game PoC providing an example of the level of detail that 3D artists and programmers require to realize effects:

- Circles appear on the ground, about 10 per second for 10 seconds. The circles grow from their starting size of about an inch diameter. Once a circle reaches a certain size, it ascends. The ascending circles rise until they reach a point on a mesh, where they stop. Once about half the circles are in place, we fade in the mesh that the points are filling out. The mesh fades in over 10 seconds.

Once all the spawned points have reached their destination, they get smaller until they are invisible. At the end of the scene (we need this to be malleable while we adjust the VO timing), circles will grow until they reach a random size, about an inch big give or take half an inch. Once a circle reaches the desired size, it holds. Once all the circles are at full size, the mesh is turned off. Once the mesh is hidden, turn each bubble to a popped bubble texture for a short time before turning the bubble off. Once every bubble is gone, the scene progresses.

Software Options

Because XR is still relatively new territory, new ideas and app concepts will require some research into the technical constraints, guidelines, and creative opportunities each immersive platform offers. However, much exploration of XR design principles can be done without the use of complex software or coding. Narrative components of the immersive experience can be tried without diving into ARKit, ARCore, Unity, or After Effects. Great stories can be told using presentation software. Software such as Figma can be used to conceptualize screen layouts nearly at the speed of a creative thought, keeping ideas fresh and spontaneous.

There are numerous software options for completing your first presentation or first draft on a tight schedule. Whatever program you are comfortable with is usually best, but if you are not sure where to start, we'll list a few programs in various categories that we have used in our project work that have proven to be very effective.

Obsidian is a useful program for keeping notes and collecting documents. It has a mind-mapping section called Canvas which you can use to create a flow chart of a script if that is how you work best. You can drag images in, write blurbs in nodes, drag connections between nodes, and create links between notes. If flowcharts are your comfort zone, it is a great place to start. Obsidian vaults (what they call their collections of notes) can be kept in your own computer, which is good when security is required and cloud storage is not an option. While Obsidian works best for individuals, it is possible to share vaults and synchronize them across multiple computers.

There are numerous note-making applications on the market. We mention Obsidian as one example among many, but there is no "best." Instead, we suggest you use what works for you and your team.

Google Slides is a good way to organize scenes of a storyboard and of having notes and comments, which makes it great for keeping lists and links to references shared across all members of the team, and read-only and comment-only links can be shared with stakeholders when required. Storyboards can be made in many traditional image editing programs, making them even more useful to developers. There are also tools like Canva or templates you can use in mind-mapping software like Miro that work well for this.

Video editing software such as Adobe Premiere Pro and After Effects is good for creating demo videos and animatics to test ideas for timing and pacing, especially at the early stages of projects. Blender, Maya, Unity, and/or Unreal can be used to create mock-ups on a tight deadline. Working closer to the final medium can make it easier to start iterating early if you are ready to do so. Simply moving a placeholder model around the scene to the right place at the right time can give you insight on the project that other methods cannot.

Outsourcing PoCs to External Developers

For both the PoCs and functional prototypes, discussed in Chapter 8, it is best to bring engineers and developers into your project as early as possible. For teams without advanced development skills and experience, hiring an outside contractor might be an option. When working with an outside developer, make sure to give an overview and a plan with at least a rough schedule to begin with. Most importantly, always be clear about the delegation of tasks.

A PoC project requires many components. To keep things well coordinated, each team member needs to know the output that they are responsible for, how their portion of the work fits into the overall project, how their portion fits locally with other surrounding parts, and the time frame that the deliverables are expected in. Making the development team members aware of these parameters as soon as possible gives them time to prepare their schedule and start thinking about the project. Cold starts on short time frames are difficult, to say the least.

Things usually change in the development of proof-of-concept prototypes and in subsequent phases of the project, because nothing can be translated with 100% accuracy between mediums. Provide what is absolutely important to the artist about what they are working on, and know where you can accept compromises. Make sure to not overload any team member with extraneous information, since this might lead to confusion. It is much more important to be clear with your desired outcome.

You need to thoughtfully include things in your direction, such as: (1) How is the asset seen by the end user? Think about it: if it is far away and doesn't rotate, then it can be created much more simply than something like a character that stands in front of the user animating and interacting. (2) Does the asset animate? Does it deform? Does it grow or shrink? The artist has to consider how the asset will look in all of its forms.

With the amount of work it can take to create digital assets for XR, it is valuable to think about a range of acceptable representations for each asset so that settling on an appropriate scope is easy. Also, having the option of purchasable assets can also help trade money for time or stress if needed.

Summary

In this chapter, we covered the process of creating proof-of-concept tests. We argued that a PoC is one of the most efficient and accurate methods of selectively testing key hypotheses and validating fundamental assumptions about the desirability and feasibility of the XR app concept. We defined a distinction between

PoCs, which are, for the most part, internally facing tests, and functional prototypes, which are more refined and elaborate simulations of the final product that give the media makers a clear view of its real-world operation and allow them to obtain realistic user feedback.

The chapter outlined the three key areas that need to be considered for proof-of-concept testing. Aside from (I) the audience hypothesis and (II) the value hypothesis, which need to be addressed at the earliest stages of the project, the PoC phase needs to focus on testing (III) the usability/feasibility hypothesis to assess how the system will perform and how immersive it will be. The accumulated learning from various PoCs can also give the design team some insight regarding (IV) the growth hypothesis, or the XR media franchise's potential to be adopted by its intended audience and to scale beyond that.

We also covered the important methods, tips, and planning processes needed for the creation of an effective PoC. Testing different aspects or components of the XR system brings the development team one step closer to the final product. But before an audience can be invited to enter the immersive space, the various bits and pieces of the experience have to be conceptualized and put together as one cohesive system. Systems thinking as applied to immersive experience design is the topic of our next chapter.

References

Cootmans, L. (2021, June 17). Get distracted by these three VR prototypes about distracted driving. *Preview Labs*. https://previewlabs.com/get-distracted-by-these-three-vr-prototypes-about-distracted-driving/

Demodern. (2017). Virtual reality Ulm experience. *Demodern*. https://demodern.com/projects/virtual-reality-ulm-experience

Eggleston, R. (with Calahan, S., & Jessup, H.). (2020). *The art of Pixar: The complete colorscripts from 25 years of feature films* (Revised & expanded). Chronicle Books.

Engage VR. (2017, February 21). *360 Storyboards for Titanic VR* [Video recording]. www.youtube.com/watch?v=QWt1hXCylL4

Fadell, T. (2022). *Build: An unorthodox guide to making things worth making* (1st ed.). Harper Business, an imprint of HarperCollins Publishers.

Javerliat, C., Villenave, S., Raimbaud, P., & Lavoué, G. (2024). PLUME: Record, replay, analyze and share user behavior in 6dof XR experiences. *IEEE Transactions on Visualization and Computer Graphics*, 30(5), 2087–2097. https://doi.org/10.1109/TVCG.2024.3372107

Nielsen, J. (1993). *Usability engineering*. Academic Press.

Nielsen, J. (2012, June 3). How many test users in a usability study? *Nielsen Norman Group*. www.nngroup.com/articles/how-many-test-users/

Oculus. (2016, January 30). Oculus Rift: Step into the game. *Kickstarter*. www.kickstarter.com/projects/1523379957/oculus-rift-step-into-the-game

Plümer, J. H., & Tatzgern, M. (2023). Towards a framework for validating XR prototyping for performance evaluations of simulated user experiences. *2023 IEEE international symposium on mixed and augmented reality (ISMAR)*, 810–819. https://doi.org/10.1109/ISMAR59233.2023.00096

Roosink, M., Robitaille, N., Jackson, P. L., Bouyer, L. J., & Mercier, C. (2016). Interactive virtual feedback improves gait motor imagery after spinal cord injury: An exploratory study. *Restorative Neurology and Neuroscience*, 34(2), 227–235. https://doi.org/10.3233/RNN-150563

Rózsa, S., Hargitai, R., Láng, A., Osváth, A., Hupuczi, E., Tamás, I., & Kállai, J. (2022). Measuring immersion, involvement, and attention focusing tendencies in the mediated environment: The applicability of the immersive tendencies questionnaire. *Frontiers in Psychology*, 13. https://doi.org/10.3389/fpsyg.2022.931955

Sen, P. (Director). (2012). *Steve Jobs: The lost interviews* [DVD]. Magnolia Pictures. www.magpictures.com/stevejobsthelostinterview/

Sutcliffe, A., & Gault, B. (2004). Heuristic evaluation of virtual reality applications. *Interacting with Computers*, 16(4), 831–849. https://doi.org/10.1016/j.intcom.2004.05.001

Usoh, M., Catena, E., Arman, S., & Slater, M. (2000). Using presence questionnaires in reality. *Presence: Teleoperators and Virtual Environments*, 9(5), 497–503. https://doi.org/10.1162/105474600566989

Section V

Planning and Building the Immersive Platform

7 Designing the System Prototype

The Serengeti Rules

Sean Carroll, molecular biologist and author of *The Serengeti Rules: The Quest to Discover How Life Works and Why It Matters*, tells the story of an aspiring Oxford graduate student named Tony Sinclair who came across a mystery while conducting research in the Serengeti region in Tanzania. The area is one of the few places left on earth that can boast an ecosystem of such great diversity of wildlife to allow researchers to study large ecosystems in the wild. Sinclair and his faculty supervisors observed that the population of buffalo in the area jumped dramatically from 16,000 in 1961 to 37,000 in 1965. Sinclair was so intrigued by how change happens within natural ecosystems that he dedicated the next dozen years of his life to untangling the problem.

After conducting aerial surveys, Sinclair was able to confirm that the buffalo population was indeed growing rapidly. He was able to count 54,000 buffalo by 1969. Since buffalo do not typically die from predation, Sinclair considered another possibility—some mysterious elimination of diseases that could either reduce the animals' ability to have babies or cause their death. One disease called cattle plague or rinderpest arrived in East Africa in 1889 and had a deadly effect on the buffalo, the wildebeest, and the livestock farmed by humans but not other animals.

A vaccine for the virus became available in the early 1960s, and a vaccination program was launched throughout the farming communities in East Africa. Sinclair hypothesized that the vaccine had a spillover effect on the buffalo. He obtained blood samples from buffalo and age-matched them to the buffalo populations to determine if their immune systems showed antibodies that could be measured in the bloodstream. That gave him his first Eureka moment. While the older buffalo showed almost a 100% exposure rate to the virus, buffalo born after 1964 showed no exposure at all. The virus that could cause massive mortality to the Serengeti buffalo was disappearing from the area. As it turns out, the virus did not originate in the wildlife but in the cows. The buffalo stopped getting sick when the cows were vaccinated (Carroll, 2016).

The other animals that got sick from the virus were wildebeest. Sinclair compared the wildebeest population between 1961 and 1977 and saw that it grew from 263,000 to 1.4 million. To confirm his hypothesis, Sinclair obtained blood samples

DOI: 10.4324/9781003442462-13

from the wildebeests to find out that they showed a similar pattern to the buffalo. No wildebeest born after 1963 had any significant exposure to the virus. Other scientists studying the Serengeti ecosystem noticed other changes taking place at the same time. Could they all be connected somehow?

In the book, Carroll takes on the challenge of connecting the dots between three other seemingly unrelated observations to create a case study of ecology in action that shows living complexity in action.

1. The Serengeti burns every year. Wildfires are a natural phenomenon, but compared to the early 1960s, the time between the mid-1960s and the early 1970s saw fewer fires, with less intensity, and covering less area.
2. Another study revealed that there were also twice as many giraffes in 1976 as in 1971.
3. Images of the Serengeti taken over a period of 30 years show more trees.

Could there be a relationship between these seemingly disparate observations of entirely different species living within the same Serengeti ecosystem over the three or so decades since Sinclair was faced with the buffalo population growth question?

The vaccines introduced into the livestock eliminated the virus, which led to a growth in the wildebeest population. How did the wildebeests react to this change? They fed on grass, and when they ate it, they mowed it down from about 30 centimeters to about 10 centimeters in height. What function does the grass play in the life of the Serengeti? It acts as fuel for the fires. With less fuel, the Serengeti experiences less frequent and less intense fires. What does fire do within the Serengeti ecosystem? It burns up young trees. With less fire, there is more tree growth. What could be the effect of more trees on the Serengeti ecosystem? Animals like giraffes who graze the tops of acacia trees can easily find more food, allowing their population to grow.

Carrol's Serengeti case study clearly illustrates how interconnected all parts of any complex, dynamic system are. The elimination of the virus set in motion a sequence of events in which more wildebeest mowed down the grass, reducing the number of fires that allowed more tree growth that fed more giraffes.

Now, what does the Serengeti have to do with creating immersive media experiences? Like all dynamic systems, the Serengeti is alive with the possibility of change that can only be understood in terms of systems thinking. Stay with us here to explore this vital aspect of XR system making.

Immersive Spaces as Systems

The creator of the *Sims* digital game series, Will Wright, argues that "A robust game experience creates an entire system of limitations and freedoms, goals and rewards, challenges and joys in the player's mind" (Wright, 2022, p. 19). We referred to the process of participants internalizing the game rules and features of the immersive environment as mental modelling in Chapter 2. Part of the allure of being immersed in play spaces is the transportation from an everyday mindset with its typical real-world concerns to a psychologically new world of gameplay (p. 19).

We are inspired by Brian Upton's definition of play as a free movement within a system of constraints to argue that mediated presence emerges out of a free movement of human agents within an immersive system of constraints put together by the design team. Like games, all applications are complex systems set up for interactions between the users and the system. Well-designed applications are like beautiful duets between the application and the users, each complementing the other. The symbiotic relationship between the XR system and its audience makes systems thinking one of the most important conceptual tools of immersive media making.

Organizational complexity theorist Jamshid Gharajedaghi (2011) suggests that without adequate concepts and tools to explain the complexity of the world around us, we will likely end up feeling overwhelmed by a world perceived as increasingly chaotic and unmanageable. Physicist Stephen Hawking referred to the 21st century as a century defined by the exploration of complexity. Thinking in systems offers XR creators a unique and necessary understanding of the emergent quality of immersive environments. Making sense of complexity requires seeing the world in terms of wholes, relationships, and ecologies rather than reducing it to its component parts.

Systems thinking is now applied to such diverse fields as science, engineering, urban planning, and organizational management. Donella Meadows, in her book *Thinking in Systems*, argued that "the systems-thinking lens" allows change makers to reclaim an intuition about the ecology and wholeness of systems they encounter, honing the ability to understand the role of parts and interconnections, to ask "what-if" questions about possible futures (Meadows, 2009, p. 7). Ultimately, thinking in systems allows us to be creative and courageous about designing and redesigning systems to serve us better.

Meadows defined a system as

> a set of things, people, cells, or molecules which are interconnected in a way that produces an observable pattern of behavior over time. Systems may be constricted or affected by outside forces, but their response to such forces is always guided by the unique architecture and dynamics of the system itself.
> (p. 2)

In contrast to wholeness, as displayed by such ecosystems as the Serengeti, Meadows brings up an example of sand scattered on the road by chance, which she describes as "a conglomeration without any particular interconnections or function" (p. 12). No matter how many grains of sand we add or remove, we always end up with just a pile of sand on the road.

Compare this with a football coach arbitrarily adding players to a well-trained team or a surgeon taking away organs from the body. The latter would result in configuring very different systems that would behave in very different ways. The field of systems thinking has developed a number of powerful concepts. Here are a few we consider to be most relevant to the design of immersive systems.

Systems Iteration Basics: Emergence

Researchers who encounter complexity in such areas as ecology, computer science, engineering, and organizational management have developed new theories, frameworks, models, and methods for exploring phenomena that cannot be understood by traditional reductionist methods. Self-organization and emergence are some of the most important concepts associated with complex systems in such diverse domains as the study of animal cells, forest ecosystems, honeybee colonies, primate brains, and entire economies. They are highly applicable to the design and iteration of sociotechnical systems such as computer games, social media, and immersive systems.

Thomas Malone, the founder of the MIT Center for Collective Intelligence, refers to groups of people acting in a purposeful manner by engaging in collective intelligence that leverages computer networks as Superminds (Malone, 2018). Social media platforms such as Facebook and YouTube and multiplayer computer games such as *Fortnite* and *Halo Infinite* are examples of Superminds where groups of individuals connected via networked computer technology dynamically co-create complex sociotechnical systems. No single participant playing *Fortnite* controls all of the action. Beyond providing the platform and the rules of the game, the game designers do not control the outcome of the game either. The *Fortnite* Supermind, its self-organized behaviors, and the outcomes of multiple players acting together in time can only be understood using the tools of systems thinking.

All complex systems, from ant colonies, flocks of Canada Geese, financial markets, and primate brains to *Fortnite*, can be characterized as being composed of many smaller parts acting together without centralized control. Global-level organization exhibited by such systems can be described as an emergent product of all the local interactions between the individual parts. The explanatory theory of self-organization draws on ideas in chemistry and physics surrounding dissipative systems, synchronization, and pattern formation emerging from the bottom up among elements governed by simple rules.

Regardless of whether we deal with flocks of birds, neurons in the brain, or connected gamers, complex systems produce order that emerges out of initially disordered states. An often-cited example of emergence is the behavior of the cellular automata in John Horton Conway's Game of Life—a grid of cells, each having one of few possible finite states evolving according to a set of four simple rules that govern the interactions of each cell with its eight neighbors. The simple rules of the Game of Life state that: (1) any live cell with fewer than two live neighbors will die by underpopulation in the next cycle of life. (2) Any live cell with two or three live neighbors will live on to the next generation. (3) Any live cell with more than three live neighbors will die of overpopulation in the next cycle. (4) Any dead cell with exactly three live neighbors comes back to life by reproduction.

Although the rules are only defined locally, the cellular automaton is capable of producing complex global-level behavior that manifests itself in visually striking patterns emerging over time. Looking at videos of the ever-changing but always beautifully organized patterns created by the Game of Life, it is hard to believe that they are not a product of a group of very talented designers.

The point we want to make in this chapter is that a team with the most talented designers and engineers cannot match what can be co-created by teams that empower the audience to exercise freedom of choice within the constraints of well-designed immersive systems.

Systems Basics: Networks

Computational social scientist Alex Pentland argues that while decisions made by social actors are a blend of personal and socially acquired information, in the instances when personal information is weak, individuals tend to rely more on their social networks. Social learning has a confidence-enhancing effect in situations involving any kind of uncertainty (Pentland, 2015). Based on this insight, Pentland suggests that the only way to understand the digitally hyperconnected world is to extend our definitions of key economic and political ideas to include the effects of millions of digital citizens and consumers engaging in exchanging ideas that influence their opinions, values, worldviews, and behavior.

Pentland is among the many computational social scientists who have demonstrated that complex sociotechnical systems can be modeled and analyzed based on data analysis that brings clarity to the messy complexity of social interactions. The main contributions to network theory have come from computer science and the graph theory area of mathematics. Network analysis can help system designers see and even plan for connections and idea flows within systems through which contagions spread. The availability of real-time dense data sets and visualizations that can be generated by immersive forms of media allow designers to apply computational social science to designing and iterating XR systems based on evidence that enriches the design team's intuitive judgment.

One of Pentland's contributions to social physics is the idea of leveraging the honest behavioral signals as digital breadcrumbs, which can be analyzed using advanced algorithms. He refers to this process as reality mining. Pentland, his students, and his colleagues have used this approach to predict which individuals might be likely to develop diabetes, secure investments, or pay back loans. Pentland rejects the idea of people as independent decision-makers. He argues for the importance of factoring in the dynamic social effects as the true driving forces behind such social phenomena as economic bubbles, political revolutions, social media memes, and the entire internet economy. These social phenomena are outcomes of billions of microinteractions occurring between individuals trading not only goods and money but, maybe even more importantly, worldview-shaping ideas that affect their individual and collective behavior.

Immersive systems can be highly persuasive environments. Behavioral nudging can be accomplished by giving users social network incentives that affect idea flows within a given network. Having the algorithmically enhanced meta-level view of the sociotechnical world gives XR designers not only an understanding of how human+computer systems work but also the tools to take steps to fix the problems inherent in their status quos. By creating the richest possible quantitative descriptions of social networks, social physics can build quantitative, predictive models of human behavior in virtual-world interactions.

Some of the most important dynamics influencing XR systems that can be leveraged by XR system design teams include the two types of feedback loops that balance the activity within the system. Balancing feedback loops (also known as negative feedback loops) are potential sources of both stability and resistance to change in systems (Meadows, 2009, p. 30). Runaway feedback loops (also known as positive feedback loops) have an opposite, amplifying, reinforcing effect on the system.

The snowballing effect of positive feedback loops can create vicious or virtuous circles that cause healthy growth or runaway destruction of the system (p. 31). For instance, as shown by the work of network theorist Yochai Benkler, popular conspiracy theories develop a life of their own with real world consequences when amplified by social media networks (Benkler et al., 2018). The challenge of designing and iterating immersive social media platforms necessitates the application of both computational social science and systems thinking to keep the system on track toward accomplishing its desirable function. The process starts with the recognition that immersive media platforms are complex dynamic systems that can be best understood as the product of Superminds that combine multiple human agents interacting over advanced computational platforms.

Phase Space: The Physics of Immersive Systems

In *The Aesthetic of Play*, game designer Brian Upton advocates the adoption of the mathematical concept of phase space as a useful abstraction representing all possible states of an interactive play space. The concept of phase space was introduced in the 19th century by Ludwig Boltzmann, Henri Poincaré, and Josiah Willard Gibbs to address the issues of dynamic complexity arising out of the emerging field of thermodynamics. The concept has now been adopted by several fields concerned with dynamical systems.

Upton describes phase space as "a hypothetical space representing all possible states of a [game] system," with any particular state of play corresponding to "a unique location" or point within the space (Upton, 2015, p. 54). Physicists dealing with mechanical systems define phase space in terms of all possible values of the position and momentum variables.

Upton argues that within interactive game spaces, the experience of play is determined by the active constraints created by the entire phase space configuration of the game (p. 54). He uses the example of the ball's position in a digital game of ping-pong to illustrate how a game system's phase space both enables and constrains players' immediate actions. At any point during play, the ball's position can be specified by the three values of its "location from side to side, its distance from the net, and its height off the table" (p. 54). During play, the system dynamically plots the coordinates as a unique point within a three-dimensional phase space. As the players hit the virtual ball back and forth, the unique coordinates of the dynamically changing position move within the phase space along a trajectory that reflects the ball's path defined by the physics of the game.

To make things more interesting for the players, game designers can add the dimensions of velocity, direction, and spin to the equation. Accurately representing the velocity and direction of the ball requires three additional coordinate numbers. Representing the spin requires three more. The dynamic state of a ping-pong ball moving between players can be represented by a set of nine numbers. As the players compete, the system tracks the state of play with nine constantly fluctuating numbers.

In terms of XR environments, phase space can be thought of as the system's process or code frozen in time. It is the operationalizable accuracy of the measurements and the ability to freeze and analyze particular system states that makes the phase space concept a natural fit for the design of XR systems. The accurate tracking of the ball's trajectory within a nine-dimensional space is a simple example of how the concept of phase space can serve as a powerful tool that allows system designers to enable and constrain any immersive experience.

But this is just the beginning of the application of the system-level phase space to the design of immersive experiences. All objects within the virtual system can be tracked using the coordinate numbers for their various states. In the following sections, we will discuss how the coordinates for more than just the position of objects can be recorded and analyzed for system calibration and iteration. Phase spaces within XR systems are full of digital breadcrumbs left behind by all of the system agents.

The Social Physics of Collective Intelligence

Digital breadcrumbs as the behavioral data points are the foundation of computational social physics. Pentland has defined a number of key social physics concepts which can help XR designers to affect the emergent behavior within phase spaces of social networks.

- "Ideas" within the social physics framework can be defined as the basic social currency of meaning-making affecting instrumental agent behavior (Pentland, 2015). Resonant, valuable ideas can spread throughout a network to become strategies for action and habits of action used in what Kahneman called "fast thinking" responses.
- "Information" is a perceived observation made by a community member that may be incorporated into a belief or used to build up a larger idea.
- "Idea flow" refers to the propagation of beliefs, values, and behaviors that diffuse through a social network by two critical processes of social learning and social pressure. An estimation of an idea flow for a social network must take into account the structure of the network, the individual susceptibility of members to new ideas, and the strength of influence between each pair of members.
- "Interaction" is an exchange between community members that can include direct means such as conversations or indirect means such as observation of behavior, overhearing a conversation, or sharing of digital content.

- "Engagement" can be defined as social learning occurring within a peer group. Engagement typically leads to the development of behavioral norms that are later reinforced through the mechanism of social pressure.
- "Society" in social physics can be described as an entity made up of networks defined by exchanges between individuals. It stands in contrast to a description of society as composed of such social science concepts as classes or markets.
- "Strategy" is a choice defined by the total potential actions an individual can take in a given situation.
- "Trust" is a core social norm that allows individuals to expect fair value when dealing with others in an exchange or a relationship. Adherence to trust assures a continuing stable exchange of value within a community.
- "Value" constitutes the extent to which an exchange will satisfy social and personal needs and goals of individuals in social exchange interactions. This can include the utility value of goods or deeper personal needs, spanning the gamut of needs from physiological to safety to love, belonging, esteem, and self-actualization.
- "Social influence" describes the likelihood of one person's behavior affecting the behavior of others.
- "Social learning" can consist of learning new strategies by observing other people's behavior. It can also involve acquiring new beliefs and behaviors through direct personal experience or observation.
- "Social network incentives" are a type of reward capable of affecting individual behavior or altering the pattern of exchanges between pairs of community members. Finally, "social norms" are sets of agreed-upon compatible strategies the community employs to produce the most desirable exchange value.

Norms are developed and spread by the process of social learning and enforced by social pressure, which is the leverage members of a community can exert upon one another (2015).

Applying the fundamental social physics concepts to the phase space of a well-designed XR system can give system designers an unprecedented opportunity to fine-tune, calibrate, and recalibrate an immersive system in play in a way that optimizes the performance of its intended function.

XR System Architecture

To be operational, the phase space a well-designed XR system needs, along with its digital breadcrumbs, must become a part of the system's essential architecture. We base our approach to designing XR system architecture on the framework Peter proposed in his previous book. The framework addresses the five critical dimensions of a well-designed system. These include (1) function (purpose), (2) structure (information architecture + hardware architecture), (3) process (code), (4) semantics (meaning), and (5) pragmatics (behavior). In addition to the five system dimensions which can be, at varying degrees, controlled by the designers, Peter identified two that might completely elude the designers' control. These include (6) context

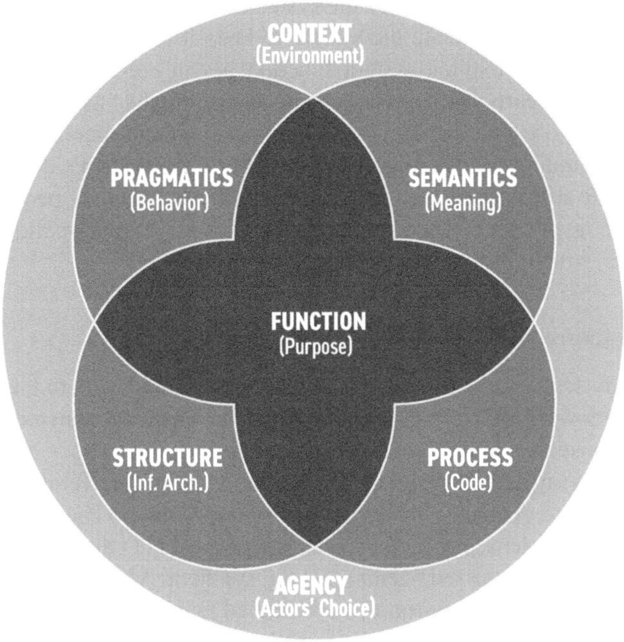

Figure 7.1 The seven dimensions of an XR system

Source: Concept and illustration by © 2022 Peter (Zak) Zakrzewski.

(system environment) and (7) the agency of the human actors within the system. Let us address each of these in detail.

System Function

Donella Meadows argued that the system's function or its purpose, even if not explicitly stated, will be apparent to an astute observer through the operation of the system (Meadows, 2009). She suggested that "the best way to deduce the system's purpose is to watch for a while to see how the system behaves" and what it accomplishes (p. 14). For example, if we observe a frog turning right and catching a fly, and then we observe the same frog turning left, around backward, and so on to catch another fly, we can infer that the purpose of the frog is not to keep turning left, right, or backward but to catch flies.

In systems not explicitly designed to accomplish a particular purpose, system outcomes will not necessarily be those intended by any single actor or a group within the system. They will be a product of emergence. As a matter of fact, one of the most pernicious aspects of dealing with complex social or sociotechnical systems is that the actions of actors within a network pursuing their own local goals may add up to an overall behavior and outcomes that no one wants (p. 14).

Savvy system designers identify the system function as the first step of system design and monitor and iterate the system to keep it on track. The foundation of choosing a clear and explicit system function lies in the answer the XR design team came up with early in the process while answering the "why" question behind the project. To those clear on the system's purpose, system function can serve as a north star—a guiding beacon to be followed during the system's design and iteration. Is the XR application meant to entertain, educate, or allow people to accomplish specific tasks? All of these different purposes might require different system architectures, structures, and code to accomplish.

System Structure: Information Architecture as Law

Meadows advises systems thinkers to "look beyond the players to the rules of the Game" (Meadows, p. 12). She argues that the elements of a system are usually easy to notice because they are often visible and tangible things.

This is not to say that the elements are not important to the overall functioning of a system. They are. As we discussed in the Context Scenario chapter, the elements of an immersive system may include such categories as (1) the audience members (avatarized or non-avatarized agents), (2) the synthetic AI-based agents (actants), (3) physical objects in the environment, (4) digital twins, (5) digital objects, and (6) enchanted objects. But the rules of the game by which the system elements interact with each other are initially defined by the overall information architecture (IA), which is a concept familiar to anyone with some experience with web development or software application projects.

Saul Wurman, one of the seminal figures in the development and promotion of information architecture (IA), argued that the goal of defining IA for a digital system is to make the information guiding the system and within the system understandable to other human beings (Saul Wurman, 2001, p. 23). Wurman defined information architects as people who map information in a way that allows others to find their personal paths to knowledge. He believed that the IA field would open the door to the emerging 21st-century professional occupation addressing the need for clarity, human understanding, and the scientifically based organization of information (Wurman, 1996). Great information architects, in essence, have the ability to cut through the necessary complexity of immersive systems to make the rules and navigation guidelines clear to the system users.

IA lies at the heart of any application design process, of organizing content of websites, games, and other digital products into categories, hierarchies, and relationships that help system designers develop the code and produce the final product. IA forms the basis of navigation that supports users in achieving their goals by organizing and arranging the relevant content in a clear and understandable way.

Finally, IA allows system designers to iterate and redesign the immersive system based on a clear and easy-to-understand structure. IA is fundamental to ensuring design consistency of the entire XR experience by standardizing the user interface (UI) conventions. If creating XR systems could be compared to building houses and offices, IA would be analogous to planning, mapping, and placement of rooms that enable intuitive flows users could expect at every turn as they navigate the immersive space.

Peter Morville, Louis Rosenfield, and Jorge Arango built on Wurman's insightful bridging of architecture and information with *Information Architecture for the World Wide Web*, commonly referred to as the "polar bear" book, which defined the field of information architecture as a community of practice dedicated to bringing principles of design and architecture to computational media design. They described the practice as the art and science of shaping digital information products and experiences that support usability, findability, and understanding (Morville & Rosenfeld, 2015).

Peter Morville, Louis Rosenfield, and their colleagues defined the scope of IA practice as the structural design of shared information environments and the structural design of information spaces to support user goals, to facilitate task completion, and to enable intuitive access to digital content. They also referred to tasks involved in IA as the synthesis of organization, labeling, and navigation systems within digital, physical, and mixed-channel information technology (IT) ecosystems (Morville & Rosenfeld, 2015).

Many of the principles of the "polar bear" book are applicable to XRXD, and we encourage those interested in learning more to read it in order to develop their own ways of applying IA to XR projects. The beauty of IA is that it can be extended to become experience architecture and developed using such simple tools as paper and pencil, sticky notes, cards, whiteboards, chalkboards, and flowcharting software to visually plan and structure the immersive experience that can become the de facto law of the immersive world.

We have divided the system structure into two essential components based on Peter's original framework. Within XR systems, information architecture must always be complemented by the physical structure on top of which the immersive experience can be offered to the audience.

System Structure: The Physical Substrate of Immersive Experiences

System structure is also the place for designers to address the physical substrate of XR experiences in a way that recognizes the indivisible nature of embodied human cognition. Over millions of years of evolution, humans have been shaped by the coexistence of other humans, objects, and natural phenomena in the environment.

Based on our evolutionary past, we have developed an embodied cognition that is, in the words of cognitive psychologist Margaret Wilson, "deeply rooted in the body's interaction with the world" (Wilson, 2002, p. 625).

In respect to embodied cognition, Wilson, whose work we already covered in Chapter 4, outlines six arguments that can serve as guides for designing the physical substrate of intelligence augmenting systems:

1. Cognition is situated, meaning that our cognitive activity takes place in the context of a specific physical environment. Cognition inherently involves perceiving objects in the environment with a singular goal of taking action.

2. Cognition is time pressured, which means that it can only be understood in relation to how it functions under the pressures of real-time interaction with other living beings and objects in the environment. Time pressure can dramatically change people's ability to think creatively and solve problems.
3. Because of limits on information-processing abilities, such as working memory and attention, humans off-load cognitive work onto the environment to reduce the cognitive workload. We have used tools such as clay tablets, papyrus, stone and chisel, and pen and paper to write notes and draw images that help us to remember and understand.
4. We are so embedded in our environments that they become part of our cognitive system. The flow of information between our nervous system and the world is so dense and so continuous that the mind cannot be studied in isolation.
5. The primary function of our cognition is to guide action. Cognitive mechanisms such as perception and memory can only be understood in terms of their contribution to situation-appropriate behavior.
6. Even when we decouple our minds from the physical environment, our cognitive activity is deeply grounded in the sensory processing and motor control mechanisms that evolved for interaction with the physical world. A metaphor such as "things are looking up" clearly shows how cognitively grounded we are in our bodily experience, even when it comes to an abstract representational system such as language.

Another cognitive scientist, Don Norman, suggested that product and experience design can be considered appealing to three distinct areas of human perception and cognition. The visceral level of cognitive processing addresses our deep aesthetic needs. A reflective level of processing extends our experience beyond the moment of engagement to a time when we contemplate the meaning of things that have happened in our past or imagine events happening in the future through personal reflection or communal storytelling. Our visceral reaction to the objects in our environment is deeply connected to the body and survival.

The third behavioral level of cognitive processing is primarily connected to the pleasure and effectiveness of using objects such as musical instruments, tools, and door handles. As Norman suggests, the experience of skillful use of well-designed products such as golf clubs, sports cars, and musical instruments rewards us with a sense of skilled accomplishment. At this level of cognitive processing, usability matters more than appearance and rationale. The solid feel of a door handle turning under our grasp and the comforting tactile feel of a physical button offer behavioral cues that give users a sense of physical quality and safety (Norman, 2013). Designers who understand the importance of the behavioral level of cognitive processing can ensure that all interactions within the immersive environment are appropriately sensory rich and pleasurable.

People infamously walking off stages while demoing VR goggles in the early days of XR is a stark reminder that XR presence is built on top of physical spaces, computer hardware, server farms, and very real and absolutely necessary XR equipment. All immersive experiences are predicated on the feasibility of XR technologies and the creativity applied to hardware and software engineering. Paying attention to the physical substrate of immersion reminds us of the need to acknowledge the complex technology behind XR immersion that leverages computational medium as an extension of the human nervous system.

Whether planning an AR experience that guides tourists on an adventure through the streets of Paris, creating an MR surgery simulator for medical students, or developing VR data visualization empowering analysts to navigate through financial data, we need to remember the XR presence is built on top of a solid foundation of a well-thought-out physical substrate.

Code as Law Enforcement

One of the earliest descriptions of the concept of serious games can be found in Clark C. Abt's 1970 book by the same title, in which the author addressed the application of board games and role play in education, science, and policy-making. Ian Bogost uses the concept to describe digital games expressly created to support the strategic interests of sponsoring organizations. Bogost argues that serious digital games have a procedural advantage over traditional ones.

Procedurality is a game design strategy aimed at rhetorically enforcing the rules of the game through code that shapes the behavior of participants within play spaces within which interaction between the player and the system takes place (Zakrzewski, 2022). Bogost argued that in the light of procedural rhetoric, digital game experiences could create aesthetic and cultural experiences that conveyed messages by compelling players to engage with and reflect on the very nature of game rules (Bogost, 2010, p. 105). Since its introduction, rhetorical procedurality has been embraced as a distinguishing feature of computer games and as a unique design affordance that allows digital game designers to introduce ethical, political, social, and aesthetic dimensions to their work (2010).

The rhetorical nature of code allows XR experience designers to enforce persuasive system rules laid out in the system architecture through expressive coding. Bogost argued that the procedural enforcement of rules needed to be understood not as a limiting factor on participant behavior but as the imposition of constraints that offer environments for creative expression. Using code to rhetorically persuade audiences focuses on how arguments can be embedded in the rules of a game and how these rules can then be understood, internalized, and expressed by the participants.

From the procedural rhetorical point of view, immersion refers to the possibility space—a phase space of immersion that itself is the product of the constantly evolving interactive process. Not all immersive experiences will aim to persuade their audiences in the areas of ethics, politics, or social issues, but using code to consistently enforce the system architecture gives artificially created virtual worlds an added sense of "reality" and solidity. The key benefit of applying procedurality in the context of XRXD is that the rules governing immersive worlds can be intentionally and consistently supported by the code created for the desired audience experience.

Procedural systems respond dynamically to human actions with consistent rules that define everything from physics (how objects move or collide) to social dynamics (how agents behave). Participants, in turn, will infuse their encounter with rhetorical meaning guided by the rules, mechanics, narrative, information architecture, and sensory affordances of the system.

Designing for Semantics and Idea Flows

Family therapists Paul Watzlawick, Jane Beavin Bavelas, and Don D. Jackson suggest that the study of human communication within complex social and sociotechnical systems could be subdivided into three areas: (1) **syntactics** addresses the structure of the communication environment to cover the problems of transmitting information, statistical properties of language, coding, channels, capacity, noise, and redundancy; (2) **pragmatics** is an area concerned with how communication and communicative actions affect behavior of agents within a system; and lastly, (3) **semantics** covers meaning-making within social systems (Watzlawick et al., 1967). Watzlawick and his collaborators suggested that while it was possible to conceptually separate the three areas for the purpose of analysis, ultimately, they needed to be understood as being highly interdependent. Syntactics, pragmatics, and semantics form the co-dependent aspects of constraints that make up robust interactive systems. We consider syntactics to be akin to the structural aspect of XR systems covered earlier. We will return to pragmatics in the next section. For now, let us focus on meaning-making within immersive systems.

As we mentioned before, XR media systems are immersive social media platforms, which means that designers can draw a lot of insights from communication patterns observed in networked communities. The rise of computational social science has given researchers an opportunity to study the contribution idea flows make toward emergent collective behavior. Pentland argues for the importance of factoring in the dynamic social effects of communication as the true driving force behind such social phenomena as economic bubbles, political revolutions, social media memes, and the entire internet economy.

All social phenomena are the emergent outcomes of multiple microinteractions occurring between individuals trading worldview-shaping ideas that affect their individual and collective actions. Decisions made by social actors are based on a blend of personal and socially acquired information. Social learning has a confidence-enhancing effect particularly in situations involving uncertainty (Pentland, 2015). Individuals also tend to rely on their social networks in instances when personal information proves to be too weak for safe decision-making.

Based on his social physics experiments, Pentland has articulated three key insights that can be applicable to managing communication within XR networks: (1) sociotechnical systems are inherently highly persuasive environments; (2) patterns of behavioral user data within digital systems such as social media platforms can be captured, identified, and used for iterating the overall system performance;

and (3) behavioral nudging can be built into the system by providing users with types of social network incentives that affect decision-making and idea flows within XR networks.

Having the algorithmically enhanced meta-level view of the sociotechnical world gives XR designers not only an understanding of how human+computer XR systems operate but also the tools to take steps to fix any calibration problems detected in their ever-evolving states. By creating the richest possible quantitative descriptions of social networks, applied social physics approaches can help us to build quantitative, predictive models of human behavior in virtual-world interactions.

Pragmatics

Watzlawick et al. argued that in an interactional setting, every behavior has message value that can influence the meanings exchanged between individuals. Our communicative behavior, including gestures, smiles, and various actions, can all serve as nonverbal communication cues that allow people to go beyond language to exchange important social signals.

Pentland argues that people are not as independent decision-makers as we commonly think. Social effects have a profound influence on shaping our worldviews and ultimately our behavior. We adapt our behavior based on learning from examples of other people in our social circle. Pentland's research showed that providing social network incentives to change behavior through idea flows was much more successful than attempts to affect change through standard economic incentives that frame people as individual, rational actors rather than as social creatures (Pentland, 2015).

As previously discussed, interactions within XR systems fall within three categories of human-to-human, human-to-machine, and machine-to-machine patterns. Each of these has a shaping effect on the overall user and system performance to the point that we can talk about system-level behavior as an emergent phenomenon tied directly to the system function. Entire sociotechnical systems such as social media platforms or online games exhibit global-level behaviors emerging out of multiple interactions. Conspiracy theories, memes, social fads, and dance crazes emerge out of social media platforms in ways that would not be possible without mediated hyperconnection.

Within sociotechnical systems, emergence originates in the cognition, affect, and behaviors of individual participants. Emergent behavior is amplified by the interactions within social systems to manifest as a higher-level collective phenomenon (Kozlowski & Klein, 2000). Social media design pioneer Khoi Vinh was one of the first media designers to argue for the necessity of a conceptual shift from the designers-as-auteurs to the designers-as-creators-of-social-spaces mode of practice. He identified the need for designers of sociotechnical environments to anticipate what could not be planned by creating a framework in which the unexpected could be expected to happen (Vinh, 2015, p. 129).

Successful design of well-calibrated immersive media environments is based not on attempts to tightly control participant behavior but on establishing conditions under which rich, immersive, and rewarding interactions can happen within the constraints created by the system designers (Zakrzewski, 2022).

The principle of emergence suggests that just as one cannot fully predict the outcome of an engaging face-to-face conversation, the emergent outcomes of XR networks will produce unanticipated creative results. This is not to say that designers cannot keep unintended consequences in check and within desirable ranges, however. A number of mechanisms can be used to ensure positive outcomes at all system levels. Chief among these is establishing the rules of play that nudge participants toward socially beneficial behavior while respecting the individual goal-seeking intentionality of each audience member. We know by now that digital networks are not immune from potential conflict arising out of intentional goal seeking by diverse groups of users.

The pioneering social media platforms such as Twitter and Reddit consistently struggled with the issue of resolving conflict between various user communities. The application of procedural persuasion as a conflict resolution mechanism might go a long way toward creating and maintaining ecological balance within an XR system, especially one involving a major social component.

XR media designers are in a unique position to nudge not just the individual participants but global-level behaviors of entire networks by strategically designing immersive environments to enable particular types of interactions and behaviors and then monitoring and iterating them to stay on course. Once the system gains an energy of its own, the designers' job is to monitor system outcomes and iterate the design of the system based on its intended function. We will cover system iteration in Section VII.

The Environmental Context

No system exists in a vacuum. Human societies are built on top of the Earth's ecosystem, which itself is part of the solar system, and so on. Immersive systems are no exception. The first question when addressing the environmental context of XR systems is to ask about the audience goals. What does the audience for this XR application want to accomplish? Where will they be located while pursuing their goal? Will they be sitting in their living room or office, or will they be walking around the streets of a busy city? This will dictate how they will need to navigate the immersive space.

Spatial computing offers unprecedented access to intelligence augmentation (IA) in the wild, but connecting contextually relevant data to real-world environments is not without its challenges. For instance, let us consider a shopper at a retail store using a mobile AR app to choose a packaged food product that matches their dietary requirements. An inability of the smart device being used to correctly identify a package on a shelf might lead to providing the shopper with a completely erroneous ingredients list.

The same can be said for street navigation. The very challenge of AR is to tightly calibrate the data presented to the user with the real-world context of the user's experience. AR's promise to overlay actionable, real-time information onto real-world environments full of products, equipment, and physical locations relies on smart data capture and two-way communication in an accurate, accessible, and intuitive manner to deliver value to the user. No matter how elaborately we can design the immersive experiences for our audience, the MR headsets they will use will have to map their physical environments to ensure that the participants do not bump into furniture, walls, and each other.

Needless to say, context of use is of paramount importance to experience designers. Some XR environments can be controlled by the design team. For instance, the design team behind a film noir–inspired immersive theater experience, Sleep No More, used a tightly controlled set of over 90 spaces in a 100,000-square-foot warehouse in the Chelsea district of Manhattan to ensure that the entire immersive environment supported the desired audience experience. This allowed the participants to immerse themselves in a story while exploring such evocative spaces as dimly lit labyrinths or the charming rooftop bar of the McKittrick Hotel.

The makers of the iOS AR app *Smash Tanks*, which turns the participants' living room floor into a cartoon-like battlefield, have to be a bit more flexible in adjusting to rooms of very different sizes and shapes. *Pokémon GO*'s impressive status as one of the most successful mobile games ever created has opened doors to location-based AR apps in such diverse media sectors as entertainment, retail, tourism, communication, and real estate.

But *Pokémon GO* is also a reminder of the environmental challenges lying ahead of XR media content creators. Location-based augmented reality apps that work by mixing geopositioning data with AR technologies to identify special geo-markers known as points of interest pose substantial environmental challenges for the design teams. The apps can only work if the smart device can figure out where it is located in the world by using data from a digital compass, accelerometers, and GPS.

Agency

The one vital aspect of XR experiences, we argue, that makers cannot and should not attempt to directly control is the agency involved in the creation of the subjective inner experience of the participants. The two critical parts of the subjective inner experience include presence, defined as a subjective user state, and agency, defined as the freedom to pursue one's own goals and actions within the system of design constraints. As Peter argued in his previous work, participants' freedom within the context of immersive systems depends on two components: the existence of a set of genuine alternatives and an opportunity to make informed choices, which refer to the participants' ability to choose their own course of action (Zakrzewski, 2022).

As XR experience designers, we can indirectly influence these factors while we do not exert direct control over participants' sense of presence and agency. We cover the controllable aspects of this audience-oriented system design dimension in different parts of the book. The rhetorical nature of immersive media allows designers to engage in a great deal of persuasion when it comes to engaging the audience and guiding their choices. Participants involved in the *Sleep No More* performances actively explore the immersive, labyrinthian environment that allows them to take their own unique paths through the experience but also share that experience by communicating with other audience members. Mediated presence is co-created by the interactions between the individual audience members and the immersive environment and the interactions between audience members as a community of participants.

Summary

In this chapter, we argue for the importance of recognizing immersive media applications as complex networks ruled by the principles of systems thinking. We outline the process of designing XR systems based on the framework of seven interdependent dimensions of a sociotechnical system: its intended function, architecture, process, semantics, pragmatics, environmental context, and participants' agency. With the ideation of the XR world as a coherent system, the team will be ready to start building the functional system prototype, which is the topic of our next chapter.

References

Benkler, Y., Faris, R., & Roberts, H. (2018). *Network propaganda: Manipulation, disinformation, and radicalization in American politics*. Oxford University Press.
Bogost, I. (2010). *Persuasive games: The expressive power of videogames*. The MIT Press.
Carroll, S. B. (2016). *The Serengeti rules: The quest to discover how life works and why it matters*. Princeton University Press.
Gharajedaghi, J. (2011). *Systems thinking: Managing chaos and complexity a platform for designing business architecture* (3rd ed). Morgan Kaufmann.
Kozlowski, S. W. J., & Klein, K. J. (2000). A multilevel approach to theory and research in organizations: Contextual, temporal, and emergent processes. In S. W. J. Kozlowski & K. J. Klein (Eds.), *Multilevel theory, research and methods in organizations: Foundations, extensions, and new directions* (pp. 3–90). Jossey-Bass.
Malone, T. W. (2018). *Superminds: The surprising power of people and computers thinking together*. Little, Brown and Company.
Meadows, D. H. (2009). *Thinking in systems: A primer*. Earthscan.
Morville, P., & Rosenfeld, L. (2015). *Information architecture: For the web and beyond* (3rd ed.). O'Reilly.
Norman, D. A. (2013). *The design of everyday things: Revised and expanded edition*. Basic Books.
Pentland, A. (2015). *Social physics: How social networks can make us smarter*. Penguin Press.
Upton, B. (2015). *The aesthetic of play*. The MIT Press.
Vinh, K. (2015). *How they got there: Interviews with digital designers about their careers*. Design Department Partners.
Watzlawick, P., Bavelas, J. B., & Jackson, D. D. (1967). *Pragmatics of human communication: A study of interactional patterns, pathologies, and paradoxes*. Norton.
Wilson, M. (2002). Six views of embodied cognition. *Psychonomic Bulletin & Review*, 9(4), 625–636. https://doi.org/10.3758/BF03196322

Wright, W. (2022). Will right teaches game design and theory. *Master Class*. www.masterclass.com/classes/will-wright-teaches-game-design-and-theory
Wurman, R. S. (1996). *Information architects*. Graphis Press.
Wurman, R. S. (2001). *Information anxiety 2* (2nd ed., expanded and updated). Que.
Zakrzewski, P. (2022). *Designing XR: A rhetorical design perspective for the ecology of human+computer systems*. Emerald Publishing Limited. http://doi.org/10.1108/9781802623659

8 Building the Functional System Prototype

What Is Prototyping and Why Do You Need It?

After the intensive thinking-to-build phase of the project, the time has come to start building the immersive application. This phase, referred to in design as the building-to-think phase, starts with the act of prototyping. In a nutshell, a prototype is a simulation of the product you are building created to test its user experience design and functionality. As mentioned in Chapter 6, full-fledged prototypes offer refined and elaborate simulations of the key features of the final product that can give the media makers a clear view of the real-world operation of the system.

Building a prototype is an essential part of following an iterative design process. After you validate your product idea with a proof of concept, you build a prototype as a first draft of your actual immersive application. Functional prototyping is even more important in immersive experience design than in interactive media because of the high level of technical complexity of XR technology needed to support the delicate balance underpinning the subjective state of mediated presence. Prototypes, by comparison to the proof of concept (PoC), focus on testing the full functionality of the key features of digital products. Prototyping is a time- and budget-friendly way to test key assumptions your team has about the general feasibility of the project. Building a prototype helps you to visualize your XR experience and get to know how it would work.

Prototypes are the best and quickest way to put the project in the hands of the end users and in front of the key stakeholders such as clients and funders. One of the key benefits of prototyping is the opportunity to get user feedback that allows the team to identify any user's pain points and to verify whether users have a positive experience navigating through your immersive content. A prototype completes the dialectical cycle between thinking to build and building to think. In XR experience design, we use imagination to explore creative possibilities and to speculate on the potential consequences of our actions. We prototype to make concepts that exist in the mind's eye, tangible and testable. Prototyping is an act of getting ideas out of our heads and into the hands of our users to get fast and inexpensive feedback that allows us to iterate, evolve, and improve our concept.

DOI: 10.4324/9781003442462-14

Thinking to Build
In XRX design, we use imagination to explore creative possibilities and to speculate on the potential consequences of our actions.

Building to Think
We prototype to make concepts that exist in the mind's eye, tangible and testable. Prototyping is an act of getting ideas out of our heads and into the hands of our users to get fast and inexpensive feedback that allows us to iterate, evolve, and improve our concept.

Succeeding by Failing Fast, Often, and Cheaply

IDEO's famous mantra of failing fast and often on the way to success is more than just a slogan; it is a core philosophy behind the iterative design approach. The secret to failing fast and often lies in understanding that prototypes and proofs of concept do not have to take months to develop or costs tens of thousands of dollars. Early-stage prototypes can often be built in as little as an hour and for little to no cost. Prototypes are incomplete models of the application that allow the design team to put the system in the hands of users and other internal and external stakeholders.

There are many ways to prototype various aspects of the immersive system, from low-fidelity paper prototypes to hacked and coded functional ones. Prototypes can be categorized based on their testing purpose. They can fall into the (1) value hypothesis or (2) usability hypothesis test types. Typically, the value-hypothesis prototypes come early in the process, and once the concept is validated, usability becomes the focus. The usability-hypothesis-testing prototypes can be categorized based on the phase of the project they best fit into. These include (1) rapid usability prototypes, (2) visual design, (3) information architecture, and (4) high-level prototyping tool buckets.

What Is a Functional Prototype?

A functional prototype is an interactive and testable model of a product that either operates or simulates the operation of the system similarly to the final immersive experience. While static and low-fi prototypes might simulate the appearance or layout of a digital product, they cannot capture its actual behavior. In game design, a fully functional prototype closely resembles the final application in terms of gameplay and technical implementation.

Functional prototypes aim to provide a comprehensive and immersive experience that captures the essence of the envisioned experience design. Functional prototypes are typically the result of the lessons learned from the technical and PoC prototypes. The main benefit of functional prototypes lies in their testability, which allows designers and engineers to identify any technical issues before final production begins. Functional prototypes offer the development team a dynamic representation of how their design might work in the real world.

Functional prototyping can give the development team a glimpse into the participant's journey through the immersive experience, mirroring what a user will see, hear, and touch. In order to achieve this fidelity, the features within the prototype have to work, and user inputs have to have consequences to replicate the flow of the real product. By letting users interact with the system, the designers and engineers get a better understanding of the product's features and flaws, helping the team to refine the design for a better user experience. Testing and debugging the system prior to production can save a lot of time, money, frustration, and possible reputational hit associated with releasing an inferior product.

According to VR developer Thomas Steffen, creating functional prototypes in VR game design serves seven important objectives.

1. First and foremost, a fully functional prototype can deliver a sense of a complete immersive experience by replicating the intended "gameplay feel" (Steffen, 2023, p. 1). This aspect of functional prototyping is critical for obtaining valuable user feedback.
2. A functional prototype can integrate all of the finalized or near-finalized digital assets across all the sensory channels to create the first representation of the immersive environment to help the design team evaluate the impact of the aesthetics on the overall participant experience.
3. A good prototype will include a more polished and functional user interface (UI), including menus and HUD elements to help the team assess the clarity, usability, and navigability of the XR environment.
4. A functional prototype can designers help to explore the entire immersive system including its mechanics, zones, levels, quests, missions, and challenges to provide the team a clearer understanding of how the experience will unfold over time once the audience enters the XR system.
5. A well-developed functional prototype needs to exhibit technical stability and performance. This means that the process of building the prototype will help to improve the system's performance by eliminating any major bugs, crashes, and other engineering limitations from the product being built.
6. The best and maybe the only reason to build a fully functional prototype is to subject it to extensive playtesting, internally within the development team and externally with a group of potential users. The valuable feedback such tests can provide will help the team to identify all critical areas that need improvement.

It will also help to validate the key design decisions made during the inscription phase to refine the overall immersive experience.

7. Finally, a fully or partially functional prototype can be invaluable in demonstrating the potential of the XR content to publishing platforms or funders.

Building Functional Prototypes

It is difficult to underestimate the importance of assembling diverse skills on an XR project team. Collaboration between content creators, designers, developers, and media distribution strategists is required. This teamwork will ensure you choose the best platform, tools, and design patterns for your project. To construct working prototypes and test your ideas, gaining hands-on experience in programming is necessary, given the technical nature of XR platforms. That said, you don't usually have to code at a low level; prototyping tools and game engines are available that abstract the hardware interfaces and facilitate your development with high-level software interfaces to the low-level array of sensors, displays, devices, etc.

Unity and Unreal Engine are the two leading industry-standard game engines widely used for prototyping and testing immersive experiences. Unity, known for its easy-to-use interface and vibrant support community, is ideal for rapid prototyping. Technically inclined designers and media makers dipping their toes in the programming waters for the first time will find Unity more accessible. On the other hand, Unreal Engine offers more advanced rendering capabilities, making it the better choice when high-performance graphics rendering is paramount. Each of these game engines has its optimal use cases, and a detailed comparison is not within the scope of this book.

For designers and media makers who don't code, understanding the function of the code behind the XR environments—such as how interactions, physics, and animations are implemented—will allow them to create better prototypes. Coders may be able to translate interaction scenarios into functioning systems. However, designers maintain the big-picture perspective, bridging the gap between creative vision, user experience, and technical feasibility. As you begin the prototyping step, focus on prototyping simple interactions and environments. The iterative design process—hypothesis, building, testing, gathering feedback, and refining—will deepen your understanding of the capabilities and limitations of your technology stack.

Functional Prototypes

Functional prototypes are typically built at the later stages of the development process to focus on representing important features of the complete product before moving to the production stage. The primary objective is to implement the baseline of functionality needed to present your project idea in a tangible form to enable end users and stakeholders to visualize and engage with key aspects of the experience to help your team assess essential questions, e.g., "Does our experience help the end user perform their job-to-be-done?"

It is important to remember that functional prototypes are not intended to be production ready. Transitioning from a functional prototype to a fully developed, market-ready product involves navigating a range of challenges that far exceed the initial scope of your prototype. While the functional prototype serves as a tool for validating early hypotheses, the path to a production-grade solution demands addressing complex technical hurdles, resolving scalability issues, and refining the user experience. What begins as a simplified model for testing ideas often results in identifying significant challenges to be addressed, requiring significantly more effort and resources to evolve into a robust, fully realized product that meets market demands.

The Journey From PoC Prototype to Functional Prototype

Moving from the PoC prototype to a functional prototype in XR experience design marks a significant transition, as the focus shifts from validating core concepts to creating a fully interactive, stable model of the intended experience. A PoC prototype primarily aims to test desirability, feasibility, and usability, often with only essential interactions, minimal assets, and limited participant pathways to demonstrate that key design and technical elements are viable.

In contrast, a functional prototype is a more advanced, integrated version that closely mirrors the final product in terms of participant experience, interactivity, and feedback. The functional prototype stage involves refining the prototype for real-world usability, incorporating finalized 3D assets, and ensuring that all interactions, functionalities, and feedback mechanisms are constructed and optimized for performance. It also enables developers to test specific features like real-time data integration, multiuser interactions, and advanced animations, which are usually outside the scope of a PoC.

The development and deployment of a functional prototype in XR require a methodical approach, typically starting with an expanded project plan that outlines finalized objectives, technical requirements, and performance goals. Teams generally begin by revisiting user feedback from the PoC phase to refine user interactions, environments, and navigation based on real-world testing insights.

Game engines like Unity or Unreal Engine are often used for functional prototypes due to their extensive support as complete development environments supporting cross-platform deployment and third-party packages for specific capabilities. Unity, for instance, has been instrumental in creating functional prototypes for educational XR applications, enabling multiplatform access (e.g., VR and mobile AR) and integrating data tracking for assessing user engagement.

Within these engines, designers will guide the creation of refined assets, while developers will focus on writing scripts that bring the experience to life. This involves writing code to manage gestures, spatial movement, and object manipulation so that the system responds to participant inputs and the appropriate feedback is provided: Unity's XR Interaction Toolkit provides prebuilt components for common XR interactions like grabbing, moving, and rotating objects, which can save development time while ensuring that participant interactions are fluid and natural.

Unreal Engine's Blueprints visual scripting system enables developers to create complex interactions without deep coding knowledge, which can accelerate the prototyping process and make it easier for your team to test and refine interactions based on user feedback.

Minimally Viable Products

Another distinction has to be made between functional prototypes and minimally viable products (MVPs). Both are critical components in the XRXD process. They serve distinct purposes and are used at different stages of the development cycle. A minimally viable product (MVP), a concept borrowed from the lean startup field, refers to product launches of an XR experience with enough features to attract early adopters and validate a product idea (value hypothesis) at a later stage of the development cycle.

The primary goals of the MVP are to go to market with a product that can test the value hypothesis beyond the lab or the studio environment. By delivering the application to a limited base of actual users, the team moves from the conceptual and development stages of the project to iterating and scaling the distribution of the final product. Only you and your team can decide where your functional prototype ends and your MVP begins, based on market readiness you sense from the early user feedback.

Four Approaches to Prototyping

UX/UI practitioners have successfully used four types of prototypes.

1. Rapid prototyping involves, as the name suggests, quick and cheap prototypes that are only used in the short term.
2. In evolutionary prototyping, you use a functional application or product that goes beyond a simple simulation. Each iteration of an evolutionary prototype is a sequel that improves on a previous version according to the feedback.
3. In the incremental prototyping process, the team divides the final project into smaller chunks and prototypes, evaluates, and refines each module individually before integrating all the parts.
4. Extreme prototyping is a common web design approach that starts with the creation of wireframes that assign content to each digital environment. We include an XRXD version of a wireframe in Figure 8.1. In the context of XR, the wireframes are brought to life with digital assets and coded behaviors in such authoring environments as Unity or Hoverlay.

Prepare for Testing

Your team needs to come prepared for testing. The design team's main task during high-fidelity functional prototyping is to observe the users in action in order to draw actionable insight from the various choices the interactions between the users and the system make. A typical hierarchy of design knowledge is built at three stages of (1) observation (evidence), (2) insight (inference), and (3) implication (consequences for the project).

Prepare a list of questions, user tasks, and flows you will use during testing. Have a list of goals you will ask the participants to try to attain during the test. With lower fidelity of testing, the design team will need to be much more involved with guiding the users through the test. High-fidelity functionality and interactivity will free the designers to focus on observing the system–user interaction instead of worrying about how to make the prototype work. Take notes of all of your observations, and be prepared to ask users for their feedback after the test.

Testing is essential throughout the functional prototyping phase, beginning with internal testing to validate interactions, asset loading, and graphics-rendering performance on the hardware platforms you plan to deploy to. Once the internal build is stable, user testing with external participants should be conducted to identify areas where further refinement is necessary. For XR experience design, testing environments on both high-end and lower-spec devices (if targeting both) will ensure that users experience smooth navigation and interaction.

For mobile AR, this means different generations of smartphones, and for MR and VR testing on each headset you plan to deploy, make sure to achieve consistent frame rates to avoid issues like motion sickness and ensure a smooth, immersive experience. This is often the moment in the project at which some scope management must come into play, and some changes may be required in terms of graphics performance expectations depending on the limitations of your target platforms. Internal testing typically involves the development team and QA specialists who assess fundamental aspects of the prototype, such as how well assets load and render, responsiveness to user input, and adherence to intended frame rates.

For XR applications, achieving a consistent frame rate—often around 60 fps for AR and at least 90 fps for VR—is essential to avoid issues like motion sickness and ensure a smooth, immersive experience. Performance testing on various device types (e.g., different generations of smartphones for AR or various VR headsets) at this stage can reveal any hardware-specific issues and give developers early insight into potential optimization needs.

Once the internal build has been validated and is reasonably stable, end-user testing is initiated to gather insights on how real users interact with the experience. This phase will bring to light usability issues that may not have been observed in the PoC prototype and will help you refine the experience to be intuitive, responsive, and comfortable. For XR projects, testing across the range of devices you intend to deploy to is essential, especially if you intend to target both high-end and lower-end hardware. High-end devices may handle larger, more detailed assets with ease, while users of lower-spec devices might encounter lag or poor image quality, which may require optimization for a more responsive approach with different assets for different levels of device performance.

In the case of mobile AR, connectivity and processing power are often variable factors that impact performance. Many AR applications rely on device cameras, sensors, and sometimes live data feeds to render digital elements in a real-world context, which can be demanding on devices with limited processing capabilities. User testing in different settings (e.g., indoor, outdoor, and in varied lighting conditions) reveals how well the application handles these contextual changes, as variations in lighting and space can impact AR tracking and object stability. For instance, an educational AR app targeted at high school students might need to display items both in a classroom and in the student's home, requiring precise depth-sensing and color matching, both of which can differ widely between high-end and budget smartphones. In addition, the lighting conditions in a classroom and in the student's home will be different. Testing under these conditions allows your team to identify constraints as early as possible, such as slower rendering speeds on lower-spec devices or instability in AR object tracking in low-light environments.

Feedback from testing not only covers technical performance but also addresses ease of use and comfort. In AR, many applications require the user to hold their phone in up front of them. However, this can become uncomfortable quickly. In VR, how you hold the controllers or the accuracy of hand tracking when not using controllers are but two examples of issues that often come up with live user testing that may not have come up during PoC testing. Participants may report disorientation if a VR experience doesn't meet performance standards or if the spatial layout does not track in real time in a mobile AR app.

Testing also must cover accessibility issues, such as whether controls can be used by people with limited mobility or adjusting visual contrast for low-vision users. Gathering detailed, diverse feedback during this phase allows your team to make user-centered adjustments to the prototype, ensuring a more seamless experience and feeling of accomplishment with their job-to-be-done.

High-Fidelity Prototypes

High-fidelity prototypes are what functional prototyping is all about. This kind of prototype not only looks like a finished product but, more importantly, performs like a finished product which gives the team an opportunity to test all aspects of the immersive experience. The main advantage of high-fidelity prototypes is that they allow all aspects of the immersive experience to be tested with real users to check whether the interface is intuitive or if the application complies with accessibility requirements for participants with different types of impairment.

High-quality prototypes in XR have to be interactive, which means that their behaviors have to be coded. Going straight to a high-fidelity prototype, while cost and skill intensive, might actually save the design team a lot of time and effort dedicated to low- and mid-fidelity versions by offering a validated user/product feedback that can be transferred to the development team for final production a lot sooner. High-fi prototypes test not only the functionality but also UI components for spatial computing that are context intelligent, are not too abstract to understand, use real physics, and allow direct manipulation of 3D elements.

High-fidelity prototypes need to be able to sense and respond to the user's environment. The system should trigger new content automatically when the user gets near it. The best immersive UI needs to offer natural interactions that take advantage of eye tracking or well-understood hand gestures like pressing a button or rotating or grasping an object.

When it comes to menus, different forms of XR will require different approaches. For instance, because of its real world context, augmented reality is not a suitable medium to flip through complicated menus. But all forms of XR ultimately tax the user's spatial memory, so it is important to show the user only what they need to know at the moment an action or decision is required. This level of testing requires a high level of coordination between software and hardware which in turn demands a lot of engineering prowess.

Choosing the Right High-Fidelity Prototyping Tools

When it comes to high-fidelity prototypes, for most projects, you will need to implement them with specialized tools such as the game engines like Unity and Unreal Engine, which offer rendering capabilities and the frameworks for interaction to deploy to most commercial VR, MR, and AR platforms. With the growth of XR, there is also a growing number of software development kits (SDKs) that allow content creators to turn their concepts into reality and to build their immersive apps. SDKs can facilitate your development process by eliminating the need for the team to code every aspect of the application from scratch. Complex functionalities such as image tracking and plane detection come prebuilt in most SDKs.

Platform SDKs are built for specific platforms and operating systems such as Google's ARCore for Android and Apple's ARKit SDK for iOS. They contain all the necessary libraries and application programming interfaces (APIs) you will need to build AR applications for the respective Android and iOS platforms. There are also existing hardware SDKs that are tailored to specific hardware components such as Apple's Vision Pro and Meta's Quest that allow developers to build applications that take full advantage of the hardware capabilities of Vision Pro and Meta Quest headsets. These capabilities can include functionalities like tracking hand, head, and body movements, managing spatial audio and video, or handling the device inputs.

Game engine SDKs are integrated into game engines such as Unity and Unreal Engine to simplify the development of VR, AR, and MR applications. These SDKs provide a wide range of development tools that enable content makers to create 3D environments, simulate physics, script gameplay, and even incorporate AI.

One of the trickiest aspects of choosing a prototyping tool involves the risk of the chosen vendor stopping the support of their platform before the team can finalize the project. This has been surprisingly common in the rapidly evolving XR space. Microsoft halting the support of popular Maquette VR and wiARframe shutting down are perfect examples of this issue. This is why we recommend workflows that involve concept development and immersive wireframing in the prototyping tool and then transitioning the work to a game development engine like Unity or Unreal Engine for the implementation phase of the prototype.

We are weary of tools that promise end-to-end solutions, since having your deployment too dependent on one specific prototyping tool can be a risky proposition. In some instances, tools such as Adobe After Effects can be great for sharing the concept and getting feedback from a limited circle of early supporters and on social media. GitHub can be a great source of code shared by a growing community of XR content creators, designers, and developers.

Prototyping High-Fidelity Mobile AR Experiences

When it comes to prototyping mobile AR experiences, some platforms stand out in terms of usability, flexibility, and ease of deployment. These include BrandXR Studio, Adobe Aero, Hoverlay, Unity with AR Foundation, and ZapWorks Studio. Each of these tools offers its own unique set of strengths and features, which should be considered based on the specific requirements of your project. It is important to investigate carefully if a tool is going to provide you with a reliable foundation for development and deployment.

All of the tools available are, in one way or another, early-stage products given the rapidly changing technology landscape. For example, as we are finishing this book, Adobe Aero is still in beta more than 4 years since it was launched, and there are still numerous serious bugs and limitations. However, this is not unique to Adobe Aero. Most XR development tools are still in their early stages of active development, which means that they can change rapidly. Today's favorite tool may be replaced with a new one in spite of its great potential and growing adoption among content makers. We will focus our discussion on tools that are currently in use, have been around for several years, and show evidence of having enough support to last.

BrandXR Studio might stand out for its designer-friendly no-code environment, making it an attractive proposition for teams without programming backgrounds. It might be particularly well suited for marketing teams looking to deploy AR campaigns quickly. The platform allows content creators to build, publish, and measure the effectiveness of 3D AR experiences using drag-and-drop interfaces. Designers using BrandXR Studio can upload their own or open-source 3D assets and integrate other digital content such as YouTube videos to enhance the AR scenes.

Adobe Aero promises to provide user-friendly platforms for prototyping mobile AR experience. It is part of the Adobe Creative Cloud suite and supports the prototyping and development of AR apps without writing code, making it ideal for rapid prototyping. You can import 3D assets from other Adobe tools like Illustrator or from third-party software such as Blender. With Aero, you can add interactivity, animations, and spatial behaviors to objects in the augmented world. Aero supports both iOS and Android devices, making it easy to preview and test prototypes in real environments and share with stakeholders for feedback.

Hoverlay is a mobile AR platform that simplifies the process of creating, managing, and publishing mobile AR experiences via a web-based editor without the need for users to download an app. Hoverlay's platform allows you to link AR content to physical markers, such as images, or directly at specific locations, enabling participants to access your experiences by scanning these markers with their smartphones.

It's valuable for applications in tourism, marketing, public art, museums, entertainment, and retail, where people can access interactive information or digital experiences in response to real-world locations or objects. For instance, a museum might use Hoverlay to create AR exhibits, allowing visitors to scan artwork and view historical context, multimedia content, or related information layered over the physical art piece. David's students have used Hoverlay to prototype mobile spatial storytelling experiences accessible through web links or triggered by walking to designated locations in the real world (Kovacev, 2018; Fox-Sowell, 2018). Hoverlay, as an augmented reality browser platform, broadens the reach of immersive experiences, allowing creators to deploy and revise prototypes with an intuitive web-based development platform to make AR experiences available on a wide range of devices.

ZapWorks Studio is another mobile AR platform currently in release that offers web-based authoring and can be used to build and deploy AR projects that can be accessed through a browser, making it platform agnostic. The authoring tools supports both image and face tracking, as well as location-based AR, which makes it adaptable for marketing, education, and entertainment applications. The platform includes a range of interactivity tools, animations, and visual effects, allowing you to build engaging AR prototypes and iterate based on participant feedback. The analytics feature also lets you monitor user engagement, which is invaluable for refining XR experiences.

Prototyping High-Fidelity MR and VR Experiences

ShapesXR is a tool specifically designed to prototype XR experiences directly in VR, making it ideal for designers and developers working on early-stage PoC and functional prototype development. 3D prototypes that are fully spatial and collaborative can be visualized and tested in true scale and depth. The platform allows you to quickly build scenes, design user interfaces, and simulate user flows without writing code, making it highly accessible for prototyping complex spatial interactions.

ShapesXR supports collaborative design; teams can work together in real time, review designs spatially, and iterate based on live feedback. The platform also includes features like field-of-view simulation to see how the experience will appear from the participant's perspective, enabling quick adjustments to layouts and object placements. The greatest value of ShapesXR for both PoC and functional prototypes is that after the initial prototyping phase, designs and layouts created in ShapesXR can be exported for further development in tools like Unity, providing a smooth transition from immersive wireframe concepts to functional prototypes.

In Unity, you can take the foundational 3D layout from ShapesXR and enhance it with custom C# scripts to create interactive behaviors, such as object manipulation, physics-driven interactions, and spatial animations. You can use ShapesXR to establish the initial placement and layout of interactive elements and then, in Unity, add functionality like grabbing and rotating objects, creating scene transitions, and enabling user-triggered actions based on input from VR controllers or hand tracking. Unity's physics engine supports realistic simulations, so you can add gravity, collisions, or material properties to objects, creating a more immersive and responsive environment that adapts to user actions in real time.

Furthermore, deploying your functional prototype in Unity opens up possibilities for data collection and analytics, an essential function for testing and refining XR experiences. Unity supports data-tracking plugins and analytics integration, enabling you to capture and analyze user interactions, such as how often participants interact with certain objects, time spent in specific scenes, and success rates for task completion. This data provides insights into user behavior, helping you refine user flows, optimize interactions, and improve overall user engagement and presence based on measurable feedback.

Unity also provides cross-platform deployment options, enabling you to build applications for a variety of VR and AR devices, including Meta Quest, HTC Vive, Apple Vision Pro, Varjo XR-4 Series, and mobile AR platforms. Additionally, working with a development environment for the functional prototype offers access to a wide range of tools, such as AI-driven avatars, environment assets, and advanced shaders, further enhancing your prototype with detailed, realistic elements that help you create presence for the human in the XR world.

Working With Experimental Tools

The tools available to prototype and test immersive experiences are changing rapidly, as this is an area of active research and development. Depending on the technical resources available on your team, you may consider using academic/experimental tool sets if there are specific features and affordances you need that commercially available tools can't provide.

ProtoAR and XRDirector are two notable examples. ProtoAR highlights the potential of rapid physical prototyping using a combination of paper and Play-Doh with mobile cross-device authoring and interactive capture tools to generate mobile screens and AR overlays from paper sketches and 360-degree captures of clay models (Nebeling et al., 2018). Immersive authoring is increasingly popular because it allows you to design and test concurrently.

XRDirector is a collaborative authoring system that enables designers to design and test interactions using AR and VR devices to manipulate virtual objects in 3D physical space and demonstrates how designers can prototype complex scenes in a Wizard of Oz style. We anticipate systems like this will become viable commercially available products in the near future (Nebeling et al., 2020).

Developing High-Fidelity XR Functional Prototypes

For high-fidelity prototyping of XR experiences, the best tools in many cases are game-development engines like Unity and Unreal Engine that provide advanced capabilities for creating realistic, interactive, and fully immersive prototypes that will closely resemble the final product and can be quickly iterated as long as you have the programming resources on your team or the budget to contract programmers.

In comparison to prototyping tools, game engines better facilitate iterative development through more extensive asset management, version control, and performance-optimization tools. As developers refine a prototype, they can adjust graphical quality, optimize assets for performance, and test on target devices to ensure a better experience. This approach to workflow is critical in XR, where frequent adjustments to spatial layouts, interactions, and environmental details will be needed. With either Unity or Unreal Engine's comprehensive suite of development tools, your teams can refine each aspect of the prototype, from visual fidelity to interactive functionality, before moving into full production, ultimately saving time and resources in the long run. We will take a look at the most widely used tools for high-fidelity MR and VR prototyping in this section.

Unity is an industry-standard development framework widely used among XR developers that provides advanced mobile AR prototyping and deployment options with two paths: AR Foundation and the XR Interaction Toolkit. AR Foundation is Unity's framework for building AR applications across both iOS (ARKit) and Android (ARCore), providing cross-platform capabilities that will streamline your prototyping process. With Unity, you can build complex interactive systems with real-time object tracking, environmental understanding, and physics-based interactions, which are difficult to achieve with more simplified platforms like Aero and Hoverlay.

Unity also supports scripting, asset importing, and integration with analytics tools, enabling you to test not only technical performance but also user engagement and interaction patterns, making it a strong contender for high-fidelity AR prototypes. XR Interaction Toolkit and AR Foundation serve different purposes within the XR development ecosystem.

Like Unity, Unreal Engine supports AR, MR, and VR cross-platform compatibility through integrations with ARKit (iOS), ARCore (Android), and various hardware platforms such as Meta Quest, Varjo XR-4 Series, Apple Vision Pro, HTC Vive, and Windows Mixed Reality. Unreal Engine also integrates seamlessly with VR and AR hardware, enabling real-world testing of prototypes in immersive conditions, and includes an asset marketplace with high-quality assets for rapid development.

While AR Foundation is a cross-platform framework specifically designed for building AR applications, XR Interaction Toolkit is a more general-purpose toolkit designed to handle interactions in various types of XR experiences, including AR and VR. Here is a breakdown of how these two frameworks differ in regard to their unique features and some suggestions for when to use each.

AR Foundation is Unity's primary framework for developing AR applications that can run on both iOS (ARKit) and Android (ARCore), making it a strong contender for mobile AR development. It provides foundational AR functionalities including Plane Detection (recognizing flat surfaces like floors, walls, and tables, allowing for object placement on real-world surfaces), Image and Object Tracking (detecting and tracking specific images or 3D objects, which can be used for marker-based AR experiences), Face Tracking (detecting and tracking facial features, useful for face filters and augmented facial effects), and Environment Understanding (estimating light intensity and spatial characteristics, allowing digital content to realistically blend with real-world lighting and conditions).

AR Foundation enables the developers on your team to write applications that can adapt to both Android and iOS environments with a single codebase, which is ideal for developing core AR functionalities like overlaying virtual objects on physical surfaces, integrating interactive markers, or creating face-tracking applications. We recommend you use AR Foundation when your project focuses specifically on AR with features like plane detection, image tracking, or environmental sensing, as it's optimized for cross-platform AR and provides the basic building blocks for these functionalities.

The XR Interaction Toolkit is a higher-level framework that focuses on adding user interactions across the XR spectrum, supporting AR, MR, and VR. Its primary role is to provide out-of-the-box interaction components that make it easier to create interactive experiences in Unity and simplify the code needed to implement these interactions. Some key features of the XR Interaction Toolkit include: Grab and Manipulate Objects (prebuilt components for picking up, rotating, scaling, and moving objects in 3D space), UI Interactions (built-in support for gaze-based or controller-based UI interactions, such as pressing buttons or toggling switches, which are essential for MR and VR but also applicable in AR), Teleportation and Locomotion (components for moving around in the augmented or virtual space through teleportation or continuous locomotion), and Cross-Platform Input Handling (input framework that supports interactions across various XR devices, simplifying input handling and interaction consistency).

The XR Interaction Toolkit is particularly useful if you need to deploy your experience across the modalities of AR and MR and even VR. Both AR Foundation and XR Interaction Toolkit can be used together in projects to combine AR Foundation's environmental sensing with XR Interaction Toolkit's interactive elements, which is handy when you need AR Foundation to detect a table surface and place an object, then use the XR Interaction Toolkit to allow the user to pick up and manipulate that object in a manner that is consistent across the AR and MR version of your application.

Unreal Engine by Epic Games is another choice for high-fidelity XR prototyping, especially for projects that demand photorealistic graphics and complex environments. Unreal's advanced rendering capabilities, including real-time ray tracing, make it ideal for high-end VR applications like architectural visualization, simulations, and gaming.

One of Unreal Engine's standout features for XR prototyping is the Blueprints visual scripting system, which empowers designers and developers to create complex interactions and behaviors without the need for deep programming knowledge. This visual scripting system makes it possible to rapidly prototype interactions, such as object manipulation, teleportation, or gesture-based controls, by connecting logic nodes within the editor. By prototyping these interactions in

Blueprints, your team can test and iterate quickly on functionality to make sure the user interactions feel intuitive and immersive.

Unreal Engine is known for its high-quality rendering and lighting systems, which provide a high level of realism in VR environments, making it an ideal choice for experiences that require photorealistic visuals. For example, in architectural visualization, Unreal's real-time ray tracing and advanced material shaders allow the user to make lifelike prototypes of buildings, interiors, and landscapes possible.

Like Unity, Unreal Engine supports AR, MR, and VR cross-platform compatibility through integrations with ARKit (iOS), ARCore (Android), and various hardware platforms such as Meta Quest, Varjo XR-4 Series, Apple Vision Pro, HTC Vive, and Windows Mixed Reality. Unreal Engine also integrates seamlessly with VR and AR hardware, enabling real-world testing of prototypes in immersive conditions, and includes an asset marketplace with high-quality assets for rapid development.

OpenXR is an open standard from the Khronos Group that enables cross-platform XR development, supporting multimodal input across various VR and MR hardware. It is compatible with Unity, Unreal Engine, and other platforms, making it a flexible option if you are seeking device interoperability. It abstracts device-specific input details, allowing you to create consistent interaction experiences that function across a range of devices.

This standard supports inputs such as hand tracking, gaze tracking, and controller-based input. For example, you can prototype a VR app in which users navigate with hand gestures and gaze across multiple headsets without changing code significantly.

The compatibility of OpenXR with a broad array of hardware increases the probability that your prototype is scalable, consistent, and easier to deploy across multiple XR devices. The OpenXR Mobile SDK provides the resources to use the OpenXR API for native development of Meta Quest and Quest 2. The Unity OpenXR Meta package allows developers to support Meta Quest devices in their AR Foundation projects. Scene functionality is currently supported on the Meta Quest 2, Meta Quest Pro, or Meta Quest 3 operating software. As of now, OpenXR simplifies VR development by allowing creative teams to reuse code across platforms while giving up some vendor-specific features. There is always a tradeoff.

Apple Xcode/Vision OS. Apple's Vision Pro is designed specifically for immersive MR experiences, and developing for this device requires using either Unity with the Apple VisionOS XR Plugin for AR Foundation to support development for the Apple Vision Pro, which also requires a Unity Pro subscription. Apple Vision Pro development is not supported in the free version of Unity.

The other option is working with Xcode, Apple's software development environment for iOS, MacOS, and VisionOS, providing full access to Apple's advanced hardware capabilities, such as high-resolution display technology, LiDAR, and spatial audio. Xcode allows developers to write MR applications in Swift or Objective-C along with access to Apple's ecosystem and enables fine-tuned optimizations for iOS and VisionOS.

One of the core components of VisionOS is the RealityKit framework, which supports real-time rendering, object placement, and interaction within an

MR environment. RealityKit integrates seamlessly with Vision Pro's hardware, providing accurate environment mapping, surface detection, and lighting estimation. Using VisionOS, you can create MR applications that blend virtual objects with real-world surfaces seamlessly, allowing users to place digital objects in physical spaces with highly accurate scaling and positioning.

VisionOS also offers multimodal input capabilities through eye tracking, hand tracking, and voice recognition. Apple Vision Pro's eye tracking can be used for gaze-based interaction, where users select options or interact with objects simply by looking at them. Combining this with hand gestures or voice commands, you can create a hands-free experience that leverages the full sensor suite of the Vision Pro. Using Xcode for Vision Pro gives you full access to Apple's customizations and optimizations. On the other hand, if your team is familiar with Unity and you need to maintain some cross-platform capabilities, then using Unity with the Apple VisionOS XR Plugin for AR Foundation might be the better way to go.

Meta Developer Tools. If you are targeting Meta's Quest platform, you may prefer to work directly with Meta's developer tools, which offer deep integration with Quest hardware features and the Presence Platform, Meta's comprehensive toolkit for creating MR and VR experiences on the Quest line. The Presence Platform includes several key components, such as Spatial Anchors, Scene Understanding, Hand Tracking, and Voice Interaction, all of which are designed to maximize the capabilities of the advanced MR functionality of Quest 3 and later models.

By developing directly with Meta's SDKs, you gain access to high-level APIs that facilitate object tracking, plane detection, and mixed reality passthrough, which are essential for building MR applications that feel natural and intuitive on the Quest 3 and later models. One notable feature of the Presence Platform is Scene Understanding, which enables the Quest to map and interpret the physical environment, allowing virtual objects to interact dynamically with real-world surfaces. For example, you can use Scene Understanding to create MR applications in which digital objects rest convincingly on tables or floors, enhancing immersion and spatial coherence. Additionally, the Passthrough API provides access to the high-resolution color passthrough on newer models, enabling you to create experiences in which participants can interact with both virtual and real elements simultaneously.

Meta's developer tools also support hand tracking, which offers more precise tracking of hand gestures and finer control over interactions. This advanced hand tracking allows developers to create MR applications that rely solely on hand gestures for input, making it possible to design completely hands-free experiences.

Meta's SDK includes integrations for Voice SDK, which allows developers to implement voice commands for actions, enhancing accessibility and ease of use. Newer Quest hardware also supports spatial audio, so you are able to program immersive audio that changes based on user position and movement, creating more engaging presence in MR experiences.

Meta developer tools are designed to integrate seamlessly with both Unity and Unreal Engine, offering an alternative to OpenXR and the XR Interaction Toolkit for applications that do not require cross-platform deployment. Using Meta's tools enables you to use Meta's specialized features for more robust device-specific experiences, but you give up the ability to easily deploy to multiple platforms. Meta regularly updates Unity SDKs and offers detailed documentation and support, making Unity a good choice for creating applications using Meta-specific MR functionalities.

Unreal Engine has slightly fewer prebuilt integrations with Meta's tools compared to Unity, but it still fully supports Meta's developer features through plugins and the OpenXR framework, allowing you to use Passthrough, Spatial Anchors, and hand tracking in Unreal projects.

If you are working with Unreal Blueprints, Meta provides plugins that integrate its features into the visual scripting environment. Unreal's high-fidelity rendering combined with Meta's tracking and passthrough capabilities makes it a good choice for visually demanding MR applications in design visualization, industrial simulation, and immersive storytelling.

High-end VR devices. Developing for high-end headsets like the Varjo XR-4 Series, for example, requires working with a platform that can leverage advanced capabilities, including ultra-high-resolution displays, precise eye tracking, and high-fidelity mixed reality passthrough.[1] These devices are designed for demanding applications in fields such as medical training, industrial design, aerospace, and automotive, where visual fidelity and immersive realism are paramount.

For functional prototype development on the Varjo XR-4 Series (and similarly with other high-end devices), there are three options: Unreal Engine, Unity, and Direct Access, Varjo's own SDK outside of a game engine. Unreal Engine is widely used for developing functional prototypes for the Varjo XR-4 due to its high-fidelity rendering capabilities. Unreal's advanced rendering engine, particularly with real-time ray tracing, can take full advantage of the capabilities of advanced headsets.

Unreal Engine's support for Varjo-specific plugins makes it possible to optimize the rendering pipeline and take advantage of Varjo's unique features, like bionic display technology, which provides super-crisp visuals at the center of the user's field of view. This is ideal for functional prototypes in which clarity and visual detail are critical, such as in medical simulations, where people need to view highly detailed anatomical structures.

Unity is also widely used for developing high-end VR applications, particularly for its large developer community, and extensive library of tools and plugins. Unity's Varjo Plugin provides direct support for Varjo-specific features like high-resolution rendering, eye tracking, and depth sensing, allowing you to create experiences that fully utilize the hardware. Unity is popular for applications that require multiplatform deployment, enabling prototypes to be easily transferred across headsets for broader testing and deployment.

Varjo's SDK and Varjo Lab Tools are designed for developers creating functional prototypes and deployments that fully leverage the range of Varjo XR-4's capabilities. The Varjo SDK provides access to device-specific features such as foveated rendering (leveraging eye tracking to increase rendering detail only where the user is looking), high-resolution color passthrough, and low-latency depth sensing. Using the Varjo SDK, you can create prototypes that adjust rendering quality dynamically based on the user's gaze direction. For example, in a medical surgical simulation, foveated rendering allows the dashboard and controls in the vehicle to appear ultra-crisp when viewed directly, while peripheral details require less rendering power, optimizing performance. This eye tracking feature is particularly useful for functional prototyping in sectors like aerospace and medicine, where visual precision in specific areas is critical.

The Five Steps in Creating a Functional Prototype

As a relatively new and interdisciplinary field, XR media design borrows methods from a multitude of disciplines such as human–computer interaction (HCI), UX/UI, software engineering, industrial design, and so on. Functional prototyping is a core component in most of these fields. Design practitioners typically follow five possible stages in creating a functional prototype. These states mirror the PoC process outlined in Chapter 6. However, the goal with the functional prototype is to build a more complete experience that can stand on its own as a step toward a marketable minimally viable product (MVP).

Step 1: Immersive Wireframing

A first common step is to use wireframing, flowcharts, and other "paper" tools to conceptualize the system and user interface.

Step 2: Virtual Prototyping

In Step 2, designers transition from a predominantly "paper" concept to software and hardware design by creating a virtual prototype. Virtual prototyping is a relatively new methodology of combining mechanical modeling and digital simulation with control design to increase the speed and efficiency of designing and prototyping complex software and hardware systems and devices. Virtual prototyping allows the development team to connect software design and control algorithms to 3D CAD mechanical models and digital assets to test the mechanics of the immersive system before building any physical prototypes.

Many XR developers use game engines such as Unity, Unreal Engine, GameMaker, or the open-source Godot Engine to create virtual prototypes of their immersive products. Virtual prototyping can lower the cost and the risk associated with immersive product design by improving the development team's understanding of user requirements, speeding up the development process, streamlining debugging, and shortening the time to market by providing tangible customer feedback.

Step 3: Building the Physical Substrate for Testing and Feedback

The third stage involves building the physical substrate and adding inputs/outputs (I/O). Adding I/O to your prototype is a major step toward creating a truly functional immersive system. While creating wireframes, flowcharts, and other forms of paper or digital design or even simulating the design in a virtual environment are largely conceptual exercises, adding sensory input and control output tests if the XR application can truly be implemented in the real world. The custom nature of sensor integration can make the I/O stage a daunting engineering task because of the difficulty in predicting the total cost of the time and resources needed to construct a custom I/O solution for a truly immersive product.

Step 4: Coding

Developing your code is the fourth step in the functional prototyping process. Transforming pencil-and-paper-system behaviors and algorithms into a robust, well-tested, and easy-to-use XR product requires specialized expertise. The world of XR is a blend of various, mostly incompatible operating system standards that include Apple's VisionOS and Meta Horizon OS, and languages, including Apple's Swift, XML, C#, and JavaScript. VisionOS is built on top of the well-known frameworks and technologies used for iOS and iPad apps.

To develop MR apps for Vision Pro, you will need to use Apple's proprietary SwiftUI, UIKit, ARKit, and RealityKit. Apple also supports developers with an integrated development environment (IDE) known as Xcode, which comprises a set of tools that help you build Apple-related apps. Apple's new Reality Composer Pro tool facilitates the development of spatial computing apps by providing an environment for designing, editing, and integrating 3D content.

Meta's Quest uses a number of software languages, including C#, JavaScript, and XML. Developers working with this platform need to be familiar with its four languages to develop MR applications for the platform, but many Quest developers opt for using Unity due to its robust features, excellent optimization, easier resource requirement, large user community offering support, and an extensive array of third-party packages for performing tasks that the core game engine does not support.

Step 5: Deployment

Progressing through the four-step functional prototyping process from concept to a paper/digital design, from a virtual prototype to a setup with fully functioning inputs and outputs, and finally developing the code to run the entire system is only a prelude to building a releasable product. Ideally, the development team designs and builds a functional prototype in a way that allows you to actually produce, deploy, and scale your XR app to a wider audience. Seasoned developers design and build prototypes with deployment in mind. The key to making a smooth transition from a functional prototype to a market-ready product lies in choosing the right tools that give you both the flexibility and capabilities necessary to effectively test your concept and that are robust enough to take to market. This means using industry standards in both software and hardware as early on in the prototyping process as possible.

Defining the Degrees of Freedom (DoF): Wireframing Your Immersive Experience

As discussed in Chapter 6, wireframes are an excellent interim step between a XR concept and the final product or experience. Creating wireframes is an equivalent to putting together a blueprint of the XR application. In UX/UI design, wireframes are commonly used to lay out content and functionality on web pages, taking into account user persona aptitudes, needs, goals, and the desired user journeys.

Peter has adapted a traditional storyboarding template used in film and video production to the wireframing of spatial computing content. The template seen in Figure 8.1 sets the scene by following the user, environment, objects, interactions framework to conceptualize the participant's journey through the immersive experience in space and time. The process involves planning the virtual space with maximum degrees of freedom (DoF), which refers to the number of ways an object can move in the virtual space. Granting users the freedom to move around a virtual space is closely linked to the system's ability to track the movements of the participant using the AR, MR, or VR headset.

Defining the Virtual Space

There are two types of motion to consider when building and prototyping virtual environments. The first of these, rotation, offers three degrees of freedom that include roll, pitch, and yaw. An airplane roll can occur when one wing of the plane is higher than the other. Pitch occurs when the nose or the tail of the airplane is higher than the other. Yaw occurs when the plane rotates around its vertical axis when it turns left or right. When a system can track roll, pitch, and yaw, it is described as having three degrees of freedom or 3DoF (IEEE Digital Reality, 2022).

Translation is the second type of motion that provides the DoF within virtual environments. Translation represents the familiar everyday movements that do not require rotation, such as moving forward or backward, left and right, or up and down. Translation in 3D space is often described in terms of an x, y, and z axis. An agent or an object that can move forward and backward along a single axis, like a train on tracks, can be described as having one degree of freedom. An object that can move forward and backward and also left and right has two degrees of freedom. Objects that can move up and down in addition to moving forward, backward, left, and right have three degrees of freedom.

A system that can track position in all three directions can be described as a 3DoF system. An object that can roll, pitch, and yaw in a virtual space, can potentially also move forward and backward, left and right, and up and down, giving it three degrees of rotational freedom and three degrees of translational freedom, or a total of six degrees of freedom (6DoF). An XR system that can track agents and objects with 6DoF can be considered to be a 6DoF system. The XRXD wireframe template accounts for the six degrees in the freedom of movement by prompting the development team to consider that the audience members can look up and down and around them as if they were inside of a sphere rather than seated in front of a computer or a movie screen.

Placing Objects in the Virtual Space

After defining the virtual space, we move on to listing and placing the objects or touchpoints the participants can interact with in the environment. As with the UX/

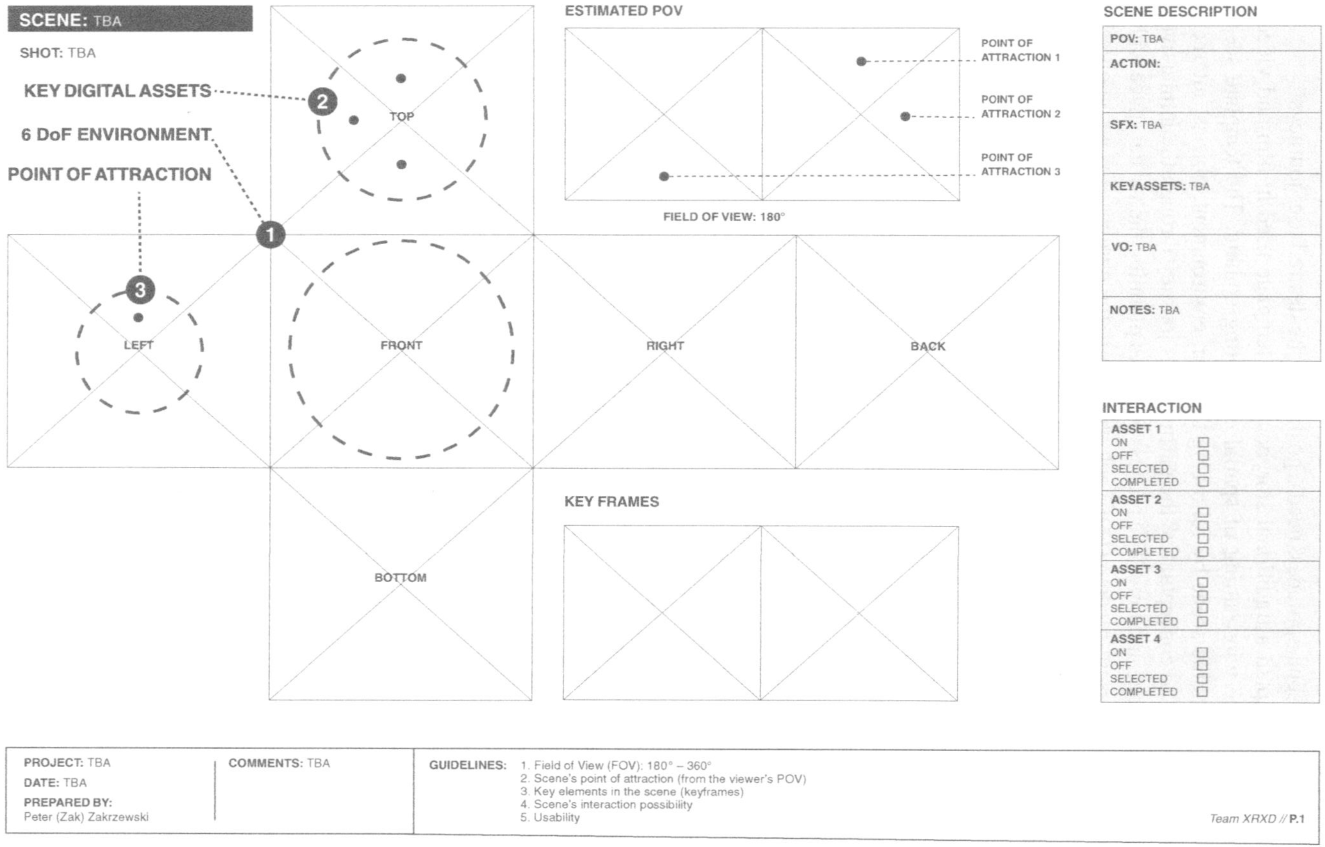

Figure 8.1 The XRXD wireframing template

UI wireframes, the same design principles apply to the placement of spatial content. All objects (touchpoints) within the immersive environment are opportunities for interaction, but some will be more important based on the user goals or the designers' narrative objectives. This means that the participants need to be guided through spatial content through an application of such design principles as focal point, hierarchy, flow, and balance. The most important touchpoints need to be placed in the places where the participants are most likely to look, unless hiding objects is part of the game mechanics of the XR experience.

Other design basics such as size, position, color, and shape will also be very useful in laying out the digital assets within the XR space. Once placed within the XRXD wireframes, the digital assets can be listed for production and acquisition.

Functional Prototype: A Case Study

Reality's Fugue: Genesis is a virtual reality experience that David co-produced and co-directed with F. Samuel Brainard. The production is inspired by Brainard's book, *Reality's Fugue: Reconciling Worldviews in Philosophy, Science, and Religion* (Brainard, 2017). Created in collaboration with a group of artists and two VR developers, the experience reimagines the Genesis creation story to explore collective consciousness and shared reality and employs VR to illustrate how individual viewpoints work together to create intersubjective reality that shapes our public world.

In the first act, the participant explores the emergence of a shared world through four distinct perspectives, each associated with a foundational element of creation: light (vs. darkness), differentiation (between light and darkness), form, and time. Guided by visual and auditory cues, the participant manipulates light to create a light–darkness boundary, builds form, and adds movement and cycles. These actions symbolize the interplay of individual and collective awareness.

In the second act, the participant engages with a four-faced entity that represents a unity of views but quickly breaks apart into just one view to illustrate the limitations of individual perspectives. The participant then must merge with other perspectives to experience the complete landscape and regain a holistic view of the world. The experience is staged in a calm, meditation-like setting in which the participant encounters interactive elements, ambient soundscapes, and shifting visuals using their hands for interaction and locomotion with gentle gestures. The narrative and interaction use the virtual environment to manifest the idea that our collective world is a product of both unity and diversity of perspectives.

Challenges Faced in Moving From Proof of Concept to First Functional Prototype

Moving from our initial PoC to a functional prototype marked a significant transition, as the focus on the project shifted from validating key interactions to creating a fully interactive and immersive version of the experience. During the PoC phase, we tested the feasibility of the locomotion, tested basic interactions, and created a baseline set of 3D assets, which together demonstrated the core elements required for a viable experience.

In contrast, our functional prototype required expanding the team to include two developers, who used the Unity game engine to create the technical components required to orchestrate the animations, interactivity, and sound elements. Working with developers with experience in optimizing system performance provided smoother interactions for participants. It also enables developers to test complex features like real-time data integration, multiuser interactions, or advanced animations, which might have been outside the scope of a PoC.

As you can imagine, this concept hit some technology brick walls as we moved the project out of the proof-of-concept phase and built the first functional prototype, which was exhibited at the New Hope Arts Center in New Hope, Pennsylvania, in January 2024, and exhibited at the New Media Exhibition at the annual conference of the University Film and Video Association in July 2024. We collected user data using a modified form of the system usability scale (mSUS) survey to assess the effectiveness of the experience and determine the level of presence participants encountered in the experience.

As this book goes to press, the *Reality's Fugue: Genesis* team is working on updating the experience based on user feedback and observation and fine-tuning the interactions, with the goal of completing the iterated MVP in 2025 in anticipation of a wider release.

Interaction Challenges

We used hand tracking as the primary input (no physical controllers). Given that this is computer vision based, the "field of view" is more limited than hand controllers, and the tracking resolution is also limited (finger data is always slightly jittery and requires filtering in code, though some platforms like the Apple Vision Pro do this for you, but that was not an option on the Meta Quest 2). Hand tracking is tricky because it (1) requires a well-lit room (or tracking will severely degrade); (2) it is more susceptible to physical occlusion (one hand covering up another hand from the view of the headset-mounted camera); and (3) it has extremely limited discrete inputs.

The primary reliable inputs we had access to were: (1) whether fingers and thumb are pinching, or (2) the transform of every joint in the hands. Aside from these, there was little else we could do for interaction. All interactions in the experience were limited to (1) finger point and hold (menus), (2) finger point—immediate trigger (light spheres being revealed in sky), (3) pinch and pull (locomotion), and (4) head gaze and hold (look down to force reset). When working with hand tracking, it is important to keep in mind that there is a lack of tactile feedback which is standard with handheld controllers.

Meeting the Challenge of Working With the Limited GPU Resources of Standalone VR

Most of the 3D assets and visual effects created for the PoC had to be reworked for the functional prototype. When working with a stand-alone headset like the

Meta Quest, you are limited by the graphics performance of a mobile GPU, and therefore, rendering performance is always a major bottleneck. It is important to remember that these are effectively mid-level mobile phones with a lot of compact optics strapped on. The baseline rendering demands for VR of any sort are intensive. Frame rates and resolutions must be kept high; otherwise, you risk delivering not merely a less-than stellar experience but one that could cause actual physical discomfort and nausea if not maintained. A Meta Quest 2 is effectively running three and a half full HDTVs at a frame rate nearly 3 times the standard acceptable rate of most major console games (72–90 fps on Quest vs 30fps for a typical console game).

Given these limitations, our developer engaged in performance profiling. This is an important step to ensure meeting rendering performance and framerate requirements. We used the Meta OVR Metrics application in-headset as the primary performance-monitoring tool. On the development headset, at nearly all times during production, the OVR Metrics dash was overlaid on top of all standalone builds with constantly updated values. This was kept in the upper field of vision at all times during development and playtesting. This was done primarily for monitoring for FPS dips, CPU and GPU usage (in percentage), and GPU level (the Quest 2 has a GPU that automatically adjusts energy consumption based on demand: 1 being least, 4 being highest). In nearly all projects, Patrick, our developer, has built or tested on this device, the Quest remains between 3 and 4 but most typically at the max output of 4.

Initial performance concerns were related directly to our inability to reproduce the original Tilt Brush sculptures that our collaborating artists had created and we had working along with the swimming locomotion in Unity using a PC-based VR prototype. We could not figure out a way to use these graphics in any way that didn't tank performance as soon as you got near a piece of a sculpture that used a lot of transparency-based shaders, so we had to create all-new graphics for the functional prototype. We had originally targeted a PC-based platform for exhibition, but when we changed our target platform to four Meta Quest 2 headsets for budgetary reasons, we had no choice but to rework the graphical elements to work within the limitations of the GPU available on the Meta Quest 2.

One of the signature graphical elements in *Reality's Fugue: Genesis* is volumetric lighting beams from hands, but we had to minimize expensive transparency draw calls, which incur what is known as "overdraw" because the same portions of the screen are drawn on repeatedly with each new transparent element. Most mobile VR experiences rely on static or "baked" lighting (if they use lighting at all).

For *Reality's Fugue: Genesis*, it was requested that the sun move and light react to this. A single distant sun is quite performant, but shadows are disabled for most objects. There are non–shadow-casting dynamic point lights attached to the mimic heads as well, which is also advised against, but they are limited such that they only cast light onto their associated mask in the animation. And then we had to deal with multiple oval mirrors that are shown in one segment and required a very "expensive" trick. These "mirrors" are each responsible for another (low-resolution) frame render from a virtual camera placed behind each mirror. It was important to make sure the volumetric hand beams were disabled by the time the mirrors were enabled, as performance would often stutter then.

Unlike many VR projects that run on the Meta Quest 2, one positive note here is that we were never bothered with approaching geometry vertex limits. Without the spatial artworks developed for the PoC in Open Brush that had high vertex counts, we specified fewer geometric objects for the functional prototype, and those that remained were all reasonably low-poly assets. We had, thankfully, taken this into account when we recreated the new graphics for the functional prototype after shifting our target platform from a desktop PC with a powerful GPU card tethered to a headset to the standalone Meta Quest 2, which sports a tiny GPU in comparison to PC gaming desktops.

One of the animations involves mimicking, and this system seemed fairly simple at first, until we started moving the mirrored Buddha heads in space so that they would turn. Maintaining a relative head-to-hand position on the duplicated heads proved to be a real challenge, but it was ultimately solved.

Locomotion Challenges

The locomotion was implemented as a form of swimming based on hand movements. The implementation was unique, and there was little access to examples of similarly implemented locomotion systems, requiring a lot of custom code to be written by our developers to get it working smoothly. It still needs some work as we work on the second functional prototype.

Reality's Fugue: Genesis was originally designed to be experienced in a swivel chair, so that users could turn freely in 360 degrees and we would only need to program the ability for the user to initiate a move "forward" in whatever direction they were facing. For many reasons, primarily logistical, this was scrapped, and a late technical addition to the project was updating the hand interaction system to enable rotating the user left and right as well as moving positionally.

The navigation code was the part that survived most intact from the PoC prototype to the functional prototype. As such, it was easy to modify to implement turning with fingers toward the end of the project. This points out the value of working in a game engine during the PoC phase.

It is important to be willing to start from scratch whenever you need to (like we did with 75% of the PoC code). However, when you write modular functions, it's often possible to repurpose portions of code, which is particularly valuable on artistic projects like *Reality's Fugue: Genesis* that have limited development budgets. Due to issues with locomotion control and precision issues with objects in the world that the participant interacts with, it was decided that locomotion would only take place on a horizontal plane, and vertical motion would be nullified. While the vertical motion we had working in the PoC prototype was interesting and well received by the users, it was the right choice to make and seemed to significantly improve user experience for the functional prototype.

Timelines

The entire experience is run from a total of two linear "timelines," which is what Unity calls its sequencing and animation system that functions similarly in some ways to a cross between a nonlinear video editor (NLE) and the dope sheet of an animation DCC app. Objects in the scene are enabled/disabled, animated, and synced with sound and narration via a series of tracks. Along with this, custom scripted functions can be called on any object (or series of objects) at certain frames in this system via "cues": place a cue at a frame and pick which functions of which objects should be triggered at that time. These cues are responsible for many things: changing skyboxes; enabling and disabling locomotion and other interactions; playing audio narration if the user has not moved after some time; instantaneously moving the player's rig; material changes on selected objects in the virtual world; and more.

Music

The first functional prototype is tied to the flow of the prerecorded music track. However, this is a limitation we plan to eliminate in the second functional prototype, and to that end, our composer delivered all of the music with individual tracks so we can remix the composition as a spatial sound field and dynamically place elements and adjust the mix based on the scenario and the participant's activities. A reactive music track will improve some sections of this work, but for the first functional prototype, it was one of many healthy scope limiters to plan for the future by having the composer deliver all of the original tracks of the musical composition along with a stereo mix but deliver on a deadline with a stereo mix.

Working Within Your Limitations

Meta does not allow developers to simply override OS functions such as disabling access to the home menu. There are enterprise solutions available for this, but our team did not have experience with them. We used the Unity XR Interaction Toolkit and OpenXR, which, for whatever reason, does not work with the Meta Quest to send an event when the headset has been removed. This is a known issue that we hope is resolved in time for our second functional prototype.

One solution would be to rework the experience entirely with the Quest SDK in Unity instead of OpenXR, which would formally lock down the experience to the Quest (which is the only device we are targeting, and we are targeting the Meta Quest 3 for the second functional prototype), but in return, it would provide reliable direct access to normal hardware events and even a number of friendly utilities for working with more sophisticated Quest APIs (gestures, spatial audio, etc.).

It's important to appreciate that any code and technology platform decisions need to be revisited at each iteration of your PoC and functional prototypes, and you must be prepared to toss components aside for new ones in order to achieve your experience goals. It is vital to budget for this and set your expectations and not get too precious about specific components. Embrace change; that is the nature of XR experience design and the technology that is constantly shifting under our feet. The takeaway is that it is important to bring developers in as early as possible in order to have them provide guidance on the creation of assets and the design of interactions and implementation to ensure they will work with the intended platform.

The transition from PoC to functional prototype on any project is the time to reassess priorities and bring the learnings from the PoC together with the realities of the functional prototype deployment platform into the process to ensure a viable project.

Summary

In this chapter, we described the process of creating functional prototypes. We argued that a functional prototype is one of the most efficient and accurate methods of testing key hypotheses and validating fundamental assumptions about the desirability and feasibility of your entire XR app concept. It is also the best way to move from the ideation to the production phase of the project.

As in Chapter 6, we made a distinction between PoCs, which, for the most part, are selective, internally facing tests, and functional prototypes, which are more refined and elaborate simulations of the final product that give the media makers a clear view of the XR systems' real-world operation and allow the team to obtain realistic user feedback. The chapter outlined various approaches to functional prototype testing that range from low-fi paper-and-Play-Doh prototypes to fully functioning hardware and software versions of the XR product.

We also shared the important methods, tips, and planning processes needed for the creation of effective functional prototypes.

Testing the XR system as a functional prototype brings the development team another step closer to the final product. But before your application can be released to the world, more members of the core audience have to be invited to put the system through its paces. Playtesting the XR system prototype is the topic of our next chapter.

Note

1 At the time of this writing, according to Varjo, their XR-4 Series headset featured the highest-resolution MR displays ever made. They are based on custom full-dome aspheric lenses capable of showcasing 28 million pixels of pure visual fidelity with industrial-grade color accuracy, 200 nits of brightness, and local dimming for high-quality contrast (Varjo.com).

References

Brainard, F. S. (2017). *Reality's fugue: Reconciling worldviews in philosophy, religion, and science*. Penn State University Press. http://doi.org/10.1515/9780271080574

Fox-Sowell, S. (2018, July 13). Northeastern students 'change the world' for nonprofits in Toronto with augmented reality. *Northeastern Global News*. https://news.northeastern.edu/2018/07/13/northeastern-students-change-the-world-for-nonprofits-in-toronto-with-augmented-reality/

IEEE Digital Reality. (2022). The differences between 3DoF and 6DoF, and why. *The 2022 IEEE 2nd international conference on intelligent reality (ICIR)*. https://digitalreality.ieee.org/publications/degrees-of-freedom

Kovacev, M. (2018, September 1). Northeastern students pioneering storytelling through AR. *Hoverlay* (Blog). www.hoverlay.com/pioneering-storytelling-through-augmented-reality/

Nebeling, M., Lewis, K., Chang, Y.-C., Zhu, L., Chung, M., Wang, P., & Nebeling, J. (2020). XRDirector: A role-based collaborative immersive authoring system. *Proceedings of the 2020 CHI conference on human factors in computing systems*, 1–12. https://doi.org/10.1145/3313831.3376637

Nebeling, M., Nebeling, J., Yu, A., & Rumble, R. (2018). ProtoAR: Rapid physical-digital prototyping of mobile augmented reality applications. *Proceedings of the 2018 CHI conference on human factors in computing systems*, 1–12. https://doi.org/10.1145/3173574.3173927

Pentland, A. (2010). *Honest signals: How they shape our world*. The MIT Press.

Steffen, T. (2023, May 17). Fully functional prototype in game design. *Medium*. https://thomassteffen.medium.com/fully-functional-prototype-in-game-design-460e9daa0ac1

9 Playtesting the Prototype

Superminds

We strongly believe that the concept of "Superminds" proposed by Thomas Malone, the director of MIT's Center for Collective Intelligence, is essential to understanding the complex reality of human+computer systems. Malone suggests that collective intelligence is an emergent property of a group of individuals acting collectively in ways that seem intelligent. This can include bees, ants, and groups of people (Guszcza & Schwartz, 2019). In the realm of XR, Superminds also include people and computers working together toward a chosen purpose.

The advanced computation, the hardware, and the software behind immersive media are such incredible human achievements that we would be justified in referring to them as being superbly impressive, but as super as they are, they are not the minds in the Superminds. The magic of XR applications that tell immersive narratives, enable gameplay, or 3D data explorations only happens once humans enter the system. Creating the seamless connection between people, and between people and advanced computational resources is the very definition of XRXD. Finding the right user–system fit can make a difference between creating a Supermind and a super flop.

The Emergence of Presence: The Four Layers of XRXD

As we argue in Section I, computational world-making is one of the most appealing aspects of XR media. XR world-making depends on the ability of the participants to fully immerse themselves in the system to the point of reaching the subjective state of mediated presence. In the context of XR experience design, mediated presence is an emergent property created at the confluence of the mindfully curated set of subjective stimuli created to elicit desired subjective states in the audience. The process of emergence within immersive sociotechnical systems occurs only after human participants enter the system and start interacting with the inscribed XR world.

As the immersive worlds become more complex and more tightly yoked with the human nervous system, it will become increasingly important to firmly keep in mind the key insight of the human-in-the-loop methodology—human actors possess an intrinsic agency of their own. This means that they navigate the immersive

DOI: 10.4324/9781003442462-15

worlds and interact with the objects in those worlds based on their own diverse mental models, skill levels, and goals and not by the way the system was originally inscribed by the design team.

The concept of emergence is frequently used to refer to the appearance of unplanned organized behavior in a complex system. Emergence has applications to the design and analysis of many kinds of systems in the social and physical sciences. The concept is based on the insight that behaviors produced by complex systems are not apparent from the analysis of their parts in isolation. Complex system behaviors result from the interactions, dependencies, and relationships that arise when the component parts are placed together in a system.

Understanding emergence is critical to the design and iteration of complex sociotechnical applications that mix human actors and nonhuman actants in ways that enhance the dynamic action potential of both. We propose that immersive experience designers need to be prepared to deal with two types of emergences.

The first one occurs at a local level for each audience member individually after they enter an immersive system. The subjective conscious experience of presence in the XR world is itself an emergent product of the participant interacting with the immersive environment. The emergence of XR presence occurs after the design team has built the XR system and after the audience has been attracted to it and immersed in it.

The second emergence, the kind of Supermind emergence Malone talks about, occurs at the global level when the XR system produces its system-level behavior, which might or might not coincide with its intended function. This emergent global behavior gives rise to the possibility of designing collective intelligence systems that create opportunities for purposeful collective action. For instance, exploration of the lunar surface is a long-term project involving scores of scientists and engineers across many organizations and countries, working with cutting-edge technologies, including XR, united by a single purpose of advancing space exploration. Advances in space exploration, like most scientific endeavors today, emerge out of microinteractions between scores of individuals leveraging all kinds of advanced technologies ranging from XR to robotics, connected via networked computers that allow them to exchange information.

Conceptually, Supermind-inducing immersive projects can go through three distinct phases. In the first, the design team lays the foundation by building the immersive system capable of supporting the emergence of presence. In the second phase, designers attract the audience to the experience and invite them to participate. The main focus of the "play" phase is to create an alignment between the system and an audience that supports the emergence of presence, leading to the collective outcome or the emergent system behavior. In the third phase, designers analyze the system behavior to compare it against the originally stated system function and iterate the system to support the emergence of a more desired state through audience participation and the dynamic of collective intelligence (see Figure 9.1).

XR systems require a conceptual framework as well as a physical substrate, but they only come to life after the human actors enter the interaction loop. In our system design, we rely on Peter's previous work, in which he described the

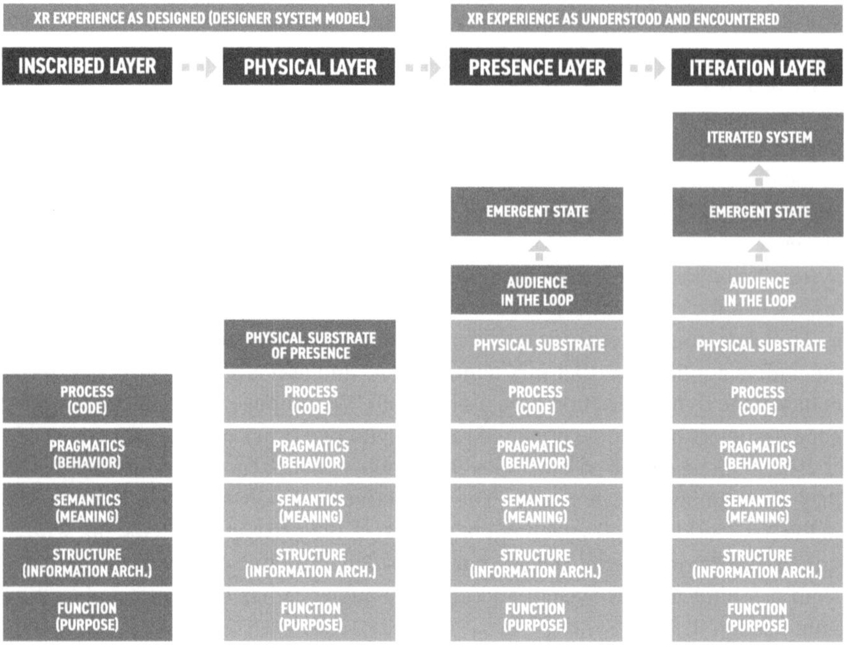

Figure 9.1 The path to Supermind emergence

Source: Illustration by © 2024 Peter (Zak) Zakrzewski.

conceptual framework for building human+computer (H+C) systems as being based on an architecture of seven elements. The first five are (1) the system's function as defined by its global behavior and stated purpose and evaluated based on the emergent outcomes produced by the system; (2) structure (information architecture + hardware architecture), (3) process (code), (4) semantics (meaning), and (5) pragmatics (behavior).

In addition to the five system dimensions, which can be, at varying degrees, controlled by the designers, Peter identified two that might completely elude the designers' control. These include (6) the context or the external environment within which the H+C system operates and (7) the agency of the audience that addresses the free and informed choice and open participation afforded to participants by the immersive system (Zakrzewski, 2022). System structure comprises both the information architecture and the hardware architecture. The physical substrate of XR systems comprises the physical space within which the immersion takes place and all of the hardware necessary for delivering immersive experiences. As much as these two layers set up the important aspects of the immersive system, XR worlds come alive during the next phase, during which the audience enters the immersive space and engages in various interactions enabled by mediated presence.

At the initial phase, XR applications represent the inscribed version of the system based on the designers' model. The first phase of presence building focuses on the immersive experience as designed. This inscribed version of the system is static. It is an amalgam of displays, sensors, and conceptually defined constraints that exist before the immersive experience begins. Life must be breathed into the inscribed system for anything interesting to happen.

In the second phase, the audience enters the immersive space. In the process of encountering the system, the audience internalizes its rules that act as a set of active constraints in each participant's mind. In encountering the XR system, the audience members mix the internalized rules with their real-world knowledge. Based on these, they invent new strategies for accomplishing their goals within the system. The system state that emerges out of the encounter between the system and its audience can be evaluated against the overall system purpose set up at the start of the process. Any discrepancies can be reconciled through iteration and tweaking of any of the conceptual and physical components of the system.

Human in the Loop: The Humane Way to Refine the User–System Fit

The mindful inclusion of human agents in the loop of XR systems is the main goal of immersive experience design. Making the immersive environments as welcoming, engaging, effective, error-tolerant, easy to use and learn, and most importantly humane are both major challenges and major opportunities facing immersive media makers. Not only can physical and mental imbalance caused by stressful external stimuli impede the attainment of immersive presence, but long-term exposure to stressful task loads can also lead to the sympathetic nervous system becoming hyperactivated, causing physical, psychological, and behavioral abnormalities in the users of XR systems (Kim et al., 2018).

One of the earliest approaches to helping users feel comfortable with complex sociotechnical systems requiring human input and interaction that has gained traction is the human-in-the-loop (HITL) method. The addition of the human actor can effectively contribute to a two-way transfer of knowledge and skills between the system designers and system users with consequences for the system outcomes in the real world. On the one hand, trainees using flight simulators learn valuable skills that prepare them to become professional pilots. On the other hand, engineers involved in the development of such systems can test, evaluate, and iterate their proposed design with human operators using immersive technologies that overcome the limitations of physical prototyping (Netti et al., 2023).

Utilizing HITL methodology allows the design team to put their prototypes in the hands of the core audience, who interact with realistic models while attempting to perform tasks they would do in an actual real-world scenario. Even the most advanced "intelligent" systems can only go so far to simulate a complex sociotechnical process. Only human designers and human users can ultimately accurately judge the success of the final design. Most designer-inscribed processes cannot fully anticipate the full range of the possible human error and creativity. Realistic HITL simulations can bring to the surface many unanticipated issues that would otherwise not become apparent until after a process has already been deployed.

As a methodology, HITL assumes that the human actors will be physically and cognitively constrained. They will need guidance in internalizing the system rules before they can fully use the system. And they will have the freedom to change the outcome of events and processes within the system.

What Constitutes Agency?

The question "what is an agent?" has always been a bit like an elephant living in the family room of the agent-based computing community in a similar way that the question "what is intelligence?" is avoided within the mainstream AI field (Wooldridge & Jennings, 1995). Although the term "agent" is widely used, it has consistently resisted attempts to produce a single universally accepted definition. Even though, on a practical level, issues of terminology might seem trivial as long as researchers develop interesting and useful applications, in any area of research, there is actually a real danger in not defining terms properly. If the concept of an "agent" were to become not much more than a background noise, it could become subject to abuse and misuse and be a potential source of unintended consequences for the entire intelligence augmentation field.

In an effort to define the term more precisely in the context of software engineering, Michael Wooldridge and Nicholas Jennings distinguish between the relatively uncontentious weak notion and the potentially more contentious strong notion of agency (1995). They suggest that in a weak form of agency, the software agents can operate without direct human intervention while having some level of control over their own actions and internal state. Agents should also be able to interact with other agents and humans via some kind of communication protocol. Agents need to be able to "sense" their environment to not only respond, in a timely fashion, to any changes occurring in it; they should also be able to exhibit goal-directed behavior by taking the initiative.

For some researchers, particularly working in the AI field, the term "agent" has a stronger and more demanding meaning. They generally consider an agent to be a computer system that, in addition to having the identified properties, can be conceptualized and/or implemented using qualities usually applied to humans. It is not uncommon in AI circles to characterize an agent using mentalistic notions, described by Gregory Bateson as "pathetic fallacies."[1] These terms include such traditionally human qualities as "belief," "knowledge," and "intention."

Other AI researchers go even further to conceptualize agents in terms of emotions, arguing that far from it being a pointless anthropomorphism, there exist good arguments in favor of designing and building agents in terms of human-like mental states (1995). In an updated version of Turing's imitation game, AI agents can be given surface-level human-like attributes by being represented visually or auditorily through realistic animation and natural language processing (NLP) capability.

While we have some criticism but no argument with the affective computing and other similar research agendas, for the sake of clarity, we like to make a clear distinction between the agency of the human actors and the agency of the synthetic AI agents. In his previous work, Peter contrasted the extrinsic agency of the nonhuman actants whose original "behavioral" code is under the control of system designers, with the intrinsic agency of the human actors constrained by their unique biology and motivated by their own goals.

Adopting a humane, participatory approach to designing immersive experiences requires designers to persuasively invite prospective users to engage with XR environments in a way that respects participants' agency and choice. The freedom enjoyed by the human participants within the context of immersive systems involves an explicit design philosophy that affords the participants an ability to make informed choices and to direct their own course of action within an immersive system based on a set of genuine alternatives.

Throughout this book, we equate the agency of human participants with deliberate conscious and informed choice. We follow Peter in proposing to divide the agents within sociotechnical systems into human actors who possess intrinsic agency of their own and nonhuman actants whose extrinsic agency is artificially inscribed into them by the system designers (Zakrzewski, 2022).

> *A discussion of social actors' freedom within the context of immersive systems involves two components: the existence of a set of genuine alternatives and an opportunity to make an informed choice which refers to the actor's ability to choose their course of action.*
>
> – PETER (ZAK) ZAKRZEWSKI (2022)

Shaping Agents Extrinsically

Having sorted the distinction between the extrinsic agency of the synthetic bots whose algorithms can be explicitly designed and iterated by the design team, we can now discuss what extrinsic agency means in the context of XR system design. As mentioned before, according to Wooldridge and Jennings, a weak notion of agency denotes hardware- and/or software-based computer systems that enjoy the properties of autonomy, social ability, reactivity, and proactiveness.

In terms of autonomy, agents need to operate without direct human intervention while having some kind of control over their actions and internal states. In terms of social ability, they must have an ability to interact and communicate with other human and nonhuman agents via some kind of language. In terms of reactivity, agents must be able to perceive and respond to their physical or virtual environment in ways that reflect, in a timely fashion, changes that occur in it. Agents' proactiveness refers to their ability to exhibit goal-directed behavior (1995). The weak notion of agency has often been used in agent-based software engineering.

Within this definition, agents can be as simple as subroutines or, more typically, larger entities with more persistent control. The quality of veracity of an agent's behavior is based on the assumption that an agent will not knowingly communicate false information. Agents' benevolence involves another assumption that they will not have goals conflicting with those of the user and that they will always try to assist the user in accomplishing their tasks. Rationality is an assumption that an agent will act in order to achieve its assigned goals, at least insofar as its policy permits.

Algorithms for AI agents can follow a number of different architectures, but they all share one common characteristic: the initial conditions driving their future behavior are set by the design team. In one of the most successful approaches to synthetic intelligence building, known as reinforcement learning (RL), AI agents are programmed to learn an optimal behavior indirectly by exploring an environment with an aim to obtain maximum rewards from it.

Seminal RL researchers Andrew Barto and Richard Sutton identified two dimensions by which AI agents' behavior could be shaped. These are known in the field by the technical terms of the policy that dictates what the agent will do and when and the value function, which sets up the parameters for expected rewards and punishments. These two design constraints have been shown to successfully mimic the behavior of people who master such cognitively demanding domains as chess playing.

As it turns out, advanced chess players do two important things very well. They develop an ability to assess how to take the right action in a given situation, and they find a way to estimate the potential future rewards that various actions may bring (Christian, 2020, pp. 137–138). Policy and function are two tools within one AI approach to extrinsically shape the behavior of AI agents.

Evaluating the Human Experience of the XRI Worlds

Immersive media straddle two quite distinct realms of research. The immersive part of XR relies on empirically verifiable advanced hardware and software engineering; presence, on the other hand, is a subjective experience of extended immersive environments. The collaborating duo of psychologist Lawrence LeShan and physicist Henry Margenau suggested that while the engineering model of the world belongs squarely to the see-touch realm of the senses and, with it, to the pre-19th-century view of science, researching the subjective conscious experience requires an entirely different guiding metaphor. They argued that while any changes in such see-touch realms as physical environment or geographical space obey one set of laws, human consciousness operates in the behavioral space affected not only by our perception of changes occurring in the physical world but also by the changes within the personal space of our consciousness (LeShan & Margenau, 1982).

Successful immersive media experiences operate within the overlapping space defined by the physical substrate of the immersive experience and by the personal space within the realm of human consciousness. For instance, two individuals immersed in a VR world standing at the edge of a massive cliff can have distinctly different subjective experiences, each shaped by the personal space they occupy, if one is an adrenaline junkie and the other has a deep phobia of heights. Understanding the fit between the XR system and the audience can only be understood by paying close attention to the system–user interactions within the broad categories of user responses, which can include biometrics, pragmatics (behavior), and semantics (meaning), recorded as user data of idea flows within the network or as subjective accounts of individual users.

Pragmatics: Observing Behavior

The concept of pragmatics comes from the field of family therapy and refers to the behaviors taking place within a social system. Observed behavioral responses can provide as many, albeit different, kinds of insights about the effectiveness, efficiency, and engagement experienced by the system users than verbal reports. The early attempts to turn the study of mind and behavior into a hard science by pursuing a mechanical model, now known as the molecular paradigm, dates back to the 19th century.

The empirically based mechanical model of behavioral phenomena aimed at scientific acceptability by avoiding any reference to such mysterious concepts as consciousness. Consequently, the approach relied on identifying basic units or discrete molecular events and specifying the laws, such as the sequential occurrence or proximity of stimulus and response causing their association in the mind, to explain behavior (Baum, 2004). Under the molecular paradigm, behavior was reduced to a movement or a sequence of movements, such as scratching, specified by the muscular contractions involved. The resulting connectionism led to the division of mental life and behavior into discrete events that could be explained by direct, immediate causes.

Psychologists use molecular analysis to examine behavioral processes in terms of elemental units analyzed in a moment-by-moment or phase-by-phase manner. Using the molecular units of behavior, researchers could measure the objectively observable responses such as gland secretion or muscle contraction, measured by electromyography (EMG).

The more recent understanding of the psychology of behavior is reflected by the molar paradigm, which analyzes behavior as complex phenomena composed of activities whose parts are closely integrated and inherently extended in time. Burrhus Frederic Skinner, the creator of operant conditioning, argued that the stimulus–response (S–R) model of behavior did not explain the behavior of complex organisms such as birds, mammals, and humans, whose responses frequently occurred without any obvious stimuli to provoke them. While a person scratching an itch might appear to follow a simple S–R pattern, a shopper hunting for a perfect pair of shoes displays a varied and prolonged behavior that allows no easy identification of direct stimuli.

Skinner proposed the concept of an operant response to represent a class of movements defined by the performance of the same operation on the environment. Building on Skinner's work but breaking off the conception of behavior as broken into discrete responses, Howard Rachlin argued molar behaviorism, based on a view that most activities are composed of several other activities. For instance, a bank teller helping a customer can engage in such activities as greeting the person, answering questions, and withdrawing and depositing money, all with a single aim of providing good service.

Psychologists use molar analysis of behavior to examine behavioral processes as holistic units extended through time. The molar paradigm postulates two important principles: (A) all activities take time and (B) all behavior constitutes choice. The principle of behavior taking time is based on the insight that identifying sequences of actions constituting molar behavior requires extended observations. Equivalence between choice and behavior stems from the fundamental assumption that every situation contains more than one alternative activity. Explanations conceived within the molar paradigm are based on concrete particulars of empirically observable and quantifiable variables, but they also rely on abstractions addressing the molar units of behavior above the reflex level whose causes are contiguous with the observed responses (Baum, 2004).

One of the most accessible forms of observing user behavior is a qualitative social and behavioral sciences research method known as ethnography. In ethnographic studies, data are collected through observations and interviews, which are then used to draw inferences about individuals or entire social system dynamics. Ethnographers use notebooks and still and video cameras to observe and record people's behavior in the natural environments, such as homes and offices, rather than trying to artificially recreate real-life scenarios in a lab. Ethnographic studies can help system designers create object, spatial, and temporal mappings of immersive spaces. They can also be used to create power mappings to understand the various forms of power dynamics occurring in sociotechnical spaces between people or between people and XR systems.

When it comes to usability research, we subscribe to the philosophical stance described by ethnographer Christena Nippert-Eng that acknowledges the existence of an objective reality outside of our minds and bodies that is individually sensed, understood, and mediated by the ways we have learned to do so (Nippert-Eng, 2015). As humans, we perceive the world through subjective and intersubjective lenses shaped by a mix of our own unique experiences and by the cognitive frameworks taught to us by our culture.

When it comes to research concerning human experience, whether we apply quantitative or qualitative methods, no matter how hard we try to pursue unmediated truth, we are not going to obtain it. The humble approach to usability research is a pragmatic blend of art and science motivated by improving the experiences of technology users. Today's usability field has access to a much-expanded repertoire of paradigms and methods since the early days of the molecular paradigm. XR usability researchers can, in addition to the evolved molar paradigm of behavior, now draw on biometric data obtained from various sensors and on subjective user reports.

Ethnographic observations can be enhanced by the application of passive sensing of digital breadcrumbs left behind by users. Such sources of more empirical behavioral data can provide system designers with a great deal of insight into various aspects of usability. For instance, collecting and quantifying a person's Facebook "likes" of articles, videos, and other media content allows data scientists to create the person's psychological profile across the five basic personality dimensions of openness (being inventive and curious vs. being consistent or cautious), conscientiousness (being efficient and organized vs. being extravagant or careless), extraversion (outgoing/energetic vs. solitary/reserved), agreeableness (being friendly and/or compassionate vs. being critical and/or rational), and neuroticism (being sensitive and nervous vs. resilient and/or confident) (Parker, 2015).

Based on this data, AI algorithms, after analyzing just 10 likes, can predict the subject's personality and likely behavior more accurately than a work colleague. With an analysis of 70 likes, they can outpredict a friend or a roommate. With 150 likes, they can outperform a family member, and with 300 likes, a spouse. Pentland argues that many types of human behavior can be reliably predicted from the analysis of biologically based "honest signals" (Pentland, 2008). Such ancient primate signaling mechanisms as synchrony and mimicry form a meta-level analog channel of communication between people that can operate alongside or separately from language.

Biometrics: Understanding Honest Signals

Smiles, hand gestures, sports cars, and designer clothes are some of the familiar signals people use to communicate who they are and how they want to be perceived. We plan and produce such displays consciously to influence our communication with others. In contrast, honest signals are produced under our conscious awareness to reveal what we are unwilling or unable to communicate about our experience. When our heads seemingly automatically move in the direction of a sudden noise, our behavior is guided by the ancient structures of our nervous system.

Immersive media offer unprecedented access to empirically based honest signals which can become a basis of assessing human responses to XR environments. XR systems can be combined with a wide array of biometric measurements of involuntary body functions such as heart rate, brain activity, breathing patterns, galvanic skin response (GSR), and muscle responses. These measurements can be obtained from such biofeedback sensors as HRV finger sensors, respiration sensors, and skin conductance sensors, or measurements of neural activity in the brain, which can be obtained with electroencephalography (EEG), functional magnetic resonance imaging (fMRI), or functional near-infrared spectroscopy (fNIRS).

Heart rate variability expresses the amount of time between heartbeats. The observed variations are usually quite small, measuring a fraction of a second between beats. Some heart rate variability may be present even in healthy individuals to indicate the presence of mental health issues such as anxiety or depression (Cleveland Clinic, 2024). Under stress, well-regulated individuals temporarily speed up their breathing, making both their breathing and heart rate irregular and temporarily out of sync as part of a normal stress response. HRV can be used as a proxy of activity of the autonomic system. As such, it is associated with executive cognitive functions, decision-making, and emotional regulation (Arakaki et al., 2023). Abnormal changes of HRV are signs of various neurological conditions that include mild cognitive impairments, dementia, mild traumatic brain injury, stroke, epilepsy, stress, anxiety, and schizophrenia.

EEG is a technology underpinning neurofeedback therapy. The technology works by measuring the varying heights (amplitude) and wavelengths (frequency) of a subject's brain wave activity. Frequency is measured in hertz (Hz) or cycles per second (cps). It refers to the number of times a measured waveform rises and falls within the time measure of 1 second. EEG systems provide objectively measured insights into the inner world of the system user's brain activity, which is directly related to the state of their arousal.

Delta waves, which exhibit the slowest frequencies of 2 to 5 Hz, are associated with states when the mind is turned inward. Researchers have found that individuals who show too much slow-wave activity in their waking state can exhibit poor judgment, poor impulse control, and unfocused thinking. Eighty percent of children diagnosed with attention-deficit hyperactivity disorder (ADHD) show excessive slow waves in their frontal lobes.

Theta frequencies in the range of 5 to 8 Hz are associated with edge-of-sleep types of experiences and hypnotic trance states. Brain states that accompany the theta frequency range support the creation of novel connections and associations by inducing a free-flowing and creative state of mind unconstrained by repeatable daily routines or logic. The frequency is also correlated with reports of being "out of it" or feeling depressed.

Alpha waves, at 8 to 12 Hz, are associated with the feeling of peace and calm familiar to people who practice mindfulness and meditation.

Beta waves, at 13 to 20 Hz, are the fastest brain wave frequencies. They orient the brain toward the outside world. On the positive side, they support focused attention while performing tasks. On the negative side, high beta waves over 20 Hz are associated with agitation, anxiety, and body tenseness (van der Kolk, 2014). Computer-aided EEG analysis, known as qEEG, can show visualizations of the subject's brain wave activity millisecond by millisecond, displaying which frequencies are highest and lowest in the monitored areas of the brain. They can also show how well various brain regions are communicating and working together. The ever-growing databases of normal and abnormal brain wave patterns enable a classification and comparison of quantitative EEG patterns between users (2014).

Neuronal activity constantly fluctuates in an active organism with tasks ranging from holding objects to using language or solving math problems. Functional magnetic resonance imaging (fMRI) is a noninvasive neuroimaging technique that can measure the fluctuation of brain activity by detecting changes in blood flow within different regions of the brain. The technology is based on the finding that cerebral blood flow is correlated with neuron activation. Blood flow increases in the active brain regions.

fMRI technology has revolutionized researchers' ability to view the human brain in action. Measuring neural fluctuations allow researchers to see which regions of the brain are engaged during various activities. On the positive side, the real-time capability of fMRI technology and the accuracy of the produced neuroimaging have been shown to be an effective basis for neurofeedback treatment methods for a variety of psychiatric illnesses (Dudek & Dodell-Feder, 2021). Real-time functional magnetic resonance imaging neurofeedback (rtfMRI-NF), built on the fMRI technology, is a biofeedback method used to help patients learn to regulate their own brain activity and to control their emotional responses in real time.

Emily Dudek and David Dodell-Feder who conducted meta-analysis of 17 studies evaluating the impact of rtfMRI-NF neurofeedback training on brain and behavioral outcomes for psychiatric disorders. They found that participants who were shown their own brain activity patterns in real time were able to regulate activity in specific regions of the brain with volitional control (Knispel, 2021).

The drawbacks of the fMRI imaging, which make fMRI difficult to apply in usability research, include such factors as the high cost, the extensive technical setup required, and various practical constraints imposed by the bulky scanner environment.

Functional near-infrared spectroscopy (fNIRS) has recently emerged as an alternative hemodynamic-based approach that is not burdened by such fMRI limitations as lack of portability and low tolerance for motion (Scarapicchia et al., 2017). fNIRS is a noninvasive imaging technique that measures changes in oxygenated and deoxygenated hemoglobin (Hb) concentrations within the brain. It does so by means of optical measurement of the characteristic absorption spectra of the wavelengths between 700 and 1000 nm (Dubreuil-Vall et al., 2022). fNIRS has shown great promise imaging the functioning of the human brain in such previously inaccessible populations as infants and in neuroimaging studies examining full-body behaviors such as exercise.

fNIRS technology has its own unique technical constraints, which include a limited depth of recording and a lower spatial resolution (2017). Despite these limitations that affect the application of fNIRS to clinical settings, the technology has emerged as part of a neuroergonomic methodology that can benefit usability assessment studies (Bhatt et al., 2019). Bhatt et al. define neuroergonomics "as the study of the human brain in relation to performance while at work and in everyday settings" (p. 181). Portable and wearable fNIRS brain imaging headsets have emerged as a practical sensor modality to capture cortical oxygenation changes in the brain that enable researchers to assess a variety of cognitive functions, such as working attention, memory, and decision-making.

Compared to fMRI scanners, fNIRS monitors are much smaller, cheaper, and easier to use, making them a cost-effective neurotechnology in exploratory and diagnostic studies that require real-time assessment of oxygen level-dependent cortical activity before, during, and after specific cognitive tasks and clinical interventions (Dubreuil-Vall et al., 2022). fNIRS headsets have been used in field studies to assess the workload of air traffic controllers and in training unmanned vehicle operators with customized flight simulator interfaces (2019). Miniaturized, battery-operated wireless versions of fNIRS have been used in outdoor settings to monitor participants using augmented spatial navigation interfaces. fNIRS headsets can be used with other data-collection methods to provide usability researchers with cognitive and affective-state-related brain activity data related to modulated task assessment and user engagement.

Semantics: Observing Idea Flows Within a Network

Semantics generally refers to the meaning-making within the immersive system. A lot of insight about system usability can be gained by paying attention to the idea flows within the XR platform. Paying attention to observable group conversations and their relation to social network dynamics is related to the concept of social physics advocated by the computational social scientist Alex Pentland. Social physics suggests that sociotechnical systems can be better understood by the observation, quantification, and assessment of patterns of not just individual but also communal communication and behavior. In Chapter 7, based on the work of Alex Pentland, we have defined a number of key social physics concepts that can help the design team to assess user responses to the XR system. These included "idea flow" which refers to the propagation of beliefs, values, and behaviors which diffuse through a social network by two critical processes of social learning and social pressure. An estimation of an idea flow for a social network must take into account the structure of the network, the individual susceptibility of members to new ideas, and the strength of influence between each pair of members. "Interaction" between community members can include direct means such as conversations or indirect means such as observation of behavior, overhearing a conversation, or sharing of digital content. Pentland defined "engagement" as social learning occurring within a peer group which typically leads to the development of behavioral norms that are later reinforced through the mechanism of social pressure.

"Society" within the framework of social physics can be described as an entity made up of networks defined by exchanges between individuals. It stands in contrast to a description of society as composed of such social science concepts as classes or markets. "Strategy" is a choice defined by the total potential actions an individual can take in a given situation. "Trust" is a core social norm that allows individuals to expect fair value when dealing with others in an exchange or a relationship. Adherence to trust assures a continuing stable exchange of value within a community. "Value" constitutes the extent to which an exchange will satisfy social and personal needs and goals of individuals in social exchange interactions. This can include the utility value of goods or deeper personal needs, spanning the gamut of needs from physiological to safety to love, belonging, esteem, and self-actualization. "Social influence" describes the likelihood of one person's behavior affecting the behavior of others. "Social learning" can consist of learning new strategies by observing other people's behavior. It can also involve acquiring new beliefs and behaviors through direct personal experience or observation. "Social network incentives" are a type of reward capable of affecting individual behavior or altering the pattern of exchanges between pairs of community members. Finally, "social norms" are sets of agreed-upon compatible strategies the community employs to produce the most desirable exchange value. Norms are developed and spread by the process of social learning and enforced by social pressure, which is the leverage members of a community can exert upon one another (Pentland, 2015).

Semantics: Listening to the Audience Members

Direct conversations with the audience members are another source of user feedback that can help the design team to assess the usability of their XR system. Media makers can learn a lot about the audience response to the immersive environmental stimuli by seeking qualitative, subjective accounts that reflect the users' conscious experience. Since presence and flow belong to the realm of subjective experience, they do not conform to the quantifiable protocols and forms of user research. Properly understood and interpreted, subjective user accounts can reveal a lot about the system's effectiveness in eliciting the state of flow and presence.

Understanding subjective audience responses to immersive environments requires digging beyond the surface of quantifiable patterns of behavior with the same tools we covered in the audience discovery chapter. These include such qualitative tools as structured and unstructured interviews, surveys, and projective techniques such as metaphor elicitation. The need to mix subjective, objective, qualitative, and qualitative methods does not need to be a cause for alarm among the positivist-minded researchers. It should rather serve as a recognition of the multidimensionality of immersive experience design and the attendant need for multimodal forms of user research.

NASA's Task Load Index (TLX)

One of the most widely used human-in-the-loop (HITL) methods is NASA's Task Load Index (TLX), initially outlined by Sandra Hart of NASA Aerospace Human Factors Research Division Research Center and Lowell Staveland of San Jose State University. TLX is a subjective, multidimensional assessment tool that allows users to rate the perceived workload imposed on them by the system to assess the system's effectiveness, efficiency, and ease of use in regard to its overall performance and function.

The index consists of six subscales that represent diverse independent clusters of variables including the mental, physical, and temporal demands the system makes on the user and the level of effort, frustration, and performance required to perform the task. The variables are rated for each task on a 100-point scale with 5-point steps. The ratings are combined to show the overall task load index.

The mental demand index assesses how much mental and perceptual activity the user required to complete the task and whether the task was perceived as being simple to execute or complex. Physical demand assesses how much physical activity was required to complete the task. The temporal demand index reflects the cognitive science finding about the decreasing ability of humans to perform tasks under time pressure. It also addresses the pace at which the task needs to be performed.

Participants are also encouraged to evaluate their own performance by answering two questions: How successful do you think you were in performing the task? And how satisfied were you with your performance? The level of mental and physical effort is measured by asking the participants how hard they had to work to accomplish the necessary level of performance. Lastly, the frustration level caused by the performance is measured by asking the participants to rate the level of irritation, stress, annoyance, relaxation, or even complacency felt during performance of the task (Hart & Staveland, 1988).

The human-in-the-loop evaluation assesses two major aspects of XR systems. It evaluates the performance of the hardware and software while being used by human operators. The measurement of hardware performance with human factors assesses the hardware's capability to perform the designated tasks and when operated by a typical user. It also assesses the impact of the hardware and software on human performance. The human performance evaluation aspect evaluates the user's physiological and psychological states elicited by the system while performing specific core tasks. These types of tests can identify system features that pose challenges for both usability and accomplishing the overall system function (Netti et al., 2023).

TLX evaluates users' subjective mental workload (MWL) by rating performance across six dimensions to determine an overall workload rating.

1. Mental demand measures how much thinking, deciding, or calculating was required to perform the task.
2. Physical demand asks the users to assess the amount and intensity of physical activity required to complete the task.
3. Temporal demand measures the amount of time pressure involved in completing the task.
4. The Effort subscale assesses how hard the participant had to work to maintain their level of performance.
5. Performance measures the level of success in completing the task.
6. The Frustration Level subscale allows the user to assess how insecure, discouraged, secure, or confident they felt during the performance of the task. The subscales are presented to the participants either during or after the experimental trial. The tasks are rated based on an interval scale ranging from 1 (low) to 20 (high).

To use the tool, the researchers first define the task(s) to be assessed. Optionally, the team can conduct a hierarchical task analysis (HTA) for the tasks under study. In the next step, researchers select participants who are chosen to represent a wide range of users. The participants are briefed on the NASA-TLX and are shown a brief run-through of the method. In the next step, the participants perform the task(s) under analysis, filling out the assessment form either during or immediately after the trial. Researchers can evaluate the overall workload scores using NASA's official Task Load Index (TLX) app, which is a free iOS download available on the Apple App Store for use on iPhones and iPads.

The System Usability Scale (SUS)

While TLX focuses on assessing the level and quality of a user's performance while operating an advanced system, the System Usability Scale (SUS) developed by John Brooke at Digital Equipment Corporation (DEC), is a tool used to assess the performance of the system from the subjective perspective of the user. The usability of a system is measured by taking into account the full context of use including such aspects as who is using the system, what they are using it for, and the context of the environment the system is being used in. The system usability scale (SUS) is a context-specific ten-item (out of 5) attitude Likert scale that measures the key aspects of the system from the user's perspective. The ten questions include:

1. I think that I would like to use this system frequently.
2. I found the system unnecessarily complex.
3. I thought the system was easy to use.
4. I think that I would need the support of a technical person to be able to use this system.
5. I found the various functions in this system were well integrated.
6. I thought there was too much inconsistency in this system.
7. I would imagine that most people would learn to use this system very quickly.
8. I found the system very cumbersome to use.
9. I felt very confident using the system.
10. I needed to learn a lot of things before I could get going with this system.

The 5Es of Usability

In Chapter 4, we discussed one of the best approaches to both design and assessment of complex sociotechnical systems. The 5Es of usability framework proposed by the co-founder of the Center for Civic Design, Whitney Quesenbery, is a framework that offers five very useful lenses for building and evaluating artificially designed systems (Quesenbery, 2004).

The 5Es can be evaluated from the designers' or the users' perspective either as an objective or subjective measure of a system's performance. Effectiveness of the system addresses the question whether the solution is useful in meeting the users' goals. Efficiency probes the system's speed and performance. How fast does the system perform or respond to the users' inputs? How accurately does it perform the intended task? Engagement focuses on the way the system is experienced by the users. How pleasant, satisfying, or interesting is it to use? Error tolerance refers to the system's ability to deal with user errors. How well does the system prevent errors? Does it help users recover from mistakes and misunderstandings? Finally, ease of learning refers to the comprehensibility and learnability of the system and its rules. This measure allows the design team to assess: How well does the system support initial orientation and deeper learning of the system model?

Multimodal Usability Studies

Vittorio Netti and his colleagues at the Sasakawa International Center for Space Architecture (SICSA) at the University of Houston are opening up a whole new area of XR-based testing environments and protocols for assessing human–system integration. Their testing system is specifically intended to help researchers working on the Lunar Surface Study (LSS) and Lunar Terrain Vehicle (LTV) programs validate their respective initial design assumptions regarding various features of space hardware.

Netti's lab comprises an 8.2 ft × 8.2 ft, 11.5-ft-tall, flexible, modular stage-lighting rig structure. The rig serves as a support for VR sensors and lighting. A control station computer operated by the experimenter sits just outside the stage area. The experimenter runs the VR simulation, controls the lighting, and collects the biofeedback data from the biosensors attached to the subjects' bodies. Subjects who enter the stage wear VR headsets and controllers and a number of biofeedback sensors measuring such biomarkers as breathing and heart rate variability (HRV). Within the stage area, subjects interact with 3D printed, low-fi physical objects paired with virtual digital twins, simulating interactions necessary for task completion. The proximity of the control station to the stage allows the experimenter to observe the actions of the test subjects and the system's operations directly.

The experimental protocol mixes the collection of subjective data from the TLX and the modified SUS (mSUS) surveys with the biofeedback data collected from the sensors worn by the subject during the performance of the assigned tasks. Actions taken by the participants within the virtual space can also be observed by the experimenter operating the control station and captured by the system. Biometric data typically needs to be triangulated with other methods such as observations of nonverbal behavior, participant retrospectives, surveys, and rating scales. In the case of immersive evaluating systems, additional sources of data can be collected based on the observed user behavior.

Subjects begin the experiment by performing an entry survey that allows the research team to collect metadata to account for the broad range of the target population covered and for the potentially large variability between participants. In the calibration phase, researchers ask test subjects to wear biofeedback sensors and perform baseline tests to establish appropriate baseline readings for each individual. Becoming familiar with sensors helps to prevent potential discomfort during the test that might influence the accuracy of the gathered data.

During the training phase, subjects are asked to wear VR sets and are introduced to the simulated environment. During this phase, each subject is to perform four tasks such as driving the Lunar Terrain Vehicle in EVA mode, accessing the habitat crew airlock from EVA mode, unloading the cargo airlock from the cargo box in IVA mode, and re-purposing the interior habitat configuration in IVA mode.

The entire test takes an average of 25 minutes, with each task taking approximately 3 minutes to complete. After completing each task, subjects are asked to complete a survey. This is done with a VR headset off but with the biofeedback sensors on. The survey comprises a mix of TLX and mSUS questions regarding

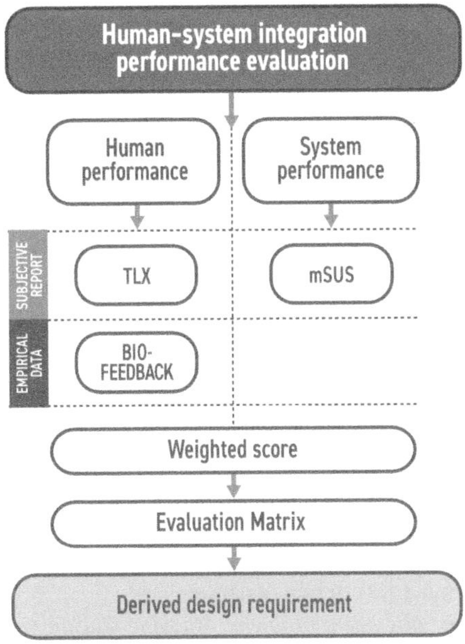

Figure 9.2 SICSA Human-System Testing Framework

Source: Illustration by Peter (Zak) Zakrzewski, based on Vittorio Netti, Albert Rajkumar, Olga Bannova (SICSA), XR Testing Framework.

the task that has just been completed. After the final survey, subjects are debriefed and asked to complete an exit survey, which collects high-level metadata on the experiment.

Summary

In this chapter, we highlighted the critical importance that the user plays in the performance of an immersive system. We discussed the concept of Superminds as a guiding principle of XR system design that combines the activities of multiple users and advanced computer systems. We outlined how the ultimate goal of XRXD—presence—emerges at a play stage that is built on top of the inscribed (conceptual) and physical layers.

We covered the human-in-the-loop concept and its importance to the design and testing of XRR systems. We distinguished between the intrinsic agency of the human actors and the extrinsic agency of AI bots and other digital objects. Finally, we discussed the various domains of reality that support multimodal usability research.

With the hard work of testing and evaluating the system prototype behind us, it is finally time to focus on inviting a broader audience to participate in the immersive system.

Note

1 In this case, the word "pathetic" does not connote the derogatory meaning in the sense of arousing pity or being miserable. It refers to its Latin roots of the word "pathēticus," meaning affecting the emotions. Pathetic fallacy is a common literary device that allows authors to attribute human qualities, thoughts, and emotions to inanimate objects. (Zakrzewski, 2022). Jerome Bruner referred to endowing animals (Bambi) and nonliving objects (Little Red Engine), cast as narrative protagonists, with intentional states as intentional state entailment.

References

Arakaki, X., Arechavala, R. J., Choy, E. H., Bautista, J., Bliss, B., Molloy, C., Wu, D.-A., Shimojo, S., Jiang, Y., Kleinman, M. T., & Kloner, R. A. (2023). The connection between heart rate variability (HRV), neurological health, and cognition: A literature review. *Frontiers in Neuroscience*, *17*, 1055445. https://doi.org/10.3389/fnins.2023.1055445

Baum, W. M. (2004). Molar and molecular views of choice. *Behavioural Processes*, 66(3), 349–359. https://doi.org/10.1016/j.beproc.2004.03.013

Bhatt, S., Agrali, A., McCarthy, K., Suri, R., & Ayaz, H. (2019). Web usability testing with concurrent fNIRS and eye tracking. In *Neuroergonomics* (pp. 181–186). Elsevier. https://doi.org/10.1016/B978-0-12-811926-6.00030-0

Christian, B. (2020). *The alignment problem: Machine learning and human values*. W.W. Norton & Company.

Cleveland Clinic. (2024). Heart rate variability (HRV): What it is and how you can track it. *Cleveland Clinic*. https://my.clevelandclinic.org/health/symptoms/21773-heart-rate-variability-hrv

Dubreuil-Vall, L., Laabs, T. L., Eyre, H. A., Smith, E., & Catuara-Solarz, S. (2022). Sex differences in invasive and noninvasive neurotechnologies. In *Sex and gender bias in technology and artificial intelligence* (pp. 133–160). Elsevier. https://doi.org/10.1016/B978-0-12-821392-6.00008-X

Dudek, E., & Dodell-Feder, D. (2021). The efficacy of real-time functional magnetic resonance imaging neurofeedback for psychiatric illness: A meta-analysis of brain and behavioral outcomes. *Neuroscience & Biobehavioral Reviews*, 121, 291–306. https://doi.org/10.1016/j.neubiorev.2020.12.020

Guszcza, J., & Schwartz, J. (2019). Superminds: How humans and machines can work together. *Deloitte Review*, 24.

Hart, S. G., & Staveland, L. E. (1988). Development of NASA-TLX (Task Load Index): Results of empirical and theoretical research. In P. A. Hancock & N. Meshkati (Eds.), *Human mental workload*. Elsevier Science Publishers B.V.

Kim, H.-G., Cheon, E.-J., Bai, D.-S., Lee, Y. H., & Koo, B.-H. (2018). Stress and heart rate variability: A meta-analysis and review of the literature. *Psychiatry Investigation*, 15(3), 235–245. https://doi.org/10.30773/pi.2017.08.17

Knispel, S. (2021, February 22). An experimental neurofeedback imaging treatment has emerged for psychiatric patients. Does it work? *University of Rochester News Center*. www.rochester.edu/newscenter/experimental-neurofeedback-imaging-treatment-for-psychiatric-patients-470142/

LeShan, L. L., & Margenau, H. (1982). *Einstein's space and Van Gogh's sky: Physical reality and beyond*. Macmillan.

Netti, V., Rajkumar, A., & Bannova, O. (2023). XR testing framework for human-system interaction design validation. *2023 international conference on environmental systems*. https://hdl.handle.net/2346/94631

Nippert-Eng, C. (2015). *Watching closely: A guide to ethnographic observation*. Oxford University Press.

Parker, C. B. (2015, January 23). Michal Kosinski: Computers are better judges of your personality than friends. *Insights by Stanford Business*. www.gsb.stanford.edu/insights/michal-kosinski-computers-are-better-judges-your-personality-friends

Pentland, A. (2015). *Social physics: How social networks can make us smarter*. Penguin Press.
Quesenbery, W. (2004). Balancing the 5Es: Usability. *Cutter IT Journal*, 17(2), 4–11. www.wqusability.com/articles/5es-citj0204.pdf
Scarapicchia, V., Brown, C., Mayo, C., & Gawryluk, J. R. (2017). Functional magnetic resonance imaging and functional near-infrared spectroscopy: Insights from combined recording studies. *Frontiers in Human Neuroscience*, 11, 419. https://doi.org/10.3389/fnhum.2017.00419
van der Kolk, B. A. (2014). *The body keeps the score: Brain, mind and body in the healing of trauma*. Viking.
Wooldridge, M., & Jennings, N. R. (1995). Intelligent agents: Theory and practice. *Knowledge Engineering Review*, 10(2), 115–152.
Zakrzewski, P. (2022). *Designing XR: A rhetorical design perspective for the ecology of human+computer systems*. Emerald Publishing Limited. http://doi.org/10.1108/9781802623659

Section VI

Launching and Learning From System in Play

10 Attracting the Audience

Inviting Your Audience to Play

User "onboarding" is one of those unfortunate, useful, but, in our opinion, ultimately misleading terms that have embedded themselves into the language of user experience design. Our focus on clear terminology here goes beyond mere semantics. Language we use as XR designers matters a great deal. Concepts and metaphors we use to describe what we do shape how we do our work.

"Onboarding" is an awkward term commonly used in reference to software products to describe a process of assisting users in acquiring the necessary knowledge, skills, and behaviors to become proficient at using apps, websites, and software services. The term originated in the field of human resources, where it described the mechanism through which new employees were introduced to institutional knowledge that would help them become effective members of the organization.

Since audience participation is an essential ingredient of immersive media design, we think that inviting an audience to participate is a much more elegant and infinitely more useful way to describe the process of bringing audiences into the XR worlds. We refer to the process of inviting an audience as the creation of a discovery path.

Attracting the Audience

Game creators, movie producers, Broadway show impresarios, theme park operators, and app and content makers all agree on one thing: there can be no show without an engaged audience. All media makers and content creators expend a considerable amount of time, effort, and money to get the prospective audiences interested in what they have to offer.

Attracting not just any audience but a right audience is a critical part of successful media making. Different audience segments will have different needs and be attracted to different types of content. Hardcore gamers are not likely to be attracted to immersive documentaries. Tourism-oriented mobile AR app users might not care a bit about fantasy game experiences, etc. This is why we consider audience discovery to be a critical early step in immersive experience design.

DOI: 10.4324/9781003442462-17

The streaming service Netflix attracts specific audience segments to its shows by creating up to eight header image variations to gauge interest. Once the header thumbnails go live, the Netflix marketing team immediately begins to track which ones inspire the most clicks and viewing responses. Eventually, the original eight invitations are narrowed down to the best-performing two or three. Within a month, one of the three thumbnail images will become the show's default representative marquee, and the rest will vanish from the home screens. Netflix's survival-of-the-clickiest attraction strategy is a form of value hypothesis testing that depends on showing the prospective viewers a number of ready options displaying potentially resonant themes and relatable characters and then counting the clicks. The clickiest options lead to the scalability of the content.

Movie theaters and streaming services are, at this point, dealing with well-established media formats. The same cannot be said for immersive media products. Like everything else having to do with the still-emerging XR media, the distribution landscape is a bit of a wild West. There are, however, some foundational concepts which can help XR media makers to find, attract, and scale up their audiences.

Prospective audience members are attracted to media content based on three highly personal factors. The first of these is an alignment with their identity and values. We refer to this attractor as relatability. Audiences who recognize themselves or identify with the people portrayed in the XR experience become open to engaging with the themes and stories that resonate with their unfulfilled cognitive and emotional needs.

Sharing resonant themes is the second audience attractor that builds up the anticipation and curiosity necessary to draw the audience deeper into the experience. As we discussed previously, relatability sets up the stage for resonance. Both of these factors blended together make conditions favorable for full audience engagement.

Audience members are also influenced by people in their peer networks. Content is often consumed and or discussed by groups of like-minded individuals. Based on the mentioned factors, media makers wanting to attract an audience must fulfill three essential criteria. (1) To attract an audience, we must appeal to their sense of

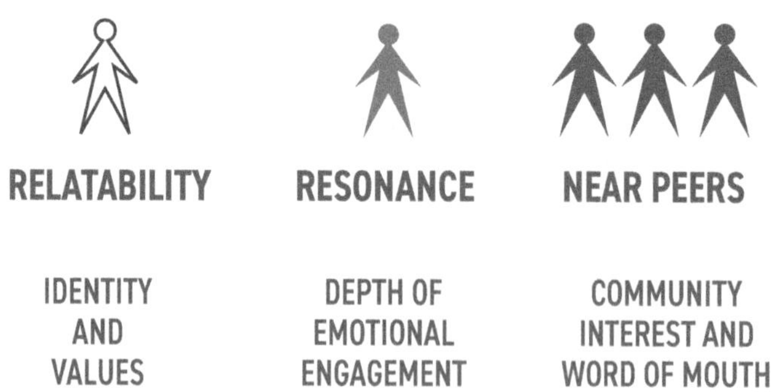

Figure 10.1 Audience attractors

Source: Illustration and concept by © 2024 Peter (Zak) Zakrzewski.

identity and values. (2) We must offer them resonant themes that address unmet cognitive and emotional needs. (3) Finally, when creating a unique XR experience, we must aim to engage not just individuals but an entire community that can make the experience a vehicle for conversation and deeper meaning-making.

Laying Down the Red Carpet of Discovery Paths for Immersive Experiences

Another term we are not very fond of is the unfortunate, useful, but also ultimately misleading marketing term "purchasing funnel." The unfortunate metaphor of the funnel conjures up images of force-feeding the audience content they might not really like or enjoy. We think that laying down a discovery path is a much more graceful and more effective metaphor and a way to think about guiding your audience to the desired immersive experience.

Laying down a discovery path is part of a bigger project of building an XR media franchise. If we view this project as yet another type of audience experience to be designed, we realize that the macro view of building an XR media franchise has the same fractal structure as the micro-level experiences within each individual application. They both share the "three steps and two transitions model" with user experience design.

As we showed in Section II, they also share a dramatic buildup with the narrative arc framework.

1. Prospective audience members have to be attracted to the immersive experience by an alignment with their identity and values. Based on this alignment, they will become open to being attracted by the themes, stories, experiences, and goals that resonate with their unfulfilled cognitive and emotional needs. Sharing themes through exposition builds up the anticipation and curiosity necessary to draw the audience deeper into the franchise.
2. Entry is a critical transition into the engagement with the franchise. It has to be designed deliberately and with maximum care, as prospective audience members are not yet committed to the franchise. A gamer might play a game for a few minutes and find it too boring or too challenging. A mobile AR user might download and try an app but decide that its navigation is too confusing. At the point of entry, prospective audience members are still not committed to the XR experience. They have yet invested enough time, money, and cognitive and emotional resources to feel like they have to work very hard at enjoying the experience. The Entry transition must raise the expectations to nudge the participant to go on while being flawless enough to effortlessly glide the users into the next stage of the engagement.
3. As discussed earlier, creating the conditions for the audience to experience the true feeling of immersive presence and flow that allow them to reach the "Aha!" and "Wow" moments are built on the foundation of design and engineering mindfulness. Presence can only happen after the deliberate execution of attraction and entry steps. Audience members who reach the immersion

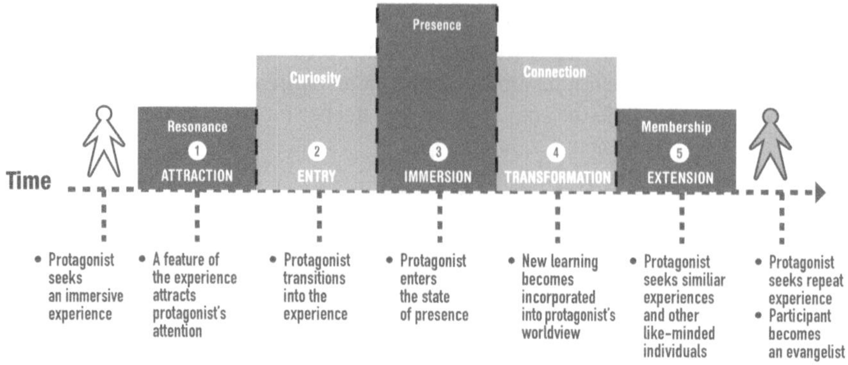

Figure 10.2 Laying down a discovery path for the audience

Source: Concept and illustration by © 2024 Peter (Zak) Zakrzewski, adapted from Compelling Experience Map by Doblin Group.

stage of the engagement with the XR media franchise are likely to become repeat users and even brand ambassadors if they continue to be guided through another critical step of transformation.

4. Just as all great narratives offer their audiences resolution of the unfolding drama, great media franchises help their audience members to exit the immersive experience and enter into a new community. Great immersive experiences change audience members and their worldview, even if it only constitutes playing a new game, finding a new way to accomplish an old task, or gaining some new knowledge about the world. XR media content has great power to transform an audience, and part of that transformation is a transition into the extension of the experience into the ongoing engagement with the franchise beyond the original immersion.
5. We cover the extension of immersive media experience in its own chapter, but in essence, extending play means connecting the audience member to the other parts and entry points of the media franchise. For instance, a gaming fan might enjoy playing *Star Wars Jedi: Survivor* without having watched the movies or the shows. They can then be invited to continue exploring the other platforms within the *Star Wars* franchise and share their experience with others. Media franchises work as entire ecosystems that tell larger stories or engage a larger community of users. Tourism-oriented mobile AR apps can encourage their users to share their unique experiences and knowledge of such places as the Roman Forum or Taj Mahal.

The Search for Resonance: The Content–Audience Fit

What attracts users to any experience is the level of resonance the proposed content offers its intended audience. Kevin Lane Keller defined resonance as the intensity or depth of the psychological bond that an audience has with the media franchise and the level of activity motivated by this loyalty (Keller, 1997). Your team's ability to attract the audience is based directly on the potential depth of the psychological bond the audience can develop with the relatability of characters and values presented by the content and the resonance of themes featured in the proposed experience.

As mentioned before, audiences enter narrative worlds they can identify with. Relatable characters and resonant stories enable presence, which can impact attitudes and beliefs in the physical world. People are attracted to stories and experiences based on their identities, needs, and values. As discussed in the Vision and Audience chapters, the discovery of resonance is the starting point of any digital product project. By defining the purpose of the project—the desired state—and the audience motivation, the team creates a "problem space" within which the clearly focused content-making work can take place.

Resonant experiences come in all shapes and sizes. Here once again, it is worth remembering Sequoia's three types of resonant problems defining the challenges and opportunities for digital products. In the case of a hair-on-fire problem, the team addresses a clear and urgent need an audience has. The demand for such a product or experience is obvious. The audience members are aware of the problem. They are actively looking for solutions and are comparing existing ones.

Hardcore gamers are constantly on the lookout for new gaming experiences. The category for such products as digital games is crowded with competitors. A team building a new game must rise above the noise. The experience they deliver must be the best-in-class solution in order to succeed. The product and its user experience cannot just be faster or cheaper; it must be truly differentiated (Team Sequoia, 2023).

In the case of a hard-fact problem, the team faces a user need or a pain point that is universally accepted as a hard fact of life. Your team, and potentially your team alone, can see that while the problem might pose a difficult challenge, it is solvable. As a consequence, your solution disrupts the status quo with an unexpected approach.

The team's challenge, besides the obvious issue of creating a new-to-the-world product, lies in the fact that potential users are resigned to live with the problem. They are not urgently trying to solve the problem or seek experiences that address the need. The experience you create for the audience must overcome the force of habit. In order to adopt the new solution, your users will have to change their current behaviors. That means that your solution must be compelling enough to be worth making a change.

In the third case, the team offers not just a product or an experience but a vision of the future. What you bring to your audience will enable a new reality—a disruptive product like an iPhone. The challenge facing a team working on a vision-of-the-future problem is that the prospective users are not aware of the problem or the opportunity to accomplish something new. Your main obstacle is disbelief. We might argue that many aspects of immersive media fall under this category. Users are not entirely convinced that they need AR glasses to shop or travel. Movie audiences are not quite rushing to buy MR goggles to watch their favorite shows.

Part of the challenge of offering vision-of-the-future products is that a solution must come with its own ecosystem. Tesla was never just a car; it was a new way to power up personal transportation devices. Apple's iPhone truly came alive with a multitude of apps available on the Apple App Store. And Vision Pro is not just a headset; it is a new way to interact with computing devices. Vision-of-the-future products succeed only when they make the new paradigm irresistible.

Managing Entry: Sparking Curiosity

There is a critical moment during the entry (trial) phase of users engaging with digital products and services when a prospective audience member suddenly sees the benefit of using the product. While the "Wow!" moments are about creating excitement for the audience and "Aha!" moments are about the transformative power of the immersive experience, the "I'm curious" moments are all about drawing the audience deeper into immersion.

Different prospective users might have different needs and goals, but a core audience is very likely to share, at least in a general sense, a key trigger that converts them into becoming an engaged participant. Imagine yourself visiting New York City and walking down Broadway with a group of friends after taking in a show looking for a suitable dinner venue. Various members of your group will be attracted by different signs, decors, and food styles. Eventually, after a brief deliberation, you decide to enter a larger Irish pub–style place.

That is a start, but the search is not quite over yet. You might be further confirmed in your choice by the pleasant music being played. You might be repelled by the loudness of the place, by the menu, by the initial service and attention you receive, and so on. The Attraction stage and the Entry transition are very closely linked. Well-designed restaurant experiences are truly immersive. They offer their patrons just the right multimodal touchpoints at each step of the decision process.

Media makers wanting to create effective entry into the franchise focus on addressing the following questions. Is our attractor strong enough to convert the prospective user into a participant? Curating the optimal "I'm curious" moment for your XR application is not always obvious. Identifying it will require some trial-and-error iteration.

The next question your team will want to ask is: How quickly can we (should we) get the prospective user to the "Wow!" moment? Can or should the path to the "Wow!" moment be shortened? You will want to keep track of how many prospective users are dropping out on their way to Wow! Where exactly are they dropping out and why?

One way to answer this question is to conduct a Netflix type of experiment by testing your entry with prospective audience members to find out what was the moment when they got excited or bored and why. Getting no response or a weak response from the audience at this early stage might likely mean a need for going back to the drawing board to discover what would delight your prospective audience members and draw them into the franchise.

This is, again, where understanding your audience's needs, goals, and motivations comes in. Is the lack of the wow response due to a faulty value hypothesis? Is your XR product really addressing a core need? Is the lack of conversion due to the incorrect feasibility/usability hypothesis? Is the product too difficult to use? Is the experience not compelling enough due to insufficient technology? All of these issues can be addressed with proof-of-concept prototypes. As painful as they might seem at the moment, they are ultimately fixable and are much easier to fix before the play phase begins.

Reflect on what attracts you to various media franchises. Analyze each touchpoint that drew you nearer to immersing yourself in the experience. Remember our definition of user experience design as the application of objective stimuli to elicit subjective states in the audience. This definition applies every bit to extending an invitation for the audience to participate in the XR experience as it does to the creation of the experience itself. The Entry stage, like the other stages and transitions of UX design, needs to be intentionally planned and executed based on the empathic understanding of the needs of the primary audience.

In the UX world, we begin the Entry design process by (1) identifying the key emotional (Wow! and Aha!) states we like to elicit in the audience. (2) We choose the appropriate touchpoints as catalysts for the attainment of these states. (3) We break down each touchpoint by the associated modality or sensory channel it needs to engage. (4) We break down each modality into submodalities such as shape,

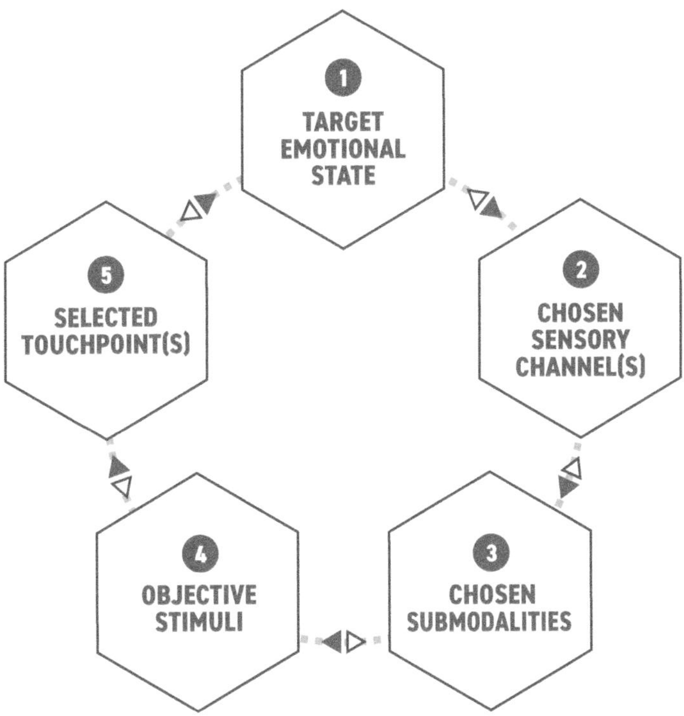

Figure 10.3 The curiosity elicitation model

Source: Concept and illustration by © 2024 Peter (Zak) Zakrzewski.

color, and texture (visual channel), or tone, pitch, and volume (auditory channel), etc. (5) Finally, we create the objective stimuli that can be experienced on the sensory level by the participant. As discussed previously, the objective stimuli are based on the formal design elements that are the foundational building blocks of multimodal composition of immersive experiences.

From the audience perspective, the entry experience process starts when (1) the audience encounters a particular touchpoint. (2) The audience members process the objective stimulus on a sensory level. Their nervous system interprets the particular sets of submodalities within the appropriate sensory channel. (3) Finally, the audience members interpret the sensory information by attaining a subjective emotional and cognitive state.

The cumulative effect of each touchpoint contributes to the creation of the subjective emotional state that affects each participant's behavior. Your team can apply this framework to the creation of any touchpoint, including the social media content promoting your XR experience. If you wanted to promote a VR game, you could start by targeting a "Wow!" moment emotional state best suited to represent the immersive fun the game offers. You would identify a particularly outstanding interaction within the game. If you were using social media platforms for the attraction, you would create a video clip leveraging the visual and auditory channels.

The submodalities you choose will depend on the medium used. In the case of social media shared on a mobile device, you could convey the excitement of playing the game by showing action and playing sound effects involved. If you had a chance to engage the prospective audience in a demo, you could add haptics and more tactile elements to convey the immersive fun and excitement of the experience.

XR Audience Segments: Do They Know It's Christmas?

By the time your team is ready to invite the audience to participate, you have already done the hardest part of promoting your XR app by investing the time to discover your audience's experiential needs and building version 1.0 of your content around it. Immersive media makers have an expanding audience in such diverse areas as gaming and entertainment, education, healthcare, manufacturing, retail, and real estate.

The global XR market has been steadily growing and is forecasted to continuously grow to reach $62.03 billion by 2029 (Alsop, 2024). But this new market is still uncharted territory for many prospective users. Even after a deep dive, an audience, especially an audience for a new medium such as an XR, can never be taken for granted. There will be some surprises along the way to wide adoption, so we recommend paying close attention to what the test users are doing and saying. This does not mean asking people what they want; it rather means listening and observing how they respond.

Like audiences for all digital products, XR users need to be segmented into the most optimal groups that can be addressed by some common offering. Different segments will require not only different content but sometimes entirely different attraction and entrance strategies. Professional users such as nurses and repair technicians will need much more training than immersive retail shoppers. The same professionals will be guided to the XR application by the organizations that employ them. Audiences for immersive entertainment will very likely, at least for the near future, be attracted to immersive experiences through traditional social media channels.

Gamers who spend considerable time in immersive worlds can be attracted to new gaming experiences from within the XR world or via social media. Attracting sports fans and moviegoers to XR will require developing novel use cases whose experience can be supported by more advanced XR technology. Most mobile phone users already have one foot in the door of immersive media when they consult the GPS apps on their devices. Immersive media makers using tablet-based AR for various use cases, from interior design to education, can follow the familiar app marketing strategy used by mobile app creators.

In order to address our core XR audience in a meaningful way, we need to segment the users into optimal groups based not only on their needs as understood by us, the media makers, but more importantly based on the awareness they have of their own needs. In the book *Breakthrough Advertising*, Eugene Schwartz introduced the framework of five levels of audience awareness, from the acceptance of the status quo and complete unawareness of the need to full solution and even specific product awareness (Schwartz, 2004).

Understanding these psychographic segments can be very helpful in shaping the attraction-stage strategy. At the most extreme end, your team might be dealing with a prospective audience who is unaware of their own pain point, problem, or need. They might not be able to articulate or be willing to admit to having a particular issue or need. Their need might seem so general and amorphous that they cannot consciously articulate what might be missing or could be improved. At the dawn of the personal computer era, most people had no idea what they might do with a computer. They had no conception of their own use cases for any type of a computational device.

XR media is, in many ways, still at this nascent stage. Bulky headsets that mess up people's hair and disconnect them from the people physically present around them do not inspire extensive use or adoption of the new technology. Anyone trying to propose a solution or even explain a potential need to people unaware of their own pain point and possibility of change will be met with indifference because the potential audience does not have the conscious experience or a mental model to help them understand it.

As a skilled impresario, Steve Jobs knew that in order to convince ordinary people to invite personal computers into their lives, Apple could not just show their products and hope for the best. They had to actively seduce their audience. While promoting their early products, and in fact the entire product category, Apple employed some of the most creative persuaders in the world. The team comprised of Steve Hayden, Brent Thomas, and Lee Clow at Chiat/Day introduced the Macintosh with one of the most talked-about marketing campaigns of all time.

At the next level, we find a problem-aware audience. This type of potential user is aware of experiencing a pain point, some kind of frustration, or an unfulfilled need. Within the context of their life project, something makes the products, services, or experiences they are currently using insufficient. This creates stress or dissatisfaction. However, the members of this group do not know what to do about the problem because they are unaware of a solution. This audience segment needs to be made aware of the existence of a solution or an available product or service.

People who move beyond the initial awareness of the problem enter the next level of those who become solution aware when they begin their search for a solution. People in this psychographic segment actively look, compare, and consider using whatever product or experience makes their life easier or more enjoyable. A solution-aware person finds what they need by either starting with an online search (individual exploration) or asking friends, coworkers, and peers within their community (communal exploration) who have relevant experience. They will investigate various existing solutions and eventually move to the product-aware level.

People who become product aware might become aware of your or someone else's product. This could mean they heard about you from a trusted friend, saw your social media posts, visited your landing page, or read some product reviews. Audiences at this level are ready to start the engagement process, which could include signing up for trials and demos. They might also access your social media content and watch product previews and tutorials. Once attracted to your specific offering, they will consider whether your product can address their need or solve their problem. They will put in an effort to become familiar with a range of brands and products to compare and contrast them against the competition.

At the very last level, we find the most aware prospects. These individuals are ready to make a decision and engage with your product, but they might need something to push them over the line. They might be unsure where or how to get your product, or they might be prioritizing other purchases or experiences.

Based on this framework, we can distinguish between four categories of immersive media users based on the level of persuasion needed to convince them to engage with immersive content.

1. Within the consumer segment, XR gamers are one of the most established XR user groups. They have embraced immersive gaming for a long time. They are members of tightly knit online communities that use word of mouth to discover and share new and interesting experiences. They are well aware of their needs for new and exciting forms of immersive media experiences. They will know what they like when they see it.
2. Professional and enterprise XR users in such well explored fields as medicine or manufacturing are very likely to be aware of the benefits of XR solutions to address their needs and problems but might not know of a specific solution or a particular product. This segment might rely on IT professionals to recommend and implement XR solutions for them.

3. Consumer sports and entertainment audiences are less convinced that their entertainment needs can be met by XR media. They still have not widely embraced XR. Addressing this audience will take a lot more seduction to draw them not just to a particular immersive experience but to the entire XR media category. They might have used mobile AR on their smartphones and tablets without even realizing it. They might be open to engaging with alternative immersive experiences at familiar places such as Disney World or visiting tourist destinations like New York City's theater district. Invite these folks with multimodal persuasion and charm and guide them to immersion gently.
4. Lastly, we have the equivalent of personal computer users in the early 1980s—people who cannot think of a reason to bother wearing VR, AR, or MR headsets. Before we can invite them to shop using smart glasses, learn brain anatomy, rehearse a complex heart surgery, study prehistoric worlds of pyramids and dinosaurs, or buy their next car or home in the immersive space, we will need to do a lot of seducing. Recognizing your audience segment based on their comfort with and understanding of the benefits of using the still-emerging world of XR can be critical to the success of your own project. It can also be a determining factor in the wider adoption of XR as a medium.

App Store Optimization (ASO) to Grow Your XR Media Franchise

For digital and immersive apps, app stores are the most important discovery channels. App store algorithms play favorites by displaying results for the most popular apps. In the app store world, the winner takes all, so becoming popular on an app store can make or break your XR content.

App store optimization (ASO) refers to the process of improving the visibility and conversion rate of an application within such places as Meta (Quest) and Apple (Vision Pro) app stores. Well-designed ASO strategy extends your app's conversion rate optimization (CRO) by focusing on increasing the number of users who download the app after encountering it directly in the respective app or by learning about it through your marketing campaign. Well-crafted ASOs are critical factors for increasing discoverability within the app stores that drive a higher volume of organic installs and impact the conversion rate of paid and earned exposure. Organic installs are typically performed by the media franchise's most engaged user base in terms of lifetime value (LTV).

Effective ASOs need to include optimized keywords strategically placed within the metadata and engaging visuals (icons, screenshots, and videos) to make sure that your XR app surfaces when users conduct a search within the app stores. Here again, understanding your prospective audience is key to knowing what they might be actively searching for. ASOs are typically localized for each relevant market. They need to be tweaked for updates and continually iterated to keep the engagement high.

The key benefits of creating a strong ASO as a part of an app market diffusion strategy include four key benefits.

1. App discoverability can be dramatically increased by assisting the prospective users in finding the app in the app stores to lead to more organic scaling of downloads. Achieving higher rankings for relevant keywords will mean not only a bigger audience but also a more relevant and engaged audience.
2. Organically attracting a core audience leads to the increase of high-quality users with the highest LTV.
3. By generating continuous, scalable growth, high app store rankings will lower your user acquisition (UA) costs. This means that your team will be able to invest the precious resources in other marketing channels or in improvements to the XR content itself.
4. Effective ASOs also lead more prospective users to your app's landing page, leading to higher page ranking, more app reviews, and overall broader exposure.

Successful ASO strategy is based on continuous improvements that keep pace with changing user expectations and evolving search patterns. Key components of a successful ASO strategy include such vital branding elements as

1. The app's name and visual identity. A unique name should ideally include a primary keyword.
2. App description needs to succinctly capture your app's core benefit.
3. To localize your app's promotion you need to invest time and effort to translate your ASO branding in your primary language to the requirements of the international listings.
4. Crafting the right visual branding that includes screenshots and videos to represent the app has proven to be essential to achieving high conversion rates. Your app's landing page also needs to be visually engaging while giving the prospective audience an accurate representation of what they can expect from using your app.

It is important to keep in mind that even though ranking factors and search algorithms differ between app stores, the fundamental goal of each platform is to offer users the best possible options according to the users' search criteria. Keyword searches based on your provided data and metadata are the backbone of your ASO strategy. Finding out which keywords users might be searching is an ongoing learning experience that will give your team invaluable insight into how your app aligns with the user needs.

Comprehensive keyword development typically follows four critical steps.

1. Start with a brainstorming session to identify keywords that closely relate to the values and identity of the core audience, address the audience's unmet need, and highlight key features and benefits.

2. Research any existing competitors to determine which important keywords they are using and which they might be missing. Both of these categories represent opportunities to expand your list and to make your keywords more unique.
3. Create separate lists for each app store, as there are differences in algorithms and user search habits between the two.
4. Read app reviews to find additional insights and information that might be relevant to supporting your app. Tools like App Radar and Gummicube can help you with your keyword research and competitor analysis.

Promoting Immersive Games

Building a strong community around your XR experience is the best way to start building a transmedia franchise. As we suggested before, the process of inviting the audience begins right at the moment the media design process starts. Like most immersive media applications, XR games are different from established media forms such as console video games which are better understood by large segments of the potential audience.

Gamers outside of avid XR game fans need to be drawn into the unique experience of playing immersive games through teasers, trailers, demo videos, landing pages, and best-yet demos. Posting game previews on YouTube and other social media channels at any stage of the development process is a great way to build curiosity around the project. Launch your dedicated Facebook, Instagram, Reddit, and YouTube accounts, and post frequently to engage with the community and to establish your credibility.

Keep in mind that it is usually more effective to maximize the opportunity of using fewer channels with higher-appealing content and messaging than to use every channel available poorly. YouTube is considered by many game developers to be one of the best video social media platforms for XR gaming promotion. YouTube dominates the long-from video landscape. The platform can help you reach audiences from every corner of the globe. It can accommodate all types of promotional content from tutorials to reviews, vlogs, animations, and demos.

Whatever channel you are using, make sure to select the most effective SEO keywords for your game in the description. Create an elevator pitch that outlines the theme, characters, story, gameplay elements, hardware requirements, and other relevant information you can share in any format. Keep it short, something you can tell someone on a one-minute ride in an elevator. It is called an elevator pitch for a reason. Create a daily blog post on your own site or on Medium, Quora, or LinkedIn during the development process to draw the prospective audience into your franchise by letting them in on your struggles and achievements.

If you have a unique story to tell, introduce the company behind the product to highlight any previous games you might have developed, to share your unique vision, or to showcase a new technology that can impact the gaming experience. Once you are ready, your team can create a simple landing page site that prompts visitors to download beta tests and demos.

A good example of stoking curiosity is Survios Studios' landing page for *Alien Rogue Incursion* available for Meta Quest, PlayStation VR, and Steam VR. Let the prospective audience know which XR platform they can access your game on. Is it PlayStation VR, Meta Quest, Steam VR, and HTC Vive, and/or Apple Vision Pro? Put your game in the hands of XR gaming influencers, podcasters, bloggers, and game reviewers who can say good stuff about your game early on. Send complimentary game keys to the most qualified XR gaming social media influencers.

Promoting Immersive Consumer Apps

At this point in the history of XR, the audience for consumer apps is both much larger and harder to attract than the generally younger, tech-savvy, dedicated audience of gamers. The key strategy with this market segment is to offer usable solutions to various unmet needs. As with XR games, posting frequently on dedicated Facebook, Instagram, Reddit, and YouTube accounts is the best way to reach and engage with a community of users, to pique interest, and to establish credibility.

To become a trusted voice within a user community, stay authentic and relevant by showing that your team is aware of the community's interests. Whenever possible, respond to community news by connecting those interests to your app's value proposition to create awareness of the benefits of using your digital product. One of the biggest decisions your team will make is choosing the platform for which your consumer XR app will be designed.

Pokémon GO became one of the biggest mobile games in history by leveraging an existing smartphone platform rather than by opting for a more exotic virtual reality (VR) format that would require users to acquire new devices and new media consumption habits. Mobile AR is and will be for the next few years one of the most widely adopted and accessible XR platforms for consumer apps. The successful launch of *Pokémon GO* holds other marketing lessons. The game deeply resonated with its intended audience by tapping into a powerful human urge to engage in treasure hunting.

Paying attention to fundamental human needs has nothing to do with technology. As a matter of fact, Brian Arthur argues that disruptive innovation is not really a parade of inventions followed by subsequent adoptions but a constant fresh reexpressing of old tasks such as travel, entertainment, or productivity accomplished within the new realms of technological feasibility (Arthur, 2009). Remember, audiences are not looking to adopt cool new technologies; they are motivated by finding solutions to jobs that need to be done in their lives.

In this respect, *Pokémon GO* offered few subtle attractors. It inserted a fictional narrative into such everyday tasks as taking a walk to give its users a feeling of a more enhanced experience of reality. The game acted as a conduit for social meet-and-greets that also got otherwise sedentary people up and moving.

Remember that Apple and Meta offer extensive marketing support to their developers in order to gain market dominance for their particular MR/VR platform. Meta offers many ways to promote your Quest app on their store. These include opportunities to showcase your images and trailers and developer posts.

For information on how to maximize your Meta Quest store presence, you can access their App Submission Guide and Store Asset guidelines.

As a VR/MR platform, Apple's Vision Pro offers a unique usability value proposition. Vision Pro brings the users closer to the ideal of natural user interfaces (NUI) by dispensing with controllers and switching to eye tracking, hand gestures, and voice commands instead. Apple's UI makes the user experience more accessible to a broader audience. Vision Pro offers other user benefits such as enhanced security and high performance that make XR experiences more flawless, richer, and more detailed. Apple's XR platform has its own dedicated App Store that offers developers a new algorithm and new ways to optimize their app for visibility and trial.

To get the most out of the Vision Pro platform, it is important to keep in mind that the App Store is redesigned for spatial computing headsets with focus on the in-app experience that changes the traditional app search and discovery. Since Vision Pro comes with no physical keyboard, users conduct their searches with air typing on a virtual keyboard, by pinching while looking at each letter on the virtual keyboard, or by voice dictation. This means that apps have to be optimized for a wide range of search methods that includes using metadata for voice dictation, which seems to be the fastest and most convenient visionOS search method (Watts, 2024).

Vision Pro App Store will recommend your app based on the metadata that includes your app's name, subtitle, and a concise, informative description with a short list of main application features. Make sure to use high-value keywords in all of these descriptions. With the focus on spatial computing, Vision Pro App Store places more emphasis on the visual and experiential aspects of the application, so invest time in the visual presentation. Optimize your app icons for Vision Pro's new spatial icon format. To learn more about visionOS best practices and creative guidelines, check out the extensive developers guidelines on Apple's website.

Promoting Enterprise Apps

Enterprise XR solutions now range from retail to design, manufacturing, and medical XR. As more and more organizations integrate extended reality into their operations, opportunities within this space will continue to grow. Unlike gaming and consumer level apps, enterprise XR applications are ruled by the same business logic that drives all enterprise-level digital products.

A clear map to success in the enterprise XR requires clear identification of a problem and use case. Solutions in search of a problem will not cut it with managers and purchasing executives responsible for XR technology acquisition. Enterprise apps need to fit organizational objectives to be even considered. Because this type of immersive application can only be developed with deep disciplinary knowledge, the best development approach follows the co-design model.

Your team will need to collaborate with the actual stakeholders and field experts, be it in finance, IT, or medicine, from the start. That dialog will continue throughout the entire process. The only difference at the attraction phase will be the progressively increasing number of potential users involved in the conversation.

Effectiveness of the solution will matter a great deal, and so will the key performance indicators, but do not forget usability.

Throughout the previewing and demoing process, pay close attention to how the actual users (doctors, nurses, technicians) respond to using your application for extended periods of time. Constantly learning from user feedback is part and parcel of organically growing your audience. Using XR for enterprise solutions is still relatively new, so your immersive application needs to be learned by the prospective users and not sold through traditional forms of marketing like promotional videos. The best way to enable potential users to learn about your app is to allow them to experience it for themselves by providing a hands-on demo. Your team can set up a permanent studio or a traveling demo in which users can try the technology with no risk and little pressure.

Promoting Immersive Experiences

Performance-oriented live immersive experiences have an advantage of coming to where the audience is physically located by the team choosing the right venue.

The very successful, pioneering Macbeth-meets-film noir immersive theatrical experience *Sleep No More* was originally staged in London at the abandoned Beaufoy Building. It moved to Brookline, Massachusetts in collaboration with Boston's American Repertory Theatre at the Old Lincoln School. The production truly expanded its audience when it moved to the off-Broadway Chelsea neighborhood at 530 West 27th Street in Manhattan. Many existing venues such as existing theaters, unused storefronts and warehouses, school gyms, and hotel ballrooms and conference rooms can potentially be adapted to work for an immersive art experience. Choosing the right venue for staging an immersive experience does not just include the venue shape, size, and technology accommodation but also the ease of access and discoverability of the physical location.

While the audience for XR events might vary from dedicated fans who seek out new experiences to novices who are just getting into the new genre, the same promotional basics similar to promoting XR games still apply. Building a strong community around your XR experience early is the best way to invite the prospective audience into the world of your XR media franchise. The process of inviting the audience can begin in parallel with the creative one. Create an elevator pitch that outlines your dramatic theme, characters, story, immersive elements, and other relevant information you can share in any traditional or social media format.

Each contact point between the XR media franchise and the audience communities it serves is as important to the success of the franchise as is the contact point with the ball to a tennis or golf player. One could argue that a contact point is all that matters. Touchpoints shape the audience perception from Attraction to Entry and beyond. This makes the choice and care of crafting the media franchise touchpoints of paramount importance. These touchpoints can range from the project's Facebook page, a TikTok video by an influencer, or an Instagram post on an opinion leader's account to mainstream media news stories about the product or the

XR experience itself. A media franchise is a sum total of the users' experience with the XR application, which makes forming relationships, creating experiences, and shaping perceptions the essential priorities for immersive media makers. Successful media makers not only brand their experiences, but they keep their franchises fresh and alive at all times.

Social media is still one of the best tools for growing a media franchise. The most successful game creators, media makers, personal brands like Beyoncé, and corporate ones like Nike have a well-coordinated and expertly curated social media presence. Social media channels are critical to spreading the word about XR media projects. They excel as vehicles for reaching entire communities of potential users.

One thing to keep in mind is that in and of itself, social media is not a strategy. Promoting a digital product takes a lot of time and resources. This is why a realistic strategy will help the team to focus energy and resources where they can do the most good. Different platforms need to be prioritized based on how aligned their audience and user experience might be to your audience and your strategic goals.

In what follows, we cover the leading platforms to help you develop your strategy. Create some drama around your production with daily blog posts on your own site, TikTok, or Instagram during the development process. Draw the prospective audience into your immersive world by letting them in on your struggles and achievements. If you have a unique story to tell, introduce the creative team behind the production to highlight any previous performances your team might have developed. The most dedicated audiences are grown over time, so start the invitation process early and build on your success steadily.

Once you are ready to show previews, expose your production art critics and other cultural influencers, podcasters, bloggers, and reviewers who can say good things about your production early on. Stage complimentary performances for the most qualified art and cultural influencers.

Summary

In this chapter, we covered the critical step of attracting the audience to your immersive media content. We shared our views on common digital marketing terms such as "onboarding" and "purchasing funnel" to highlight the importance of mindfulness in how we approach our audience. We outlined the impact of relatability, resonance, and near peers to the process of media content adoption. We introduced the concept of a discovery path as a guiding principle behind content promotion. We highlighted the role of attractors and touchpoints in this process.

We also discussed the uniqueness of XR media and its audience-building strategies with regard to the more established forms of digital content making. Finally, we covered the specific tactics used to promote immersive media content, which included the use of social media and app store optimization (ASO) strategy.

Having attracted the interest of your prospective audience, you are now ready to immerse them in your XR world. This is the topic of our next chapter.

References

Alsop, T. (2024, January 10). Topic: XR: AR, VR, and the metaverse. *Statista*. https://www.statista.com/topics/6072/extended-reality-xr/

Arthur, W. B. (2009). *The nature of technology: What it is and how it evolves*. Free Press.

Keller, K. L. (1997). *Strategic brand management*. Pearson.

Schwartz, E. M. (2004). *Breakthrough advertising*. Bottom Line Books.

Team Sequoia. (2023). The Arc product market fit framework. *Sequoia Capital*. www.sequoiacap.com/article/pmf-framework/

Watts, D. (2024, March 12). App Store optimization for the Apple Vision Pro. *Yodel Mobile*. https://yodelmobile.com/app-store-optimization-apple-vision-pro/

11 Immersing the Audience

The Privileged Places and Moments and the Four Degrees of Immersive Transportation

In his 1958 book about architecture titled *The Poetics of Space*, French philosopher Gaston Bachelard expressed the powerful impact our environments have on us by arguing that by leaving the space of our usual surroundings and walking into new ones, we enter into new forms of communication with spaces that can be psychically transformative (Bachelard, 1994). Bachelard suggested that by changing places, we create opportunities to change our nature.

It is not just the environments that we immerse ourselves in that affect us; it is also the timing that makes a difference to how we respond to any form of rhetorical communication. Immersive experiences unfold in four dimensions of space and time. Psychologist Robert Cialdini addresses the time factor of rhetoric when he discusses the concept of privileged moments as an aspect of persuasion. Cialdini found in his research that whatever concepts are elevated in a person's attention at the time of decision-making will be the most critical factors in determining a person's choice in a given situation (Cialdini, 2016, p. 24).

In other words, people have to be ready for the message. The individual audience member's state of mind lays the foundation for the potential of persuasion and transformation. Audience members have to be ready or prepared for the experience and ready, in turn, for each phase of it.

For instance, marketing researchers Bolkan and Andersen demonstrated that they could significantly increase the chance of recruiting people to participate in surveys by beginning a conversation with a particularly "pre-suasive" opener such as asking people if they considered themselves to be "helpful" (p. 26). Once the concept of helpfulness was elevated in the respondents' attention, they were more likely to be well disposed toward being helpful to the researchers.

Something similar happens with immersive presence—the conditions for the participant to be immersively present must be just right. Audience members must be attracted to the XR experience before they can enter into it. They have to enter before they can immerse themselves in it, And they have to be immersed before they can experience a mediated presence.

DOI: 10.4324/9781003442462-18

The word "privileged" refers to something being given a special, elevated status. The word "moment" evokes two different but equally useful meanings. The first one connotes a time-limited period of opportunity. The second relates to the concept of momentum in physics, which refers to a unique leveraging force that can bring about unprecedented change. Cialdini argues that putting together the two dimensions of privileged moments, one temporal and one physical, can have the capacity to bring about extraordinary transformation in the domain of human psychology. Supporting audience members in achieving a state of mediated presence is the holy grail of immersive media design.

With this in mind, we have adapted Marie-Laure Ryan's framework of four degrees of narrative transportation (2015) to help XR design teams assess the states of presence their experiences offer.

1. In the Absence mode, the immersive world offers either insufficient engagement or such a barrier to entry that the participants remain absent from the experience, vulnerable to the distracting stimuli of external reality, or shift their focus to reflective mental states. In XR, these undesirable states are referred to as the breaks in presence.
2. Imaginative Involvement is a state in which participants who enter the immersive world are not fully immersed and can contemplate the XR world with sensory, aesthetic, or epistemological detachment.
3. Presence refers to the sweet spot of XR experience design, a pleasure stemming from a nonreflexive engagement enjoyed by participants who are so completely caught up in the XR world that they lose sight of anything external. This unique cognitive state must align all three levels of consciousness: the proto, the core, and the extended layers, which we will discuss in detail later in this chapter. Ryan acknowledges that audiences can be pushed too far in their narrative or immersive enjoyment.
4. On the extreme end of the scale, we find an XR-based Addiction, which carries a very similar connotation as its nondigital counterpart. Immersive addiction involves the loss of the capacity on the part of the participant to distinguish immersive worlds, especially those of fiction and fantasy, from the actual world. It can also motivate participants to continue the experience at any cost (Ryan, 2015). XR experience designers must be aware of this potential danger and build plenty of guardrails into their content against such extremes.

The immersive transportation framework illustrates that the magic of XR experiences is found deep inside the four-dimensional space that ranges between absence and addiction. Finding this elusive sweet spot of mediated presence requires careful planning and mindful attention to crafting compelling immersive spaces revealed at the privileged moments. The state of presence cannot be forced. Participants need to be guided toward the state of presence. The role of XR media designers is to participate in the process of immersive emergence we discussed in the previous chapters.

Managing Entry: From XR System as Designed to XR Experience as Encountered and Understood

The Immersion layer, which undergirds the state of presence, starts building with the entry, which will transport the audience from their ordinary surroundings to the virtual world with its own function, structure, coded laws, semantics, and behaviors. Audience members need to be carefully guided through this intricate process.

As XR designers, we need to remember that we only indirectly affect the audience's experience by creating the initial conditions and the shell of the system. We can inscribe rules-based environments that the audiences will inhabit, explore, and manipulate, but before presence can emerge, the system must come alive with the audience joining in. The joining-in process fractally resembles the diachronic view of XR experiences we outlined in Section II.

Whether on a micro or macro level, immersion starts with attracting the audience to the experience and is followed by the audience entering the XR system. If done right, it finally culminates in the emergent state of presence. Each phase is its own privileged moment, so before audience members can immerse themselves in the XR experience, they must be skillfully guided along the entry.

There is an intriguing analogy between narrative and immersive forms of transportation. The history of media seems like an ongoing process of mediating the audience's reality at ever-deeper levels of consciousness. Each step brings us closer to the core consciousness and even the proto-self. A printed book can only transport the reader in terms of extended consciousness. Magazines add more sensory stimulation with visual imagery and color. Electronic media such as radio and TV add sound and motion. Finally, XR technology aims to address all eight sensory channels. These media transport their audiences along a spectrum of dual reality stretched between the natural world and the collectively imagined fictions of games, stories, and ways of living, working, and accomplishing tasks.

Richard Gerrig outlined six stages of narrative immersion in his Folk Theory of Immersion. We extensively adapt Gerrig's framework (2018) to XRXD to outline how an audience becomes immersed in an experience.

1. A participant is attracted to an immersive world that promises to temporarily reshape their identity and situation.
2. Entry into this world is accomplished through immersive "transportation." During this phase, the participant is introduced to the rules of the new world. Since the new world (to varying degrees) differs from the laws of the participant's native reality, the participant must successfully turn system directions into an internal representation of a virtual world. The holy grail of immersive media design—XR presence—can only occur after effective immersion and once the participant internalizes the idiosyncratic laws of the immersive world, which overrule the principles of their native reality.
3. During the flow of XR presence, the participant travels some distance from their world of origin.

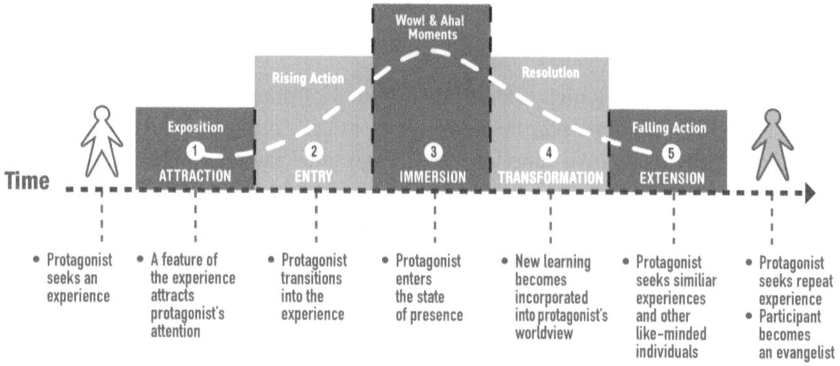

Figure 11.1 The diachronic view of XR experience

Source: Illustration by © 2024 Peter (Zak) Zakrzewski, adapted from Doblin Group's Compelling Experiences model.

4. As all good things must end, the participant exits the immersive world and reenters their native reality. The participant returns to the world of origin, changed by the journey.
5. In the extension phase, the participant wants to return to the XR reality to relive the experience, is motivated to connect with others who have experienced the immersive reality, and feels compelled to tell others to try it.

In *The Aesthetic of Play*, Brian Upton describes the task of designing compelling game experiences as a challenge of managing four broad categories of a game's constraints: the game as designed, the game as encountered by the participant, the game as understood by the participant, and the participant's conceptual background (existing worldview). He defines the game as designed as part of the initial inscribed system shell of static constraints that designers put in place before the experience begins (Upton, 2015).

In the context of the XRXD framework, we would like to change this order to suggest that audience research is the starting point of immersion.

1. The participant's conceptual background and goals form the broad collection of previously learned behaviors and strategies the participants bring to the immersive experience. For instance, the "do not be seen" strategy in many digital games relies on the player's prior experience playing such childhood games as peek-a-boo, hide-and-seek, and cops and robbers. Audiences bring various strategies for interpreting information and acting in the virtual world from their day-to-day experiences. This conceptual background can act as a resource for bridging the gap between the designed experience and the understood experience.
2. This is followed by the constraints of the designed immersive experience, which includes the structure of the information architecture, including the behaviors of the AI actants and the code that reinforces and enables that structure.

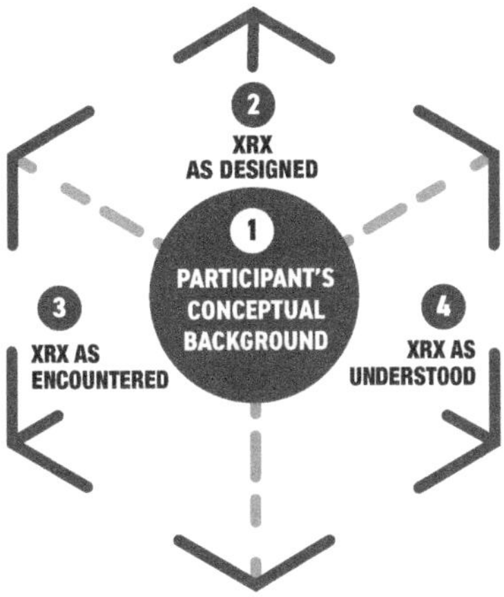

Figure 11.2 Immersive media constraints

Source: Concept and illustration by © 2024 Peter (Zak) Zakrzewski, based on Brian Upton, Varieties of Constraints.

3. The immersive experience as encountered comprises all of the active external constraints that will influence the participant's immediate actions. We can think of them as the rules that matter in each immediate moment of immersion. Finally, the goal of the entry phase is to address the experience as understood by the participants. This dimension of immersion represents the set of active constraints in the participant's mind. This mindset combines real-world knowledge, internalized system rules, and invented strategies. Fostering the experience as understood is critical to immersing the audience in an XR system. Achieving presence and flow depends on matching the framework created by the design team and the conceptual background with which the audience members start their journey.

The ability to traverse and bridge the conceptual gaps between immersive experiences as designed, as encountered, and as understood by the audience members is a fundamental component of becoming a skilled XR experience designer. Upton argues that if participants need a lot of help understanding a game, the designers need to do a better job of guiding them through the inscribed system (Upton, 2015). This means the rules might need to be simplified, or the learning curve needs to be lowered. He reminds immersive designers that the goal is not for the participant to play the game as designed but rather for the system "to guide the player toward a successful internal play space—the game as understood" (p. 50).

Like game developers, XR designers have direct control over only the static framework of the system's potential constraints, restrictions, and affordances as designed. The magic of presence happens only when this is matched by the dynamic responses participants make to the inscribed framework as they encounter it in the XR system.

Moving Freely Within an Immersive System of Constraints: Agency, Intrinsic Motivation, and Goal-Directed Immersion

The philosopher Marcus Aurelius shared an insight that might be of great value to XR system designers in his *Meditations*. He cautioned people in power to set their sights on realistic goals by advising them to realize that improving people and their circumstances, even to some small degree, is a worthy goal. He also suggested that the best instrument of influence is persuasion, because without the ability to change sentiment, those in power would be surrounded by reluctant, enslaved people and hypocrites (Marcus Aurelius & Hays, 2002).

As outlined throughout this book, the rhetorical power of immersive media gives the platform owners and system designers a lot of influence. There will be times when it might be tempting for immersive media makers to decide to procedurally coerce the audience into adopting behaviors and beliefs. We argue that the whole idea of designing ecologically sound human+computer systems is founded on the premises of participatory culture in which individual human agency is an integral part of collective emergence.

Real influence in immersive systems design comes from helping users achieve their goals, whether in entertainment, shopping, learning, or work. It logically follows that understanding audience goals and motivations is not only the best starting point but a constant compass for all design activities. The state of presence can only be approached when the audience moves freely within a system of well-designed immersive constraints. We rely here on Katie Salen and Eric Zimmerman's definition of play to argue that like the experience of play, immersive presence is not something that XR media makers directly create or control (Salen & Zimmerman, 2004, p. 316).

Presence and flow, whether in games or immersive environments, are emergent states that arise from the interaction between the system and the participant. Steve Kozlowski and Katherine Klein, who discuss emergence in the context of organizational systems, argue that system-level emergence originates in the cognition, affect, and behaviors of individuals, all of which are amplified by their interactions, to be finally manifested as a higher-level collective phenomenon (Kozlowski & Klein, 2000, p. 55).

The concept of emergence is highly relevant to understanding how individuals contribute to the overall system effectiveness. The freedom the participants enjoy within immersive systems involves explicit design decisions regarding their ability to choose their course of action based on genuine alternatives. We equate the agency of human participants with deliberate, conscious, and informed choice.

One of the most important factors contributing to the dynamics of emergence within XR systems is the goal-directed intrinsic motivation of the participants. Autonomous or intrinsic motivation involves the experience of volition and choice as opposed to controlled or extrinsic motivation, which involves the experience of being pressured and coerced (Delle Fave et al., 2011). Understanding users' motivations in terms of relatability, resonance, and goals is one of the best investments an XR design team can make toward achieving engagement, presence, and flow.

Optimal experiences are closely correlated with intrinsic rewards based on the users' lifelong construction of personal interests and goals. Assisting participants in achieving flow states "can trigger the active investment of time and effort in the practice and cultivation of the associated activities," progressively leading to increased skills, competencies, and search for higher challenges (p. viii).

The Three System Models and the Participant's Conceptual Background

Cooper et al. suggest that digital products can be represented by at least three distinct mental models. These include an implementation model, a user's model, and a represented model that bridges the two (Cooper et al., 2014, p. 18). Immersive media makers start by devising the system's implementation model, which outlines the architecture of software and hardware that allow the system to perform.

As critical as this model is to the design team, it is only partially and strategically revealed to the audience. Managing the audience entry into the XR experience is based on bridging the gap between the user's expectations based on their model of not just how the system should work but how they believe the world should work and the represented model of the system or the way designers choose to represent the working of the immersive system to the audience. This last shared model is referred to, in design, as a basis of a user interface (UI). The most important challenge for the immersive system designer is to make the represented model of the system match the user's mental model as closely as possible and to allow the participants to internalize that model as effortlessly and efficiently as possible.

Cooper et al. use an example of a mechanical movie projector to show how any machine can be described by the mechanism by which it accomplishes its function. A projector is an assemblage of intricately connected parts that create an illusion of motion on a movie screen. It shines a beam of bright light through a series of miniature still images embedded in a translucent film base flickering 24 times per second. The mechanism makes the illusion of motion possible, which supports the higher goal of narrative transportation the film audience goes through while watching a movie.

Media makers and theater owners of the analog era went out of their way to have the moviegoing public forget the mechanical substrate of gears, pulleys, sprocket holes, light sources, etc. (Cooper et al., 2014). Instead, they focused the audience's attention on an absorbing drama unfolding on the screen. Most moviegoers have little idea nor do they really care about how a projector works or how projection differs from the display of a movie on a computer screen. From the audience's perspective, technology is indistinguishable from magic as long as it delivers the desired experience.

We all use complex technologies involving smartphones, toaster ovens, and electric cars without knowing how these products work. We create convenient cognitive shortcuts to explain to ourselves how our gadgets work. Those explanations are powerful enough, argues Cooper, to allow us to interact with the objects without truly reflecting their inner mechanics. Immersive experiences are built on a substrate of cameras, sensors, advanced computer chips, algorithms, and modules of code that communicate with each other.

The system model is the engineers' representation of how such systems work. It is much easier for developers to design and explain XR products in ways that reflect their system model. While system models accurately reflect the infrastructure of engineering efforts, they do not provide a coherent user-centered schema to help the participants achieve their goals (in cognitive science, a schema is a mental framework people use to organize general knowledge, which helps to organize categories of information and their relationships). Applying system models as the driving system representations can alienate and confuse the users (p. 1). With immersive media, as in most computational media, any discrepancies between the user's mental and system models can make a massive difference in how the immersive application is experienced.

The participants' conceptual background—their model of the world—is the real starting point for the entry phase of immersive experience design. We start the design process by discovering who the prospective audience is likely to relate to, what themes might resonate with them, their goals, what technologies they are comfortable with, etc. Understanding the users' mental models is essential for creating compelling, immersive experiences, attracting the audience, managing their entry into the experience, immersing them in the experience, managing their exit, and supporting them during the extension phase in which they become active members and ambassadors for the community curated by the media makers around resonant themes, interests, and needs of the audience members.

Virtual worlds can leverage analogs to the physical world much better than traditional media forms. The ability to represent the functioning of digital media independent of the underlying operation of software and hardware allows skillful XR experience designers to create usability by masking the technical aspects of how digital products get their job done. Immersive experiences succeed by building bridges over the troubled water of XR technology.

Guiding Toward Understanding

Brian Christian, who investigated human, animal, and machine learning as part of his research into the development of AI, discusses the concept of shaping as critical to effective learning. Burrhus Frederic Skinner, the often misunderstood but now frequently quoted (in the AI community) social philosopher, psychologist, and advocate of extrinsic rewards, discovered the effectiveness of the gradual shaping of behavior by approximations of the final goal instead of waiting for the perfect response (Christian, 2020). Shaping[1] can be a useful technique for managing entry into an immersive system.

Users entering an immersive system learn how to navigate it through trial and error. They will likely first stumble onto what works and then gradually build up a tendency to do more of whatever appears to have worked. Machine learning researchers have perfected our understanding of trial-and-error approaches we humans go through when exploring new environments. They call their process RL.

In humanistic psychology, the process of helping people to progressively acquire new beneficial behaviors is known as guided mastery. In the context of XRXD, we will refer to this method as assisted learning.

There are four different approaches to assisted learning: (1) Model-based assisted learning refers to an agent explicitly trying to understand how the environment works to navigate it successfully. (2) An agent can hone their instincts in model-free learning. (3) Value learning approach relies on the agent learning how much reward certain actions can bring them. (4) In policy-based learning, the agent is guided by knowing which strategies can be more successful than others (Christian, 2020).

Guiding system users toward understanding the system by offering them rewards for accomplishing their goals is the most obvious usability strategy, but there are better ones available. RL researchers have identified a pernicious problem associated with environments that offer sparse rewards. They refer to this problem simply as a sparsity problem. Creating sparse reward environments by guiding learning through rewarding only the end goals can lead to much user frustration and very long wait times.

One of Christian's informants, Brown University's Michael Littman, illustrates the shortcomings of rewarding only high-level end goals by jokingly suggesting to his wife not to reward their infant son with food until he learns to fluently speak Mandarin (2020). Creating environments that offer users dense rewards on their way to internalizing the system rules is the best place to start, but what strategies are available to designers whose user goals or systems might be too complex for simple end-goal rewards? Fortunately, two other strategies are available if that is not an option. One is referred to in the RL community as incentives-based learning, and the other is known as curriculum-based learning.

Guiding Through Incentives

The first strategy for overcoming sparsity is to present the learners with a full-scale version of the system while adding some bonus rewards that point them in the right direction or encourage them toward behaviors that will lead them toward success. The intermediate incentives known as "pseudo-rewards" can be very helpful at giving the new users some sense of getting "hotter" or "colder" instead of flailing at random to accomplish their goals by pure chance or giving up altogether (2020).

The well-documented challenges of setting up a system of shaping rewards known to economists and management consultants comes down to the design problem of perverse incentives—rewarding behavior X while hoping for behavior Y. University of Toronto economist Joshua Gans found it out the hard way when recruiting his older daughter to help potty train her younger brother. She was offered a piece of candy every time she helped her brother go to the bathroom. The daughter quickly found a loophole in the reward system, which consisted of feeding her brother buckets of water to increase the frequency of his bathroom visits (2020).

In cases of incentives going wrong, it is important not to blame the user but rather to iterate the incentive-and-reward system by seeking deeper insights into what kinds of Xs are needed to get behavior Y. B.F. Skinner summed up this critical insight in his address at the Eastern Psychological Association meetings in Philadelphia in April 1955, titled *A Case History in Scientific Method*, by admitting that he had a hard time admitting to the design failures of his experiments. Eventually, he realized that when the experimental predictions went awry, the subjects behaved as they ought to have behaved. The subjects were always right. It was Skinner, as the designer of the experiment, who had made bad predictions and set up faulty experiments (Skinner, 1956).

The critical insight is that reward system design needs to balance the system's overall function with the rewards and penalties for moving away from the goal. Jane McGonigal, the author of *Reality Is Broken*, argues that one of the most compelling aspects of computer games is how clear they always make it what the player needs to do and how achievable a well-designed game always makes those actions seem. Game levels frame a perfect curriculum, while points constantly offer pseudo-rewards to move the player on their journey (McGonigal, 2011).

UC Berkeley AI scientist Stuart Russell argues that the reward-shaping system can be best understood as a "conservative field" in which pseudo-rewards are seen as potential energy nudges guiding the agent toward new learning (Christian, 2020, p. 169). Russell suggests that system designers must create performance measures based on what state they want in the environment rather than micromanaging how an agent should behave.

Christian's insight is to shift the focus of incentive structures of immersive spaces from actions to states. A child helping her brother go to the bathroom needs to be equally incentivized for the positive behavior and disincentivized for incessantly feeding him water to get more candy by the parents praising the desired state rather than rewarding individual actions.

Guiding Users Through a Well-Designed Curriculum

The expression "You must walk before you can run" succinctly captures the key insight of assisted learning. To encourage users to navigate complex systems, we first need to reward them strategically for simpler behaviors. We humans spend years and even decades moving through life with both literal, as in learning how to ride a bicycle, and figurative, as in acquiring education and life skills, training wheels.

Unlike nature, we do not always appreciate how the human world is frequently elaborately designed to be more learnable. Educators scaffold knowledge to be more accessible. Human resource specialists and UX/UI designers create elaborate onboarding experiences to help employees transition into the workplace. With digital products, onboarding refers to introducing users to new applications and helping them understand how to use them more effectively and efficiently.

We prefer to use the metaphor of *entrance transition*, but the concept is the same. Managing entrance before immersion is critical in ensuring that users have a positive experience with the XR application, continue to engage with it, and become ambassadors who spread the word. The typical digital product onboarding process can include creating an account, going through a tutorial on the product's features, and offering support and troubleshooting. Laying the path to successful entry has led to increased user engagement and improved retention rates for many digital products. Slack helps its users to enter by educating them with empty states and a friendly bot. Grammarly helps the users with an explicit learn-by-doing approach. Duolingo transitions their users with a guided product experience.

Great game experiences can shape not only our play but even our worldview. Legendary chess world champion Bobby Fischer used a shaping curriculum to instruct novice players first to recognize opportunities to end the game. His best-selling chess book starts with examples of checkmates accomplished in a single move and progresses to two-, three-, and four-move mating combinations. Fischer left the discussion of such complex topics as opening play, middle game, and longer-term strategies for later and more advanced books (Christian, 2020).

Nintendo's *Super Mario Bros.* is one of the best examples of carefully and brilliantly designed curriculum-based game entries. The game's legendary designer, Shigeru Miyamoto, wanted to teach the players in the most natural way possible the game's most important lessons: mushroom-looking characters were bad, and players should avoid them by jumping over them. In the first few seconds of playing, a new player is confronted by an enemy Goomba approaching them from the right. Not responding to this threat results in dying.

The lesson about dodging the "bad" mushrooms was relatively easy. The more challenging part was to teach players not to dodge but to seek the game's "good" mushrooms. The designers accomplished that feat elegantly by having the good mushrooms initially approach players in areas with too little headroom to jump over them easily. The players would naturally brace for impact upon seeing a mushroom character approach them, but in this case, instead of killing the player, the good mushrooms doubled the player in size. Once the game's mechanics have been communicated to and internalized by the players, the designers would let the players loose to immerse themselves in the experience.

The players were introduced to the game seemingly simply by playing, but they were carefully and inconspicuously trained by the system. Miyamoto's recipe was to help the users (1) learn the rules, (2) learn the exceptions and basic mechanics, and then (3) give them free rein. Complex behaviors can be instilled through successive approximations, whether aimed at helping the users navigate the system or achieving the system's overall function. Constructing an effective, learnable curriculum is one of the best ways to help users enter and succeed within complex immersive worlds. *SuperHot VR* and *Beat Saber* are good examples.

Consciousness and Presence

Vinod D. Deshmukh suggests that presence can be defined, in terms of human neurology, as a state of being fully consciously present and receptive to the current moment or situation (Deshmukh, 2022, p. 1). When we are present, we experience a sense of spontaneous existential unity while being aware of our embodiment and our embeddedness in the social and the natural world (2022). Presence is a critical evolutionary feature of the central nervous system and a defining feature of selfhood. Its function is to embed sensory information in the internal functional space of the nervous system for action in the environment and to maintain the balance, separation, and communication between the external and the internal worlds (Riva & Waterworth, 2003).

While experiencing the sense of presence is a unitary feeling, from the perspective of immersive experience design, it can be conceptually divided into three different layers based on the layered view of consciousness proposed by neuroscientist Antonio Damasio: proto-consciousness, core consciousness, and extended consciousness (Damasio, 2003).

Giuseppe Riva and John Waterworth use a visit to the Colosseum in Rome as an example of the three layers of consciousness underpinning presence. All experiences are underpinned by the proto-self receiving sensory signals from the five channels of the eyes, ears, nose, tongue, and skin. Within a few microseconds, this activity is monitored by the core self and becomes the content of one's core consciousness. A few microseconds more are required for the activation of extended consciousness. Some milliseconds later, it adds dispositional records of the place (or similar places) coming from our autobiographical memory.

We may recognize the place because we studied it in architectural history. We may have emotional ties to it because we associate it with special memories. Perhaps our most recent experience of the Colosseum was watching the movie *Gladiator* with our parents. The result of all of this activity is a single conscious experience integrating our perceptions, emotions, and feelings. Once the experience ends, it is restored in dispositional space with new data about our most recent experience and ready for the next (Riva & Waterworth, 2003).

Damasio's theory is based on a hierarchy of stages, each building upon the previous one. The Proto-self is the most basic representation of the organism that many species share. It is based on an unconscious process arising from the nervous system's constant interaction with the body to produce a collection of neural

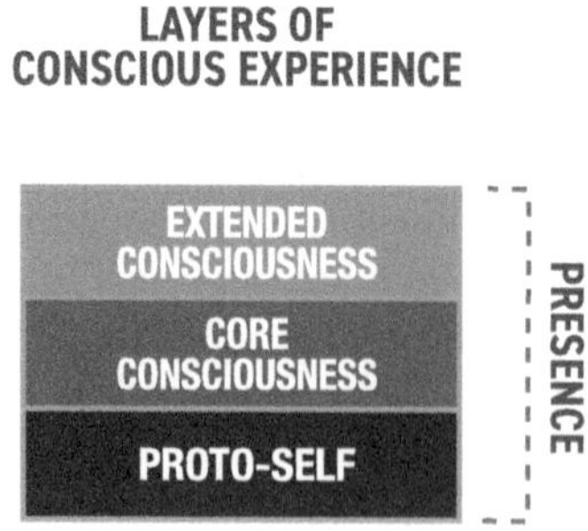

Figure 11.3 Three layers of conscious experience

Source: Illustration by © 2024 Peter (Zak) Zakrzewski, based on Damasio, Theory of Consciousness.

patterns representative of its internal state. The constantly updated mapping of the body's physiological state created by the Proto-self forms the basis for the brain to generate higher levels of conscious experience (Damasio, 1999).

Damasio suggests multiple brain areas are required to function properly in this preconscious state. The hypothalamus, the brain stem, and the insular cortex support the Proto-self. The functioning of the Proto-self does not require language. It communicates using neural objects or mental images whose patterns contribute to the rise of emotions, manifesting as the emergence of the second layer of self, known as core consciousness.

The emergent process Damasio identifies as his second conceptual layer is Core Consciousness. It arises when an organism develops a momentary sense of self and becomes consciously aware, through feelings, of the changes occurring within its internal bodily state. Core consciousness, like proto-selfhood, is not exclusive to human beings. All sentient beings experience mental patterns caused by interactions with internal or external stimuli as the nervous system continuously generates images representing the organism's experience of what philosophers of mind call qualia.

Damasio describes consciousness emerging from the layer of Core Consciousness as the feeling of knowing a feeling (Damasio, 1999). Core Consciousness emerges when the organism becomes aware of its Proto-self or the feeling of its bodily state. It concerns the present moment of here and now, not the reflection on past experiences or projections into the future. Those higher-level states are the domain of the Extended Consciousness.

Damasio's third layer emerges when consciousness moves beyond the here and now to reflect on past experiences, to plan future courses of action, and to imagine things not present to the senses. This highest level of consciousness is built firmly on top of its predecessors. Unlike the other layers, it is dependent on the use of all types of conventional memory, from working to short- and long-term memory. In Damasio's terms, the autobiographical self emerging out of Extended Consciousness draws on the memory of past experiences to generate higher levels of thought. Extended Consciousness also activates linguistic areas of the brain to enhance the organism's ability to reason, plan, and communicate.

Immersion, Presence, and Mediated Presence

One of the main reasons so much attention is paid to presence within the XR community is that for the first time in history, presence may be mediated or elicited by media technology as much as by the natural world. The possibility of immersive presence, or the mediation of the sense of "being there," is critical for our ability to design immersive environments as a flawless extension of the natural conscious state of "being here and now."

The promise of XR lies in the discovery that the neurological processes that allow people to make sense of the richness of stimuli supplied by the nervous and the sensory systems in the natural world can be leveraged in the realm of immersive media to augment cognition and decision-making within both the natural and the virtual worlds. Traditional forms of mediation, from books to film and television, affect only a limited number of sensory channels and layers of consciousness, frequently fragmenting the senses. This advanced form of mediation gives XR a unique status as a medium capable of delivering rich multimodal experiences that fully engage our embodied cognition.

One of the unique characteristics of XR is that it affects the human sensorimotor system in a way that tricks the brain into creating internal psychological states that transport participants beyond the immediate physical environment and into the virtual world of digital media. Virtual reality researchers refer to the simulated sense of "being there" and "acting there" as presence. In the context of XR, presence can be understood as a form of visceral communication (Jerald, 2016).

Researchers working in VR have distinguished between two possible forms of extended presence: telepresence, in which the human participants experience the feeling of being present at a location other than the actual, physical, and immediate one, and virtual presence, in which the participants feel themselves being transported to a computer-generated location supported by various immersive technologies (Sheridan, 1994).

Virtualized presence can take many forms. We can experience mental and embodied presence. We can also experience social and emotional presence when virtual environments simulate social interaction. Presence can be defined in at least two ways. One research perspective views presence as a subjective mental state in which a user "feels physically present within the computer-mediated environment" (Sanchez-Vives & Slater, 2005, p. 4).

Researchers who consider reality to be formed through action rather than mental filters define presence as the ability of the XR system to support the participant's actions in the virtual environment successfully. Under this definition, the perceived reality of the user experience is relative to the functionality of digital objects and environments rather than their appearances (Sanchez-Vives & Slater, 2005). While the experience-centered approach focuses on creating virtual environments that support the sense of "being there," the action-centered approach aims to create environments that support a targeted range of user actions and their agency to "act there."

Slater reserved the term "immersion" for the technology aspect of the XR experience objectively controlled by the system designers, arguing that the immersiveness of a system can be measured by the number of sensory channels it can affect while preserving fidelity concerning their real-world sensory modality equivalents (Slater, 2003). The immersiveness of the system can be objectively assessed by the design team based on its fidelity in such controllable submodalities of the visual channel as shape, color, and value or the submodalities of the auditory channel including volume, pitch, and tempo of the output delivered to the audience.

In explaining the difference between immersion and presence, Slater uses the analogy to color science, where a color can be described objectively as a wavelength distribution. This objective measurement differs from the subjective perception of color we all have when, for instance, viewing a sunset. He suggests that while immersion is analogous to an objectively controlled description of color, such as wavelength distribution, presence is analogous to how the audience perceives color.

The designers cannot directly control presence because it is a subjective state whose emergence can only be teased out or elicited by the media makers. The subjective state itself is a sentient human reaction to immersion, whether that immersion is in the natural world or is mediated via XR technology. Different participants may experience the same immersive system with varying levels of presence.

Different immersive systems may give rise to very similar levels of presence in different individuals (Slater, 2003). Sanchez-Vives and Slater (2005) clarify the distinction by suggesting that immersion is a product of technology, while presence is an internal physiological and psychological state experienced by participants immersed in XR environments.

For immersive media creators, it is important to remember that presence is an emergent phenomenon, a product of immersive technology eliciting a *subjective* state in the user. Your ability as the system designer to support audience members in attaining a sense of XR presence is the core of immersive transportation.

One of the most interesting features of presence is that participants in immersive media appear to experience at least partial inattentional blindness to the real world and the technical substrate of the virtual experience. Slater argues that mediated presence is based on two primary sensory illusions: the illusion of *place* and the illusion of *plausibility*. The place illusion addresses the question of location, while the plausibility illusion convinces the nervous system that whatever is happening is real.

Technologically driven factors such as low-latency head tracking, accurate 1:1 body tracking, and fast frame rate fool our perceptual system into believing that we are present in a place other than our real physical location. In the same way, a virtual world that accurately meets our expectations in the way it behaves and reacts to our actions makes us perceive the immersive environment as being coherent in its construction (Bye, 2017).

Immersing the Audience With Layers of Presence

Riva and Waterworth propose a particularly useful conceptual framework for immersing the audience by leveraging Damasio's three layers of consciousness and combining them with a three-dimensional model of the experience consisting of Focus, Locus, and Sensus. By applying Damasio's layers-of-consciousness model, Riva and Waterworth address the fundamental aspect of XR mediation that aims to simulate our conscious experience of the natural world.

To briefly review Damasio's model, the proto presence in the natural world is based on interoception, or the information about the body's internal state. Proprioception is a way of knowing bodily orientation in the environment, both of which can be subconscious. The proto layer's spatial presence aspect requires tracking body parts matched with appropriate and timely updating of displays. The physical substrate of XR matters greatly because our evolutionary survival depends on our ability to be always present in the real world. Whether we are engrossed in reading a book or playing games in virtual spaces, we are never completely divorced from the internal state of our bodies and our current physical situation.

Core presence, which is the pivot point for judgments about the outside world impacting the self, will be affected by the vividness of displays and other sensory inputs. We can think of the core layer in terms of sensory presence. Since core consciousness is integrated with proto consciousness, proto presence must be addressed by the XR system at all times to support the overall sense of presence.

The integration of these two levels produces a degree of immersion. The further we move down the chain from the extended consciousness, the higher the technological demands that will be placed on the medium. The perceptual models and predictions generated by the proto and core consciousness layers are evolutionarily designed to be generated rapidly to cope with interactions in the natural world. Mimicking this natural interactivity demands highly accurate and rapid response times between the XR system and its users.

Information addressing the proto and the core layers must be displayed in concrete rather than abstract ways to be accepted as realistic. Simulating proto presence presents XR media designers with the most demanding technological requirements, and at this point, our ability to meet this demand is far from complete.

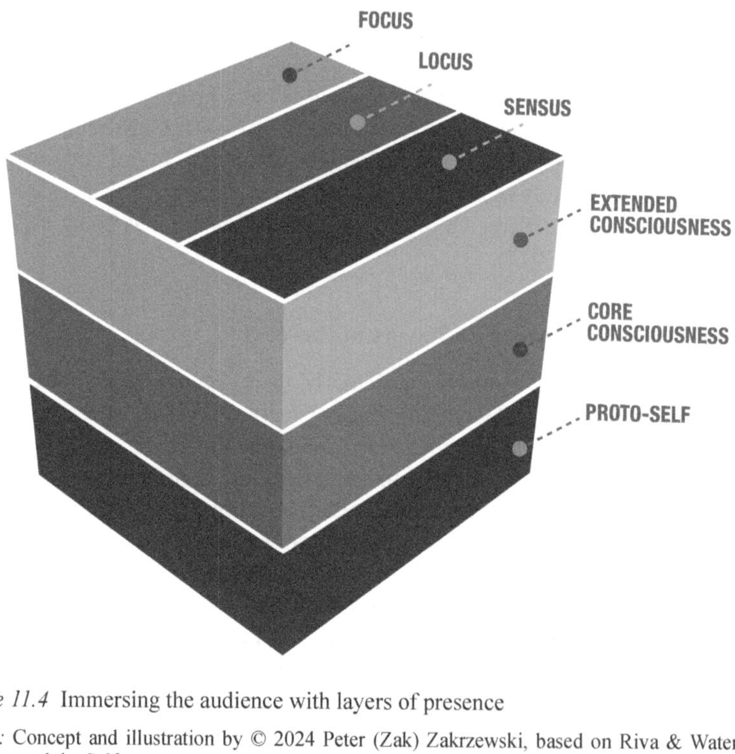

Figure 11.4 Immersing the audience with layers of presence

Source: Concept and illustration by © 2024 Peter (Zak) Zakrzewski, based on Riva & Waterworth, Presence and the Self.

Extended presence is built on top of the other two layers. Integrating and aligning the three layers of consciousness amounts to convincing the participant's nervous system to believe that something worth engaging with is happening to the self in the here and now. Absence of mind arises when extended consciousness is either minimally concerned or not involved with the current situation in which the other two layers might be involved.

The opposite might also be true. The extended self, engrossed in activity, might shift focus outside of the XR experience if it receives a sudden signal from the lower two levels that a break in presence has occurred. A great novel can deeply engage extended consciousness while one is sitting comfortably on a train. Still, we will be quickly jarred out of the narrative transportation experience if the train stops abruptly.

In the virtual and natural worlds, both the proto and the core consciousness are prerequisites for the privileged moment of extended presence. Only once the privileged moment of the extended immersive presence is achieved will the audience be ready for the intellectually and emotionally resonant content processed by the extended self (Riva & Waterworth, 2003).

The Focus of Mediated Presence

The Focus, Locus, and Sensus model was initially introduced by Eva Waterworth and John Waterworth. Combined with Damasio's three layers of consciousness, it gives the design team a powerful conceptual framework for designing the process of immersing the audience in the XR world, but even on its own, it can bring a lot of clarity to the design and evaluation of immersive experiences.

The focus aspect of the model addresses the nature of the participant's attention. They suggest that the more the participant's attention is directed toward present events, be it in the natural or virtual worlds, the higher the degree of presence that will be achieved than if attention is directed toward internally generated information that is not currently present in the environment (Riva & Waterworth, 2003).

In the context of presence, "focus can be seen as the degree to which the three layers of presence are integrated towards a particular situation" (Riva & Waterworth, 2003, p. 8). The more the XR environment supports the participant in integrating the three layers of consciousness, the higher the degree of presence will be experienced. Absence or break-in-presence arises when the highest layer of extended consciousness is less concerned with the current situation.

The proto-self exists moment by moment by monitoring the internal and external environment. The proprioception and interoception processes necessary to maintain our bodies from within and in the natural world are largely unconscious and automatic. From a survival perspective, our bodies existing in the natural world always serve as the background against which cognition is framed and have priority over reflective or virtual aspects of experience.

A break in presence spells the breakdown of the virtual illusion. Glitches at the physical and technological substrate level, such as loss of headset tracking or bumping into objects in a physical environment, are detrimental to XR experiences.

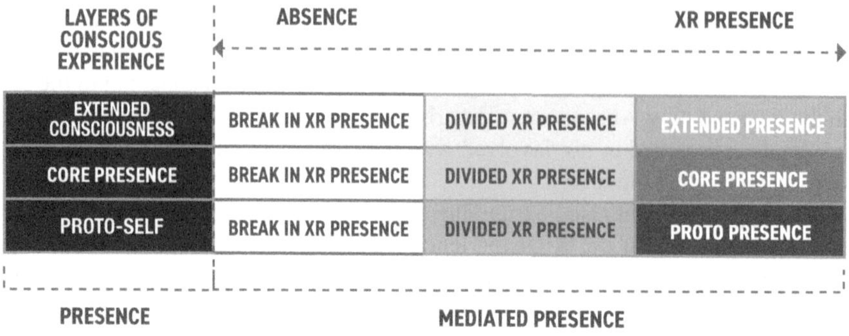

Figure 11.5 The focus aspect of mediated presence

Source: Illustration by © 2024 Peter (Zak) Zakrzewski, based on Waterworth & Waterworth, The Three Dimensions of Virtual Experience; Damasio, Theory of Consciousness.

The Locus of Mediated Presence

While the focus aspect of the model helps designers consider the degree of attention paid by the participant to events happening within the immersive experience, the locus addresses where the participant is focusing their attention, be it the physical world, the immersive world conveyed through media, or somewhere in between.

The goal of the concept of locus is to support the users in attaining their goals while avoiding any undesired breaks in presence. XR designers need to appreciate that the state of immersive presence they are trying to help their audiences achieve is in direct competition with presence in the natural world. The biological purpose of presence dictates that attention will be always dominated by the current state of the body (interoception) and the perceptions of the current position and balance of the body (proprioception and vestibular system) in relation to the physical environment in which the person is located.

As Slater and Steed argue, breaks in immersive presence are a natural outcome of rapid shifts of attention between the real and immersive worlds (Slater & Steed, 2000). Such shifts can be caused by the misalignment of Damasio's conceptual layers of consciousness. They are temporary changes in the locus of the participants' experience that can happen when the participant receives mixed messages from their proto, core, and extended conscious selves.

Some immersive experiences can possibly also, as Spagnolli and Gamberini argue, lead audiences to temporarily distribute their attention over a wider locus that spans both the real and the virtual worlds (Gamberini & Spagnolli, 2003). This is not necessarily an undesirable outcome when it comes to AR experiences when designers want the participants to be co-present in the real world and the world of information.

The locus aspect of immersive experience design is where narrative and immersive transportation concepts overlap. As Marie Laure Ryan (2015) suggests, traditional nonimmersive media such as books can transport readers to situations and places unrelated to their immediate situation or current position in the real world by engaging their extended consciousness without seriously affecting their core or

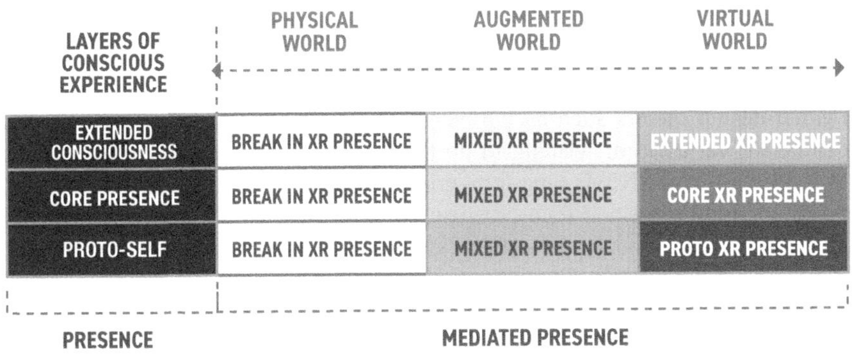

Figure 11.6 The locus aspect of mediated presence

Source: Concept and illustration by © 2024 Peter (Zak) Zakrzewski, based on Waterworth & Waterworth, The Three Dimensions of Virtual Experience; Damasio, Theory of Consciousness.

proto presence. When reading a book, we can easily temporarily shift our attention to adjust the position of our bodies in the chair we are sitting in only to return to the story without missing a beat.

The three layers of presence might be affected to a higher degree by electronic media such as television, personal computers, and gaming consoles. There are well-documented cases of gamers playing on to the point of serious dehydration. The entire furniture category of gaming chairs is a testament to the power of computational media to make us forget our physical location. But by addressing the eight senses discussed earlier in this book, no media has allowed media makers to affect and align all three layers of the audience's consciousness as much as XR. The question is, to paraphrase the 1979 hit song: Now that we can mediate presence, what are we gonna do with it?

The Sensus of Mediated Presence

Sensus, the third component of the model, refers to the level of attentional arousal engendered in the participant by the XR experience. Arousal directly affects the degree of engagement and activity exhibited by the participant during the experience.

We view Sensus in the context of the state of consciousness known as flow, initially investigated by Mihaly Csikszentmihalyi in the mid-1970s, and the wider concept of optimal experience discussed by Delle Fave, Massimini, and Bassi. Delle Fave et al. argue that flow lies at the core of psychological selection made freely by active agents who make choices based on personal meanings, goals, and experiences that are "partially constrained by biological and cultural inheritance" (Delle Fave et al., 2011, p. viii). In flow, Csikszentmihalyi identified a "particularly complex and positive state of consciousness" that was "characterized by deep involvement, absorption, and enjoyment in challenging tasks" in which individuals were able to perfectly match the levels of challenges presented by the tasks with their resources and skills (p. viii).

One of the key aspects of flow in the context of eliciting presence is the complete absorption and focus of attention exhibited by the individuals experiencing the flow-like situation. Optimal experiences are characterized by important experiential design factors such as intrinsic motivation, which are assessed through qualitative variables of wishing to do the activity. Other factors include a clear awareness of the activity's long-term goals and short-term stakes. Optimal experiences elicit effortless and focused attention, a sense of control over the situation, deep involvement, and positive affect. They transform participants' temporal experience, so time is perceived to be passing faster than usual (2011).

We base our interpretation of the Sensus aspect of the XRXD model partly on the experience fluctuation model from Delle Fave et al., which assesses the quality of the experience based on the balance of challenges and skills required. Low challenges matched by low skill levels produce worry, apathy, and boredom. High challenge levels not matched by similar skill levels produce anxiety. Lower challenge levels matched by high skill levels produce a state of relaxation and control. The state of flow emerges when the challenges presented by the experience are matched by the equally high levels of participants' skills and resources.

We augment Delle Fave et al.'s model of the skills/challenges continuum with two additional continua of positive/negative valence and high and low energy to

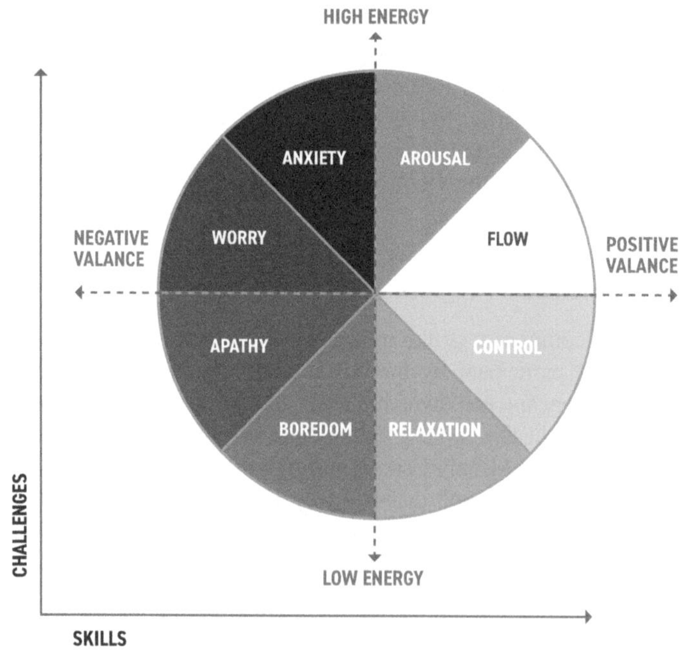

Figure 11.7 The Sensus aspect of mediated presence

Source: Illustration by © 2024 Peter (Zak) Zakrzewski, based on Riva & Waterworth, Presence and the Self. Delle Fave et al. Experience Fluctuation Model.

address the levels of affect experienced by the participants during the experience. Together, the matrix of Damasio's three levels of consciousness combined with Waterworth and Waterworth's Focus, Locus, and Sensus (2001) aspects of XR mediation can be a powerful tool that allows the design team to immerse the audience in the XR world with humane grace and creative skill required for the emergence of mediated presence.

Summary

In this chapter, we examined the dynamics involved in the process of immersing the audience XR media environments. We revisited the four-dimensionality of immersion in the dimensions of space and time by discussing Robert Cialdini's concept of "privileged moments," highlighting the impact of heightened attention on persuasion during critical decision-making moments (2016).

Expanding on Marie-Laure Ryan's theory of narrative transportation (2015), we categorize user engagement into four progressive stages: Absence, Imaginative Involvement, Presence, and Addiction. Each stage describes the depth of user immersion, from mere distraction to deep engagement, where the distinction between virtual and actual realities blurs, potentially leading to addictive behaviors. These stages underscore the gradual intensification of user engagement necessary for successful immersive experiences.

This immersive intensification process is crucial for achieving a state of presence, where users can lose (where appropriate) consciousness of their external surroundings and become fully absorbed in the XR content. We emphasized the importance of seamless integration into the virtual environment to foster this sense of presence, suggesting that technological setup and psychological impact play pivotal roles in this process.

Furthermore, we discussed the importance of finding the right balance between the freedom involved in respecting the user's agency and the requirements of system architecture. Effective immersive experiences require that users feel in control and can influence their experience within the environment. This balance is achieved through careful design that intuitively supports user interactions without leading to confusion or cognitive overload. We highlighted the critical understanding of presence as an emergent property that is a product of the dynamic interaction between the user and the system. This vital insight acknowledges the unpredictable nature of user responses and interactions with the immersive system.

Finally, we discussed Riva and Waterworth's conceptual framework that blends Damasio's layers-of-consciousness hypothesis with the Focus, Locus, and Sensus model of the experience as a way to immerse the audience in the XR world.

Having covered the critical step of mindfully immersing the audience in the immersive system, we are now ready to consider the final stage of the XR experience design process—the act of extending play beyond the initial immersive experience.

12 Extending Play

Spreadable Media: The Extended Play

Extension is the final and often overlooked phase of the experience design process. After audiences are attracted to the media content, after they enter and immerse themselves in the experience, they invariably exit to return to their everyday lives.

But great experiences transform the audience. Participants emerge changed, and frequently, they wish to continue to be involved in the life of the franchise. Millions of *Star Wars* and *Star Trek* fans group together into communities, collect memorabilia, produce fan media, and organize festivals centered around the resonant themes introduced by these media franchises. This spontaneously occurring dynamic creates a tremendous opportunity for media makers to extend the play beyond the original experience. Audience activities increase the potential cultural and economic value of media experiences by making individual and collective emotional investments in extending their shelf life.

Extending play is based on the definition of cultural production that is not limited to activities involving only media production skills. It includes such modes of participation as evaluation, appraisal, critique, and recirculation of content. Daniel Dayan has articulated an important distinction between audiences seen as collections of individuals and publics as coherent entities. We can draw parallels in contrasting individual fans with fandoms whose members identify with larger communities and their resonant themes. The changing media landscape has created a shift in collective power that comes from the transformation of audiences as separate, individual members to the connected, socially active communities of fans. The labor performed by the audience in extending play generates both cultural and commercial value (Jenkins et al., 2013).

Henry Jenkins, Sam Ford, and Joshua Green have introduced the concept of spreadable media to address the active participation of media audiences and their impact on media franchises. Spreadable media is part of the larger theme of participatory culture. The approach challenges media makers to consider and factor in the cultural desires and the ingenuity of their audience.

In his previous book, *Convergence Culture*, Jenkins argued that social media simply leveraged Web 2.0 technologies to retrieve the long-standing folk traditions of cultural production at the grassroots level (2006). While the interactive

DOI: 10.4324/9781003442462-19

platforms themselves were prestructured and controlled by media companies such as Facebook (Meta), Twitter (X), and YouTube, the participatory practices that went on within them were shaped by cultural and social protocols. Jenkins and his co-authors argue that the plurality of how media content spreads in culture, from grassroots to commercial, from top down to bottom up, has a single rule: "If it doesn't spread, it's dead" (Jenkins et al., 2013, p. 1).

All forms of cultural idea flows, including media franchises, replicate and emerge through the power of human agency. Media circulation is based on a hybrid of both top-down forces of media producers and bottom-up media consumption practices of socially connected audiences (2013). Networked connectedness of the audience is not a new phenomenon created by an array of technological tools but a continuation of word-of-mouth practices of sharing stories "by fire," as Jenkins and his colleagues write, or by "cloud" (p. 3).

Media content designed for spreadability have been leveraged by such diverse stakeholder groups as the Madison Avenue executives selling Old Spice deodorant, civic groups who work toward erasing social injustice, and independent content producers seeking to involve an audience beyond the point of initial or immediate immersive experience. Content creators can design media experiences that audience communities want to reengage with and promote by paying close attention to both the desirability and the mechanics that propel the grassroots acts of circulation. Best practices include making different parts of media content available where and when audiences want it and making it portable, easily reusable, resharable, and relevant to multiple audiences.

As mentioned in Chapter 10, the secret to inviting an audience lies in creating a regular stream of related material. The most impactful immersive experiences spark social exchanges by offering resonant content and then leaving open space for audience participation while providing resources for shared expression. The pioneering immersive theater production *Sleep No More* offers its audience individualized singular journeys within the play. It is designed to be an individual experience.

But while the audience members are actively encouraged to explore the McKittrick Hotel alone, that actually is meant to stimulate them to trade stories afterward. The co-director and choreographer of the show, Maxine Doyle, told *Interview* magazine that the creative team was very mindful of encouraging an extended dialog in which the audience members would share what has happened to them, where they have been, what they have done, and what they have seen (Symonds, 2011). The shared dialog around the experience of the McKittrick Hotel only grew over time. During its run of over a decade in NYC, the show has grown a dedicated fan base of people actively engaging online, comparing notes, and drawing new recruits to the franchise.

Jenkins et al. suggest that independent media makers need to expand access and build more direct and richer relationships with their audiences by following the gift-economy and creative-commons approaches to media co-creation. The gift-giving aspect of Web 2.0-based media was never about giving away free content because of the openly acknowledged requirement for reciprocity, which includes the well-established social media tropes such as "like," "subscribe," and "share". As author Cory Doctorow has found out, "free" content has great publicity and relationship-building value. Doctorow has increased sales for all of his books as a result of his decision to freely share digital versions of his works online.

Spreadability: If It Doesn't Spread, It's Dead

Jenkins, along with media industry colleagues Sam Ford and Joshua Green, proposed an alternative way of understanding media circulation they refer to as "spreadability." Jenkins et al. described this transmedia franchise co-creation model as a hybrid of top-down forces of media producers and media makers promoting content and bottom-up media consumption practices of socially connected audiences.

In their research, they identified the necessity of shifting from the distribution model and mindset of promoting media content to the circulation model that views the audience not as simply consumers of preconstructed media experiences but as active participants who can shape, share, reframe, and remix content in ways the creators might not have anticipated (Jenkins et al., 2013). The spreadable diffusion strategy is based on understanding audiences as engaged communities and networks that can embrace, reshape, and spread content beyond the boundaries reachable by the content makers alone. Spreadable strategy recognizes that networked communities can play a major role in shaping how media circulates and how media franchises are built.

The networked connectedness of the audience is not a new phenomenon created by an array of social media platforms. It is a cloud-based amplification of word-of-mouth practices and ancient motivation of sharing stories and ideas while sitting around a fire. Spreadability strategy needs to take into account audience members retrofitting pre-produced content to better serve the interests of their own communities. As media content is shared, it will get remixed and remade, leading to an erosion of perceived barriers between production and consumption.

By Jenkins's own definition, spreadability refers to (1) the technical resources that enable the circulation of content, (2) the economic structures that support (or potentially restrict) circulation, (3) the unique attributes of media content that appeals to a community's motivation for sharing, and (4) the social networks that link people through the exchange platforms of meaningful bytes (Jenkins et al., 2013). The core idea behind the spreadability strategy for attracting and engaging the audience is the same as the one behind our recommended approach to immersive media design—participation and co-creation. Media makers looking to spread their XR media products must imbue them with unique attributes strategically designed to appeal to their core audience's motivation for sharing. Attraction and circulation through participation has been successfully utilized by makers of such digital media as games, apps, and immersive experiences.

Immersive Experiences as Media Franchises

Computational media that emerged from the mass adoption of personal computers combined with the networking power of the internet and mobile computing devices such as smartphones and tablets, culminating in a phenomenon known as media convergence. The defining process of turning all media content from photographs to audio files and videos into ones and zeros turned computers into a medium. Media convergence has brought previously separated traditional forms of media such as newspapers, magazines, radio, TV, and telephones together into a user's single computerized device.

Jenkins identified a unique opportunity computational media has created for digital media makers by defining the concept of a transmedia franchise as a product of a coordinated effort to brand, market, and distribute diverse but well-coordinated content under the conditions of media convergence. Transmedia franchises tell stories across multiple platforms. Each platform makes its own contribution to a unified story world and creates a unique entrance point into the franchise.

For instance, Beyoncé fans can find out about the star's newest album on Instagram. They can see a video for a new song on Tik Tok. This can lead them to Beyoncé's website, where they can find out where and when to see a live concert. Fans of the *Star Wars* franchise might be familiar with the central narrative from the feature films and continue their engagement by playing *Star Wars Jedi: Survivor* or by watching TV shows like *The Mandalorian.*

Since XR media consumption is not yet well established, the transmedia approach can be particularly beneficial for immersive media makers as a way to draw the audience via existing media channels such as social media. This requires a way of looking at the XR media offerings as single media franchises that build synergy between different platform offerings while expanding and solidifying the story world and the brand and user experience.

Attracting the audience and keeping the audience engaged in the franchise needs to be a part of a single transmedia strategy that matches audience awareness with well-planned distribution. This alternative approach, used by the entertainment industry to attract and mobilize audiences, has allowed cult shows and movies such as *Breaking Bad* and *Star Wars* to not only stay culturally relevant but also be financially successful.

Jenkins et al. suggest that core fans empowered to become brand ambassadors can bring new audiences to the media franchises. The XR media brand as a media franchise model leverages the engagement-based logic of media viewing, which places a premium on viewers as a community of active agents whose creative labor can generate alternative forms of both cultural and market value. This is quite contrary to the traditional paid, scheduled, and tightly controlled advertising-driven engagement and attention-getting models. The focus of the media franchise is on the labor performed by the audience as generative of both cultural and commercial value. Transmedia engagement, which factors in multiple touchpoints, taps the collective intelligence-gathering and meaning-making activities of the audience as a core component of brand experience.

Building a media franchise shares the "three steps and two transitions model," which should by now be familiar. Let us review the steps in the context of building XR media franchises. Prospective audience members have to be (1) attracted to the immersive experience by an alignment with their identity and values. Based

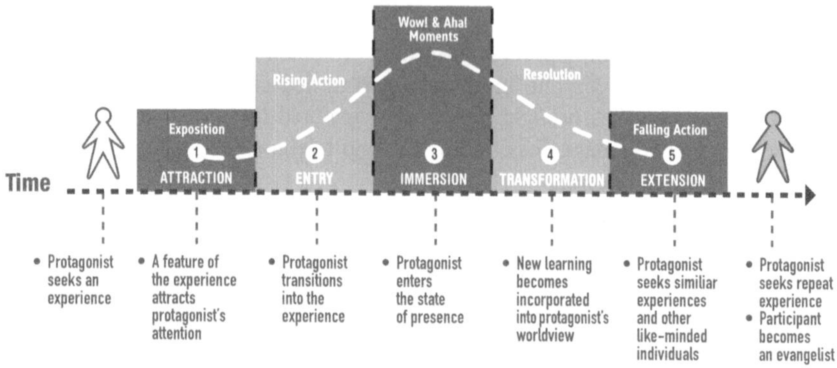

Figure 12.1 The diachronic view of XR experience

Source: Illustration by © 2024 Peter (Zak) Zakrzewski, adapted from Doblin Group's Compelling Experiences model.

on this alignment, they will become open to being attracted by themes and stories that resonate with their unfulfilled cognitive and emotional needs. Sharing themes through exposition builds up the anticipation and curiosity necessary to draw the audience deeper into the franchise.

(2) Entry is a critical transition into the engagement with the franchise. Entry into the XR media franchise has to be designed deliberately and with maximum care, as prospective audience members are not yet committed to being engaged with the new experience. A gamer might play a game for a few minutes and find it too boring or too challenging. A mobile AR user might download and try an app but decide that its navigation is too confusing.

At the point of entry, prospective audience members are still not committed to the XR experience. They have not yet invested enough time, money, and cognitive and emotional resources to feel like they have to work very hard at enjoying the experience. The entry transition must raise the expectations to nudge the participant to go on while effortlessly gliding the users into the next stage of the engagement.

(3) As discussed earlier, creating the conditions for the audience to experience the true feeling of immersive presence and flow that allows them to reach the "aha" and wow moments are built on the foundation of design and engineering mindfulness. Presence can only happen after the deliberate execution of attraction and entry steps. Audience members who reach the immersion stage of the engagement with the XR media franchise are likely to become repeat users and even brand ambassadors if they continue to be guided through another critical step of transformation.

(4) Just as all great narratives offer their audiences resolution of the unfolding drama, great immersive media franchises help their audience members to gain new knowledge that allows them to enter into the new community of the committed. *Sleep No More* audiences could all share a common language and set of experiences that allow conversation that could go beyond the confines of the McKittrick Hotel. Great immersive experiences change their audiences' world and points of

view. They transform them, and part of that transformation is a transition into the extension of the experience into the ongoing engagement with the franchise beyond the original immersion.

(5) As we argue throughout this chapter, extending play means connecting the audience member to the other parts and entry points of the XR media franchise. For instance, a gaming fan might enjoy playing *Star Wars Jedi: Survivor* without having watched the movies or the shows. But once they do, they can then be invited to continue exploring the other platforms within the *Star Wars* franchise and share their experience with others. Media franchises work as entire ecosystems that tell larger stories or engage a larger community of users. Tourism-oriented mobile AR apps can encourage their users to share their unique experiences and knowledge of such physical places as the Roman Forum or Taj Mahal. Conversely, people visiting those locations can be drawn into the AR apps through social media channels that discuss immersive guided tours.

Transmedia Storytelling

In a blog post titled Transmedia Storytelling 101, Jenkins outlined a process that systematically disperses various integral elements of a single narrative across multiple delivery channels with an explicit purpose of creating a unified entertainment experience (Jenkins, 2007). Each channel of the transmedia story makes its own unique contribution. The core bits of information in *The Matrix* franchise are conveyed through the live-action films, animated shorts, two collections of comic book stories, and a number of video games. No single source offers the audience all of the information needed to navigate the *Matrix* universe.

Transmedia narratives are based on complex virtual worlds and intersubjective realities that sustain multiple locations, interrelated characters, and their unique stories. The world-building process associated with transmedia narratives necessitates and encourages Wikipedia-like information-gathering and meaning-making impulses shared by media makers and their audiences. Communities who form around transmedia franchises are driven by deep curiosity. They feel compelled to master what can be known about virtual worlds that continue to expand beyond their grasp. Jenkins argues that this narrative strategy offers a very different experience from the immediate closure that can be found in classically constructed stories that allow the audience to walk away knowing all that is required to make sense of a particular narrative.

Media franchise extensions may take a variety of forms and serve a variety of functions. Social media offers one of the greatest opportunities for extending the audience. Specially prepared or fan-made media clips bring new audience members by circulating within niche communities of interest and on wide-reaching platforms such as YouTube or Facebook. James Cameron created *Titanic*-related documentary content to attract history buffs and to provide additional historical context for his eponymous movie.

Extensions can keep the audience engaged with small chunks of content while more elaborate content is being produced. BBC presented a series of radio dramas for its *Doctor Who* franchise while the television episodes were absent. The James Bond franchise uses its fan base to create buzz around new actor searches and new movie releases. Disney aired *The Clone Wars* on the Cartoon Network to provide additional bits of information that bridged the time lapse between *Star Wars II* and *III*. DC comics entertained the superhero universe audience with the web version of the *Daily Planet*.

Transmedia storytelling practices can not only expand the XR media's audience with additional content, they offer an added benefit of expanding the potential market for a media franchise by creating multiple entry points for different audience segments. Marvel comic books such as *Spider-Man Loves Mary Jane*, the coloring book or picture book versions, tell the Spider-Man story in ways that appeal to female, romance-oriented, younger readers. The same strategy may draw the audience of gamers to the franchise by telling part of the narrative in an immersive game format. It may also compel audiences who are comfortable in a particular medium, like comic books, to experiment with novel media platforms such as XR.

Transmedia narratives are modular in nature. Each individual story component, regardless of its delivery channel, needs to be accessible on its own terms while also making a contribution to the entire narrative universe as a whole. Game designer Neil Young refers to this media-making strategy as additive comprehension. As audiences traverse the virtual universe of their particular media franchise, they encounter new texts that add new pieces of information that forces them to revise their understanding of the system as a whole. For instance, the origami unicorn dropped to the ground by one of the characters in the director's cut edition of the movie *Bladerunner*, raised questions in the minds of the audience about whether agent Rick Deckard, played by Harrison Ford, might have been a replicant himself. This kind of information "bit" offers all kinds of potential creative avenues for transmedia storytellers.

Transmedia storytelling requires a high level of coordination across the different media channels, which means that they need to be considered by the creative team throughout the production process. XR media franchises thrive on co-creation between the media makers, creative partners, and the audience. They work best when the same creative team can shape the story and its universe at all audience touch points allowing co-creators to tap each other's expertise, pool information, and spread the word as they work together to create new narrative meaning.

Jenkins argues that media makers interested in transmedia storytelling need to understand both the social and institutional factors that shape the nature of media circulation within the culture. Any friction between the social logic, which motivates the noncommercial media fan practices, and the commodity logic of commercial mass media will create obstacles to transmedia storytelling. Sharing and remixing of media content erodes any perceived barriers between production and consumption.

Stories in an age of collective intelligence function as cultural attractors. They bring together like-minded people who form dedicated knowledge communities around common issues. Transmedia narratives can function as social activators, spurring the community to collectively assess, archive, and produce information (2007).

The phenomenon of widely circulating conspiracy theories on social media platforms is the (im)perfect example of this very human urge to find and share meaning in all kinds of places. Communities chasing UFOs and deep-state whistleblowers follow the same co-creation and meaning-making pattern outlined by Jenkins based on the ABC television drama *Lost*. When the show's creators flashed a dense map on the screen during one of the second-season episodes, the fans responded by digitizing a freeze-frame of the image, which they posted on the still relatively new medium of the Internet. This digital artifact started a heated debate about what the map might reveal about the fictional Hanso Foundation's activities on the show's island. Transmedia storytelling engages audiences in the ancient task of collective hunting and gathering as they move back and forth across the various narrative bits of information, attempting to build some sort of narrative accrual from content dispersed across different touchpoints.

Narrative accrual is one of the ten core qualities Bruner assigned to narrative meaning-making. He argued that by cobbling together various bits of information, we collectively create history, tradition, and culture. Our own accounts of happenings in our private social worlds eventually lead to more coherent autobiographies. As families, we create a corpus of shared stories that unite and bind us (Bruner, 1991). Brands like Nike and Apple invent traditions and myths of the founders out of previously ordinary events to endow them with mythical status. Bruner contrasted narrative and scientific meaning-making, arguing that sciences also achieve the accrual of commonly accepted knowledge but by very different means.

Scientists use procedures that, first and foremost, assure verification, not verisimilitude. They derive understanding from general principles and relate particular research findings to central paradigms. Narratives, on the other hand, employ such explanatory strategies as the imposition of historical-causal entailment. For instance, the assassination of Archduke Ferdinand is still seen by some as the cause of the outbreak of WWI.

Transmedia stories do not simply diffuse information; they provide audiences with a set of roles, goals, and guiding principles that they can enact in their everyday lives. People dress like Ken and Barbie, wear James Bond watches, and proudly display *Star Wars* action figures in their homes and offices. Audiences immerse themselves in fictional worlds by playing related games, constructing their own stories, dressing up like fictional characters, and role-playing.

Building Immersive Media Franchises

Computational media has given us the era of media convergence, with all mediated content flowing into our mobile devices from across multiple media channels. Immersive media will take media convergence even further. The Vision Pro platform has shown how digital assets from across all media can be brought into XR experiences.

Immersive media makers are standing at the entry point to a world in which we can, for the first time in history, immerse our audiences in virtual and augmented worlds dedicated to powerfully resonant themes. XR worlds, more than any other media, can sustain imaginative locations and rich casts of characters with their own stories. Narrative world-making mixed with the added sensory richness of XR experiences offer media makers a unique opportunity to develop rich universes that can support a variety of relatable characters who can sustain multiple narratives that extend beyond the initial experience to keep the audience co-creating and co-imagining.

Audiences who have grown up around converged media will expect their stories to be told across multiple channels. Kids who ran around parks looking for virtual Pokémon or playing *Game of Thrones* on their Xbox 360 will expect the same kind of multi-channel experience from the next binge-watching blockbuster as they get older. Pokémon makers unfolded their franchise across such platforms as games, TV programs, films, and books by design, with no single media privileged over any other. Younger media audiences have grown up as information hunters and gatherers, solving mysteries of salient plot points, tracking down character backgrounds, and actively making connections between different media experiences within the same franchise. Immersive media makers can expand on the transmedia landscape, shifting between fully virtual, augmented, and traditional media worlds.

XR is laying down a technological infrastructure for the superconversion and transmediation. The audiences are primed, and the economic prospects are very encouraging for media makers who are able not just to design stand-alone experiences but to create compelling transmedia universes. Building XR media franchise experiences is governed as much by the team's artistic vision and community-building practices as it is by the economic logic of social media marketing.

A transmedia franchise mindset is a unique approach that requires a common language and a common vision to unify media projects crossing boundaries between designers, programmers, production units, licensing arrangements, and other vital components of the franchise. The new co-creation model must make room for audience participation, fan input, and fan collaboration. In the immersive media world, audiences will increasingly desire media franchises that offer them expanded experiences within virtual and augmented worlds. Media makers who can satisfy this demand for greater investment and mastery will be rewarded with engaged audiences who will follow, support, and promote their franchises.

Summary

In this concluding chapter, we explored the vital importance of extending play beyond the moment of the initial experience. We covered the related concepts of convergent media and spreadability and their application to building XR media franchises. We argued that XR media makers can benefit from creating multiple entry points to their immersive experiences by leveraging the more established media forms, such as books, live events, online vlogs, and various other popular social media channels.

Throughout the chapter, we drew an insightful (we hope) comparison between transmedia and immersive forms of world-making, both of which give content creators an expanded view of ways to increase audience engagement. By actively promoting extended play, XR media makers leverage the powerful human desire to collectively seek understanding and co-create meaning. The point of extension is, in fact, not the final step on the immersive journey; on one hand, it is a point of reengagement for the audience, and on the other, it is a point of evaluation and reflection for the design team. This is the topic of the last section of the book.

References

Bruner, J. (1991). The narrative construction of reality. *Critical Inquiry*, 18, 1–21.

Jenkins, H. (2006). *Convergence culture: Where old and new media collide*. New York University Press.

Jenkins, H. (2007, March 21). Transmedia storytelling 101. *Henry Jenkins* (Blog). http://henryjenkins.org/blog/2007/03/transmedia_storytelling_101.html

Jenkins, H., Ford, S., & Green, J. (2013). *Spreadable media: Creating value and meaning in a networked culture*. New York University Press.

Symonds, A. (2011). *Art not without ambition: Sleep No More's Maxine Doyle*. Interview Magazine.

Section VII

Iterating Immersive Systems

Iterative Design

As should be pretty apparent by this point, going from a great idea to a great product takes a tremendous amount of dedication, care, and craftsmanship. As XR content creators, we might not be able to arrive at the perfect solution in one fell swoop. Let's remember that the original iPhone came with no official way to install native third-party apps until the App Store was launched a few months later in July 2008. The device had no front camera, and users could not copy and paste text. So even the masters of usability at Apple did not have to and did not try to aim for instant perfection.

The process of immersing and delighting audiences takes many smaller steps. XR experience design is best served by a strategy of progressively getting closer to the optimal user experience. As mentioned throughout the book, the process by which your creative intuition and your delivered application eventually converge is known as iterative design. The approach has been applied to scientific research, educational projects, and agile product development. Iterative design views XR experience design applications as living projects that are regularly tweaked and improved upon as users' needs and expectations change, and competitive products catch up to the leaders.

Iterative design is based on a continuous cycle of (1) forming a hypothesis, (2) building a prototype, (3) conducting an experiment, (4) gaining insights, (5) and restating the original hypothesis. In this process, the team, rather than building and perfecting an entire digital product from start to finish, takes time to seek user feedback and conduct desirability, feasibility, and usability testing. Each new iteration is built on top of previous versions.

The iterative design approach, as applied to individual phases of development or to the entire project, has four major benefits. One of the major ones is that it provides the team with continuous robust real-world and user feedback. Putting your XR product through its paces and into the hands of the prospective audience can help you understand which elements of the experience are working and which need further development. User testing allows the team to catch any problems earlier. The cycle of prototyping and testing the project can save the team a lot of time and effort by raising red flags for any potential problems earlier rather than later in the design process.

DOI: 10.4324/9781003442462-20

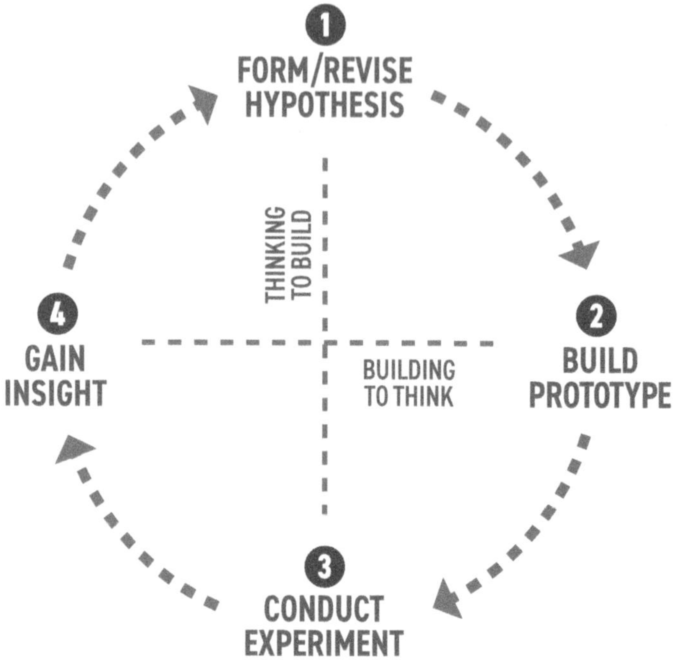

Figure VII.1 The hypothesis-testing feedback loop

Source: Concept and illustration by © 2024 Peter (Zak) Zakrzewski, based on Eric Ries', Build-Measure-Learn Feedback Loop.

There are many places where XR experiences can fall short of supporting presence. In addition to catching problems early on, ongoing commitment to user testing improves usability by giving the team a clear path to improving any rough spots identified in the experience tests.

Finally, the iterative design approach, despite seeming cumbersome and expensive, is in fact the most efficient and cost-effective way to arrive at a truly immersive product. It allows the team to spend less time creating documents describing the design and more time working on the actual product.

Now that we have taken all of the smaller iterative steps that have allowed us to invite the audience to participate in the immersive experience, let us look at applying the iterative process on a larger scale to iterate the XR system toward its next, better-optimized, and more compelling version.

Transforming and Iterating Sociotechnical Systems From the Meta-Level Position

Seeing immersive media environments not simply as apps but as complex sociotechnical systems gives XR experience designers a powerful framework to both design and iterate the entire XR media ecosystem. Practicing family therapists Paul Watzlawick, John H. Weakland, and Richard Fisch applied two theories drawn from the field of mathematical logic to problems of ecology, conflict, and dysfunction concerning a variety of social systems (Zakrzewski, 2022).

The theory of groups, introduced in 1832 by the 20-year-old French mathematician Evariste Galois, is a unique framework for thinking about the peculiar interdependence between parts of a closed system and a wonderful case study in ecologically sound rule writing. Galois's theory concerns the relationships between elements and wholes by outlining four unique properties that guide the relationships within a closed system.

1. A group is composed of members that are all alike in one common characteristic.
2. The second property of a group is that while one may combine its members in varying sequences, the outcome of each combination will remain the same. For example, adding 2 + 8 + 5 + 10 will be the same as adding 2 + 10 + 5 + 8.
3. A group contains an identity member whose combination with any other member gives that other member. This type of member may act without making a difference. Combining with the identity member actually maintains or even confirms the other member's identity. For example, in groups with additive combination rules, the identity member is a zero (e.g., 5 + 0 = 5). In groups whose combination rule is multiplication, the identity member would be 1, since any member multiplied by one remains itself. William Ross Ashby showed the importance of this principle for cybernetic systems by highlighting the importance the null function of the group of parametric changes plays in the maintenance of the stability of such systems.
4. Lastly, in a system satisfying Galois's group theory conditions, one is able to find for every member its reciprocal or opposite. The combination of any member with its opposite gives us the identity member mentioned earlier. For example, in the groups with addition as the combination rule, 5 + (−5) = 0. While this combination produces a marked change from 5 to 0, the result is itself a member of the group of positive and negative integers, including 0 (Watzlawick et al., 1967).

Watzlawick and his colleagues saw the theory of groups as a useful framework for thinking about the peculiar interdependence between persistence and change in systems within which the more things changed, the more they stayed the same. They argued that the theory was not a useful model for the types of change that required the problem-solver to transcend the system's boundaries and its frame of reference in order to transform it (Zakrzewski, 2022).

For the class of system design problems involving transformation, they found a complementary systems thinking tool in the theory of logical types proposed by Bertrand Russell. The theory was originally aimed at dealing with the emergence of antinomies and paradoxes in logic.[1] One of the most famous of these was proposed by Thomas Fowler, who offered the following puzzle: If Epimenides the Cretan says that "all the Cretans are liars," is Epimenides, as a Cretan, himself lying? As a logician, Russell attempted to rule out any such self-references in logic by clarifying a distinction between a class of objects and its individual members. By considering a class to be of a higher logical type than its members, system (re)designers avoid the confusion of any self-referential statements.

Being able to be clear about logical types leads to an understanding that a class cannot contain itself as a member. This led Russell further to propose the law of metapropositions, or propositions about propositions. Under this law, self-referential statements are not considered to be true or false but rather invalid due to their logical meaninglessness. Russell's law finds applications in all kinds of contexts. For instance, linguists and semanticists have developed a metalanguage in order to talk about a language to address such issues as semantics and syntax.

So what are the insights XR system designers can draw from Watzlawick and his colleagues? There are at least four. The first two address the importance of the need for system designers to be able to attain a meta-level position in relation to the system. XR system designers can only write system rules from outside of the conceptual phase space where immersion occurs. When Will Wright creates rules for the new edition to the *Sims* franchise, he does not do it as an individual player from within the game, so to speak. As a system designer, he defines the rules that constrain and define the play within the system from outside the game.

So our first insight is that (1) system designers must keep such logical levels as being within a system or outside of the system, strictly apart, in order to prevent paradoxes and confusion from emerging. (2) System designers must keep in mind that going from one logical level to the next-higher level, as in moving from a member to a class or from a participant in a system to a system designer, entails a discontinuity—not a simple change but a fundamental transformation of position.

When it comes to XR system iteration, there are two entirely different types of change (3) A first-order change occurs within a given system while leaving the system itself unchanged. Reshuffling a group of integers 2, 8, 5, and 10 without changing the combination rules from addition to multiplication, for instance, will not change the outcomes of the equation. The free-for-all, conflict-ridden nature of play emerging out of putting together highly competitive and system-gaming-minded players in a game with no fairness rules or enforcement by code procedural rhetoric will not change by shuffling the participants around the same game environment.

To transform such a system, designers will need to engage in (4) a second-order change, whose occurrence transforms the system rules and the system itself. Groups as systems are invariant on the first-order-change level—the change from one member to another—but they are open to change on the second-order level, spurred by the changes in the body of rules governing their structure and internal order governing their relationships. Second-order change will always be characterized by a discontinuity or a seemingly (il)logical conceptual jump, similar to what Thomas Kuhn (1996) referred to as paradigm shifts in science.

So as long as the system design team maintains their outside-of-the-system meta-level position, they can iterate and XR environments in a way that transform them toward their intended function. These functions, as we will argue later in this section, can range from the pure bliss of mediated presence meant for playful discovery, entertainment, and fun to optimizing the XR systems for collective intelligence or for high levels of social goals involving purposeful emergence.

Auditing Systems in Play

Let us now return to some of the key concepts discussed in Chapter 7 regarding the design of XR applications as sociotechnical systems. The same five critical system dimensions used for designing XR systems, can be used to guide the system iteration process. Let us restate these in the context of system evaluation and redesign.

1. System function expresses its purpose. What the system is intended to do guides the questions behind evaluation and iteration. The contrast between the professed goal of connecting people and the practiced reality of social media platforms optimizing algorithms for maximizing engagement and advertising revenue at the cost of everything else showcases Donella Meadows's argument that a system's function, even if not explicitly stated, will always be apparent (to an astute observer) through the operation of the system and the outcomes it creates (Meadows, 2009). System functions for XR experiences are rooted in the original vision for the project we discussed in Chapter 1. For the issues concerning whether the system is fulfilling its function, we need to ask: Is the overall system state, its behavior, and its outcome consistent with the original vision set for it? In the case of a system performing its function and yet not satisfying the intended vision for the project, the design team might consider a pivot or the rethink of the stated function.
2. The system's structure comprises its information and its hardware architecture. System-level changes need to be addressed, at least in part, at this higher level of leverage. The system's structure, with its rules of the game, will be the first place to start when iterating its ability to attain the intended function. As discussed earlier, system rules, be it a game or an app, are the invisible hand guiding the system's overall performance. Let us also keep in mind the critical importance of the structure of the XR system's physical substrate to the overall user experience. Cramped spaces, uncomfortable HUDs, and glitchy software–hardware integration can destroy the feeling of presence faster than any imperfect rules.
3. Process is all about the code giving form and enforcement to the structure of the immersive system. Going down the ladder, the code behind the experience can fall short of properly supporting the desired user experience and the system's inscribed structure. The procedural maxim stating that code is law clarifies the importance of iterating the system's process as another high-level leverage point of course correction. The doctrine of procedural rhetoric sheds another light on the contribution process and code make to the overall system function. Procedural rhetoric is commonly used by game makers to craft laws and rules within a game to convey a particular ethos or philosophical

stance. Coded rule-based representations and interactions can be as persuasive as images or spoken and written words.

4. Semantics address communication and the meaning-making within the system. Social bonding and social polarization happen according to what stories we tell ourselves within the intersubjective realities we create as social groups and communities. Because of the power of communication as a binding and coordinating force, semantics offer system designers an important area of iteration. When it comes to auditing semantics, we are very familiar with the importance of the conversations within the network from the two decades of social media. Some of them support positive user experiences, and occasionally, some become runaway trains. Social media algorithms have shown us the power of directing the flow of messages and their contribution to the overall system function.
5. Pragmatics addresses the behavior of both the individual participants and the system as a whole. Behavior within XR systems appears to be transparent. It is easy to observe and assess. Participants leave plenty of digital breadcrumbs wherever they go and whatever they do. In this respect, we can also gain much insight from tightly monitored social media platforms such as Facebook. According to Stanford Department of Computer Science researcher Michal Kosinski, mining a user's Facebook preferences based on the articles, videos, artists, and other content the person has "liked" allows data scientists to create a very accurate psychological profile based on the big five personality dimensions of openness (inventive/curious vs. consistent/cautious), conscientiousness (efficient/organized vs. extravagant/careless), extraversion (outgoing/energetic vs. solitary/reserved), agreeableness (friendly/compassionate vs. critical/rational), and neuroticism (sensitive/nervous vs. resilient/confident).

As astonishing as it may seem, based on these criteria, by analyzing just ten likes, social media data science teams using AI-based computer algorithms can more accurately assess the subject's personality and predict their likely future behavior than a work colleague can. By analyzing just 70 likes, they can outperform a friend or a roommate. With an analysis of 150 likes, they can do better than a family member, and with 300 likes, better than a spouse (Parker, 2015). But being able to observe a behavior should not be confused with understanding what motivates that behavior. The higher levels of system dimensions, including semantics, might be a place to look to understand the causes of why participants do what they do.

In addition to the five system dimensions that can be, at varying degrees, controlled by the designers, in his previous work, Peter identified two that have to be acknowledged as being outside of the designers' control. The first of these is (6) context, or the system's external environment. Every system operates within a larger context just as the Earth's ecosystem exists within the context of the solar system, which, in turn, is a part of the Milky Way galaxy. Walking on planet Earth, we sometimes forget that we are subject to the same mysterious laws as the distant stars and planets and that our bodies are made of cosmic dust.

In the same way that Earth's ecosystem is influenced by its position within the solar system, XR environments are influenced by the context of their physical, social, and cultural environments. The immersive world of *Sleep No More*'s McKittrick Hotel is physically located within the 100,000 square feet of warehouse space in Manhattan. It is culturally located within the solar system of an audience who know Shakespeare's Macbeth, Hitchcock's suspenseful dramas, and film noir movies. The *Sleep No More* immersive media franchise also lives within the larger galaxy of people who read *The Guardian* and the *New York Times*'s theater reviews and who communicate with each other about their experience using social media platforms.

Needless to say, all of these external environments exert an elusive but yet powerful influence over the immersive world of the show. Physical environments are particularly critical to MR and AR applications because any misalignment between the participants' phenomenologically experienced physical world and the overlaid data will lead to poor user experience. While the external environment cannot be directly controlled by the design team, it can be audited based on user feedback and addressed by giving the XR system more of what in cybernetics was referred to as requisite variety.

The cybernetic principle known as Ashby's Law suggests that a successful functioning of a system, be it a living organism, a machine, or an entire human society, depends on its ability to change its state to match changes occurring in the environment. The XR system's requisite variety is based on the application being able to actively process the flow of data from feedback loops created by imaging sensors such as time-of-flight (ToF) cameras, vertical-cavity surface-emitting laser (VCSEL), light detection and ranging (LiDAR), and binocular depth sensing to help the user alter their behavior to maintain a well aligned system-environment coupling.

Finally, we come to perhaps the most challenging part of the XR system design—(7) the agency of the human actors within the network. When Marshall McLuhan poetically suggested that the medium was the message, he attempted to redirect our gaze away from the content shown on media platforms and toward the impact any technological medium has on its audience.

In a less well-known aphorism, he suggested that while we have the power to first shape our tools, after that, our tools start shaping us. A similar sentiment about the importance of paying attention to human agents within sociotechnical systems was expressed by the VR pioneer Jaron Lanier, who argued that the most important or interesting thing about any technology is how it changes people (2017).

Throughout this book, we argue that the XR system's true function emerges out of an interaction between the designers' manifested vision and the audience's subjectively and intersubjectively experienced journey through the XR environment. Based on the audit of the semantics and pragmatics within the system, the design team will have a wealth of information about the participants' contribution to the system outcome.

The knowledge about user communication and behavior should not be equated with coercive control. We make a point of drawing a distinction between AI-based actants' extrinsic, synthetic agency within the system and the human agents' intrinsic, organic agency and motivation. We define immersion, like play, as involving free movement within the system of constraints, which means that designers can only rhetorically persuade users toward attaining presence or engaging in particular behavior. Using multimodal rhetoric to persuade the participants toward adopting new behaviors more consistent with the intended function is very different from explicitly redesigning the algorithm driving the behavior of AI-based actants.

The design team's most elegant option to affect the audience behavior within the immersive system is the XR's unique and unprecedented capacity for multimodal rhetoric. While XR audiences cannot be designed and redesigned like AI-based actants, they can be persuaded. The good news is that there has never been, in the history of media making, a more persuasive medium than XR. When it comes to system iteration, the actants can be redesigned, while the human participants need to be guided.

The above five-plus-two system design dimensions can serve as powerful guides and useful guidelines for auditing and iterating immersive systems, be it VR games or mobile AR apps. Systems that do not perform their intended function can be tweaked based on the mentioned dimensions, but first, they need to be understood

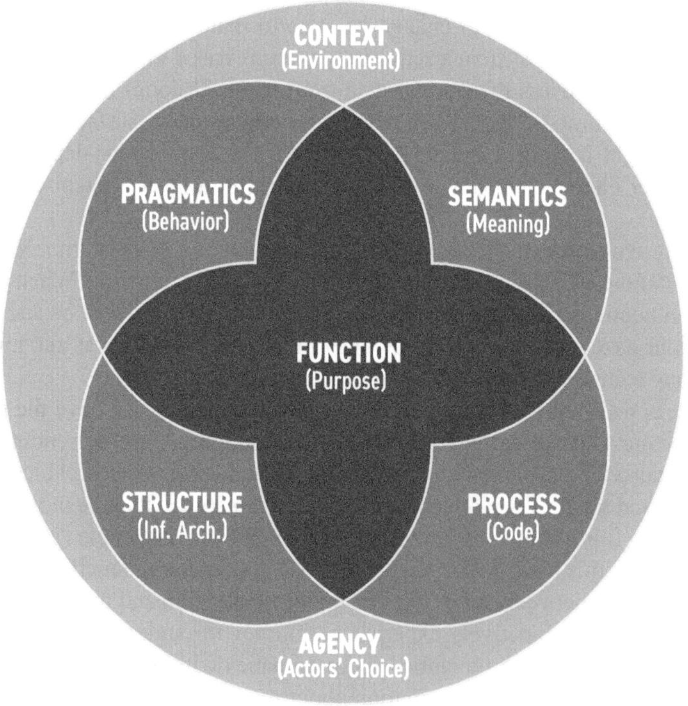

Figure VII.2 The five + two dimensions of immersive systems

Source: Concept and illustration by © 2022 Peter (Zak) Zakrzewski.

in the context of the 12 leverage points outlined by environmental scientist and proponent of systems thinking Donella Meadows.

Donella Meadows's 12 Leverage Points: Places to Intervene in a System

Donella Meadows spent an entire career investigating complexity, ecology, and systems thinking, but her epiphany about the relative importance of leverage points occurred during a meeting of the World Trade Organization policy wonks who were discussing new global trade regimes known as NAFTA and GATT. The more she listened to the various proposals on how to make world trade work better, the more she became convinced that the people who were inventing a huge new system had not the slightest idea how their complex structure would actually behave *in situ*. More importantly, she realized that the control measures being discussed, such as small parameter adjustments and weak negative feedback loops, were "puny," ineffective, and capable of potentially "cranking the system" in the entirely "wrong direction" (Meadows, 1997, p. 1). According to her own account, at one point during the meeting, she spontaneously rose up, marched to the flip chart, and, tossing over to a clean page, outlined 12 leverage points system designers need to consider when attempting to course-correct systems not performing their intended function.

Leverage points are places within complex systems such as entire economies, corporations, cities, biological ecosystems, and our own bodies, where small shifts in one thing can produce big changes in the entire system (Meadows, 1997). If the idea of leverage points sounds a bit too good to be true, like silver bullets or miracle cures, it is because they are, but mainly in one aspect—their counterintuitiveness.

When systems science pioneer Jay Forrester was asked by the Club of Rome to show how major global problems of poverty, hunger, unemployment, environmental destruction, and urban deterioration might be solved, he created a computer model that pointed to a clear leverage point—growth—both population growth and economic growth. No one was surprised by Forrester's answer because that answer had already been intuitively understood.

What was counterintuitive about the results shown by Forrester's model was that the answer to the grand challenges involved a growth strategy portfolio that involved different kinds of growth from faster to much slower, in some cases no growth at all, or even negative growth in some areas of the economy. This counterintuitive understanding of the identified leverage point was supported by the key insight from the model's output: growth has its costs as well as benefits.

When we become enamored with intuitively appealing solutions, we typically do not count the costs, which, in the case of Forrester's study, included compounded poverty, hunger, and environmental degradation that decreased the overall quality of life. In the case of digital media, rapidly growing an online platform will force the team to invest in scaling their servers or cloud services, taking valuable resources away from product development. It will also mean having to satisfy new users who are not part of the well-understood core audience segment. Attempting to please too many different segments may end up diluting the experience that attracted the original audience, possibly leading to that segment abandoning the platform.

If there is one insight about leverage points, it is that they are not intuitive. When we use them intuitively, we seem to use them backward, systematically worsening the problems we are trying to solve. Meadows's epiphany about the relative effectiveness of leverage points came from attending a policy meeting about global trade. She might have been the only person in the room to realize that the proposed trade regime was, in fact, a system aimed at managing complexity in the global economy. All of the proposed rules represented various types of leverage points, and leverage points were not only not equal in their effectiveness, but they could also be counterproductive if applied incorrectly. She considered her list to be not a definitive recipe for dealing with leverage points but an invitation to think more broadly about system change.

Meadows put the list in the reverse order to show how powerless the intuitively appealing easy fixes are in the face of complexity, starting on the lowest end of the spectrum.

- At number 12, she placed system elements (constants, parameters, numbers), exemplified in policy making by subsidies or taxes. Taxing fossil fuels at specific tax rates, for instance, has not caused the majority of people to stop driving cars with internal combustion engines.

In mathematical system modeling, parameters are used to quantify or to assign numerical values to aspects of state variables. Think of the price you charge as one parameter that will affect the financial success of your XR media venture. You can lower the price to hope for more volume of sales, but lowering prices can also lead to lower profits. Meadows put numbers expressing the sizes of flows dead last on her list of powerful system interventions, arguing that fiddling with the details is tantamount to arranging the deck chairs on the *Titanic*.

- At number 11, Meadows placed the system dimension known as sizes of buffers and other stabilizing stocks relative to their flows. To understand buffers, we can compare an example of a lake that is like a huge bathtub with slow in and out flows to a river that is like a much narrower bathtub with rapid flows. There are a lot more catastrophic river floods than lake floods because systems with big stocks relative to their flows are more stable than small ones. Buffers are known in chemistry and biology as stabilizing stock. They have the power to stabilize systems. It is the reason retail stores hold inventory instead of ordering new stock every time customers walk out the door with their groceries (Meadows, 2009). In banking, capital buffers are additional layers of capital that banks hold to support ongoing lending and cover any losses in order to smooth the impact of economic downturns of business or credit cycles.
- At number 10 sits the structure of material stocks and flows such as transport networks and population age structures. The plumbing structure and its

physical arrangement can have an enormous effect on the stocks and flows that determine how the system operates. To illustrate this point, Meadows uses the case of the Hungarian road system, which was laid out to direct all traffic from one side of the country to the other by passing through the capital city of Budapest. This decision had a lot more impact on air pollution and commuting delays than fuel taxes, pollution control devices, traffic lights, or speed limits. The only way to fix a poorly laid-out system structure is to rebuild it. Rebuilding system structures is typically excessively time-consuming or prohibitively expensive. This is why designing them properly in the first place is critically important.

- At number 9, systems are internally and externally guided by feedback loops that occur when a system output is routed back as input in a chain of control. The term "feedback loop" is commonly used in biology and chemistry to signify how the output of a reaction either increases or decreases the rate of the reaction. Feedback loops are critical to shaping player experiences in immersive games. In a game such as *Mario*, the system takes the input of how well a player is doing in order to create a response that makes the game either more rewarding or more challenging.

One of the most common feedback loop strategies in game design is based on motivating players to keep playing with the reward and progression loops. Players are rewarded with points, achievements, and unlocks for completing tasks. A delay in the reward and progression loops can disrupt a sense of accomplishment, satisfaction, and curiosity that drive the player onward in the game experience.

Meadows addressed the lengths of delays within the system relative to the rate of system change as number 9 on the leverage point list. She argued that she would list delay length higher on the list except for the fact that delays are usually not easily changeable. Things simply take as long as they take. The first proofs of concept for immersive media, done as far back as the 1950s and 1960s, and the functional prototypes of VR created in the 1990s have not led to a massive adoption of XR media throughout the economy for decades.

Ideas injected into social media conversations might take months or even years to manifest themselves as social movements. Delays in the feedback process are critical in relation to the rates of change in the stocks that they try to control (Meadows, 1997). Delays that are too short can cause overreaction and system oscillations amplified by the jumpiness of the response. Delays that are too long can push the system over a threshold past which irreversible damage, such as overshoot or collapse, can occur.

- At number 8, feedback loops can be either positive or negative. Positive feedback loops spur the player–system interaction. A winning player interacting with a system providing them with a positive feedback loop will keep getting more emboldened to keep engaging with the system, keep playing, and keep winning. Negative feedback loops tell the system to slow down.

Sweating, for instance, is a good example of a negative feedback loop. We sweat as a consequence of an increased body temperature. As the sweat evaporates, it cools our skin through the process of evaporative cooling. The negative feedback of sweating reduces the speed of the reaction of our bodies heating up. Meadows placed "the strength of negative feedback loops, relative to the impacts they are trying to correct against" as number 8 on her leverage point list (p. 153).

Dealing with feedback loops marks the demarcation between physical and more tangible aspects of the system and the information-and-control ones, where Meadows argued more leverage can be found. Complex systems have numerous balancing feedback loops they can bring into play to self-correct under different environmental conditions. Balancing feedback loops are those that keep important stocks such as energy, temperature, and cash reserves within safe bounds. Balancing feedback loops within a thermostat keep the system stock commonly known as the room temperature fairly constant near a desired level.

Like heating and cooling systems, immersive systems need settings establishing optimal goal ranges. They need monitoring and signaling devices to detect deviation from the goal and a response mechanism (a furnace and/or air conditioner) to correct any imbalance. Just as healthy nutrition, exercise, and stress reduction can bolster the body's ability to fight disease, optimizing algorithms to keep the spread of unsubstantiated news stories or bullying behavior on social media networks within a desired range can help to keep the conversations within them advancing the desirable system function.

- At number 7, while balancing feedback loops have a self-correcting effect, reinforcing feedback loops self-reinforce the ongoing trend. The more they work, the more they gain power to work some more, driving system behavior in a particular direction. The more people use and talk about an XR application, the more they spread the word about it to other people in their social networks.

Meadows designated the gain around driving positive feedback loops as number 7 on the leverage point list. Reinforcing feedback loops can be the sources of healthy growth but also of uncontrolled explosions and erosions, potentially leading to a system's collapse. Systems with unchecked reinforcing loops ultimately destroy themselves through such phenomena as economic bubbles. Balancing loops are needed sooner or later to maintain the system's health. Just as epidemics run out of infectible people, apps run out of core audiences. That is not a bad thing as far as media franchise creation is concerned. Happy core audiences give media franchises longevity. Controlling the gain around a reinforcing loop to balance growth is usually a more powerful leverage point than strengthening balancing loops.

- At number 6, Meadows illustrates the importance of the structure of information flows by recounting a story of a Dutch housing development whose management wanted to control electricity use. In some of the houses, the meter was installed in the basement, and in others, it was given a prominent place in the home by being installed in the front hall. With no other differences in the house design, electricity consumption ended up being 30% lower in the houses where the meter was placed in the highly visible location.

Leverage points affecting the information structure of the system form their own new loops, delivering feedback to impactful places where vital information was not available before. We will return to this point when discussing Alex Pentland's strategies for opening up information ghettos. The structure of information flows affects who does and does not have access to what information.

Missing information flows are some of the most common causes of system malfunction. Adding or restoring information can be more powerful and usually much easier and cheaper intervention than rebuilding physical infrastructure. The tragedy of the commons we know as overfishing happens in no small part because there is very little feedback from the state of the fish population in our lakes, rivers, and oceans to affect the decisions of fishing companies to bring more fish to the market.

You might ask: What does the price of fish have to do with iterating XR media systems? The incredible power of systems thinking lies in its rules applying equally to oceans, seafood markets, and XR applications. In the case of overfishing, Meadows argues that in contrast to the information about stock levels, prices can actually reinforce the overfishing trend. As seafood becomes scarcer, it becomes more expensive and more profitable for those willing to go and catch the last few.

The same abstract concepts of scarcity messaging and strategic structuring of information flows have the power to change the course of XR systems. Informing users about relative scarcity of digital assets, just like giving feedback to the consumers about the price of seafood, will have some effect on their behavior, but the question is: Is it going to resolve your overfishing problem? Directing the wrong information at the least impactful system participants can take immersive systems, be it social VR apps or medical training modules, farther away from their intended function.

There will always be some people willing to pay exorbitant prices to eat the last few fish on the planet. Using the structure of information flows as a leverage point means restoring the missing feedback to the right place in the system and doing it in a compelling form. If your goal is to save the oceans from overfishing, you will need to deliver information about the shrinking fish population to those whose livelihood depends on fishing in a way that compels them to manage the fish stock more wisely.

- At number 5, we have already covered the critical importance of the system rules that define incentives, punishments, and constraints. The best gaming experiences are based on free movement within a system of creatively designed constraints. The rules of baseball that stipulate nine players on a team, having to touch every base, and getting the batter out after three strikes make the game more fun to play than a free-for-all with anyone doing what they please. A system's rules define its scope, its boundaries, and the degrees of freedom allowed to the participants. Meadows placed the structure of rules, incentives, punishments, and constraints at the relatively high number 5 on her list of leverage points because changing the rules by which institutions, games, and apps work changes the way people behave. Rules form a set of extrinsic rewards for every community.

- At number 4, in our system framework, rules are part of the system structure because of their importance in structuring intangible digital products. In biology, system structure relates to the physical aspect of living systems. One of the unique properties of living systems and social systems is their ability to spontaneously change themselves by creating whole new structures and behaviors. In systems thinking, this is known as self-organization.

The entire concept of evolution is based on the ability of living organisms to add completely new physical structures, such as brains or wings. In the case of humans, we have not stopped at evolving our brains. We have added whole new artificial ones in the form of computers. The ability to self-organize is the strongest form of system resilience.

Systems capable of evolving themselves can survive almost any change. The reason Meadows put the power to add, change, or evolve system structure at the high number 4 position is because the ability to evolve the system's structure means changing any aspect of a system lower on the list.

Meadows argued that biologists worship biodiversity because they, more than any other group of scientists, have understood the power of system self-organization. Encouraging agency, variability, and experimentation goes hand in hand with losing some control over the system. From the pragmatic design perspective, managing trade-offs is part and parcel of dealing with social complexity and the core skill necessary for designing and iterating information-bonded, multiminded systems. We encourage XR design teams to appreciate the leveraging power of adding, changing, and evolving system structure while letting a thousand flowers bloom and seeing what happens.

- At number 3, as mathematicians interested in the functioning of economic and other social systems, John von Neumann and Oskar Morgenstern proposed a system design axiom that posits that one cannot maximize two (or more) functions at once. Just as it is not logically possible to maximize for more than one variable at the same time, it is not possible for any system to perform two functions equally well (Zakrzewski, 2022). There can only be one core function a system can perform.

As a system designer, your question is: What will the function be? As publicly traded companies, social media platforms are, by default, optimized for user engagement leading to advertising revenue and shareholder profit. There is plenty of evidence available to show that their professed goal of connecting people routinely takes a backseat to the practiced function of profit maximization whenever a conflict between the two arises.

Meadows put the system function at the number 3 spot on her list. As we mentioned before, she argued that a system's function, even if not explicitly stated, will be apparent to an astute observer through the operation of the system (Meadows, 2009). The way to infer the system's function is to watch it for a while to see how it behaves and what it accomplishes. To illustrate this point, she used an example of a frog turning right and catching a fly and then turning left, around backward, and so on to catch another fly. After a few minutes of observation, it will be pretty apparent that the purpose of the frog is not to keep turning left, right, or backward but to catch flies.

In systems not expressly designed to accomplish a particular purpose, system outcomes will not necessarily be those intended by any single actor or group within the system. They will be a product of emergence. One of the most pernicious aspects of dealing with systems and especially social systems is that the actions of individual actors within a system pursuing their own local goals may add up to an overall global behavior and outcome that no one wants.

Here comes an important distinction between spontaneously occurring system function and a deliberately designed purposeful system. Immersive media allow design teams to be purposeful in both the inscription of key system dimensions and in system iteration. Meadows argued that the goal of a system is a superior leverage point to all the others below it on the list, including the system's ability to self-organize. An explicitly enforced goal can override the system's self-organization. Immersive system designers have an unprecedented power to affect the XR system's purpose, but this meta-level position needs to be matched by the humility to allow the participants to exercise their agency to self-organize in response. By observing the system in play, designers can nudge the whole system toward a self-organization that fulfills the system's desired function, but they cannot control the entire process, nor should they attempt to engage in coercive technopaternalism just because they think they can.

Choosing the system function needs to be done with great care and foresight because from that point on, it will be the beacon guiding the team in the design and iteration process. A useful conceptual tool to brainstorm system function can be found in Roger Martin's process for thinking and deciding. As mentioned before, system design always involves managing some kind of trade-offs. Managing those trade-offs is critical to ecological system design. Martin's process involves four important steps.

- The first one requires the design team to choose the most critical variables defining the project. These can be such potentially competing goals as high engagement, ease of learning, error tolerance, and efficiency levels of systems, products, or services. These key building blocks need each other for mutual support to make the system perform its main function. The blocks work best when they work together, but they also compete for resources.
- The second step of the process entails establishing and clarifying the causal links between the constituent parts and their subsystem features. In building an overall function for an XR system, designers need to reconcile such important considerations as the desired user experience, the availability of appropriate technology, the cost of implementing the technology, the internal processes

required, and the external partners needed to deliver the final product. The nonlinear causality of aligning such multifaceted aspects of a complex system requires what Martin calls the integrative thinking approach. This means considering any unsatisfactory trade-offs as possible thesis/antithesis pairs for which superior alternative syntheses can be developed. Critical examination of seemingly incompatible trade-offs exposes opportunities for the type of quantum-leap thinking that allowed innovators such as Apple and Netflix to leapfrog their competition.
- The third step in the model deals with the architecture of the entire system. At this stage, the team will need to develop an overarching architecture that encompasses all of the moving parts. The process of building XR systems is one of zooming in and out from the details of each block to the big picture of the entire system and back again.
- The fourth and final step focuses on the creative resolution of tension within the proposed system. This last stage of the alignment process involves evaluating the system function in the light of the transformative vision set out by the team at the start of the project. Following the model as a systems-level conceptual tool allows problem-solvers to envision and evaluate the function of the system before implementing any of the individual steps.
- At number 2, Meadows put system users in the first and second place on her list of leverage points by identifying the mindset or paradigm out of which the system's goals, structure, rules, delays, and parameters arise as one of the two top ways to transform systems. She argued that implicitly shared ideas in the minds of a community or a society—the great big unstated assumptions and cultural codes—constitute that group's paradigm, their mental model of how the world should work. She argued that "paradigms are the sources of systems" (p. 163).

System goals, information flows, feedback loops, stocks, flows, and all of the other system dimensions arise out of shared social agreements about the nature of reality. Digital game designers use cultural codes shared by their audiences as a foundation of their games. Specific social contexts affect game playing practices and experiences. For instance, the fictional world, its monsters, and many of the quests of the role-playing game (RPG) *The Witcher 3* draw their unique character from espousing a mindset based on Eastern European folklore. The game's entire system is built on top of a particular Slavic mindset. Just as the ancient Egyptians built pyramids because they believed in an afterlife, gamers playing *The Witcher 3* go on quests because they believe in the concept of slaying monsters. Without this underlying belief, the game would represent an entirely different system of rules.

For the audiences of the 2002 film *Bend It Like Beckham*, the rules of that world are uniquely based on a blend that would shape the identity of a South Asian Sikh teenage female soccer player raised in Great Britain. To understand the film, one has to immerse oneself in the value system and the logic of this particular worldview. To the people living within a particular social system, the mindsets are unstated because it is unnecessary to state them. In a world where one can "own" land, communal land ownership does not make much "common" sense.

- At number 1, paradigmatic assumptions of one's culture can utterly dumbfound other cultures, who do not think of them as being the least bit obvious. This is the reason different cultures or subcultures create very different social systems and outcomes. Changing cultural codes is one of the most impactful system leverage points and one of the hardest things to accomplish. Meadows put this particular skill at the top of her list.

Systems thinkers and system designers who can gain mastery over paradigms can affect the most powerful form of transformative change within sociotechnical systems. Meadows put the power to transcend paradigms at the top of the list of leverage points, even above that of changing a paradigm. What she meant by the power to transcend paradigms was the ability to keep oneself unattached to any one paradigm, to stay flexible.

As discussed in the introduction, paradigms are extremely useful constructs in the realms of scientific discovery, but no matter how appealing they might seem, they always offer "a tremendously limited understanding of an immense and amazing universe that is far beyond human comprehension" (p. 165). To see a map as a map and not as a territory itself is to liberate oneself to construct new solutions with a mind open to separate what is ordinarily confused and to unite what is ordinarily separated. In the words of the Nobel Prize winner in Medicine Albert Szent-Györgyi, creative problem-solving means seeing what everybody has seen and thinking what nobody has thought.

Redesigning for Presence

The attainment of immersive presence is one of the key objectives of the XRXD framework. Ensuring that the conditions for the emergence of presence are set correctly within the XR system is one of the first goals of system evaluation and iteration. In Chapter 11, we described the core framework for immersing the audience

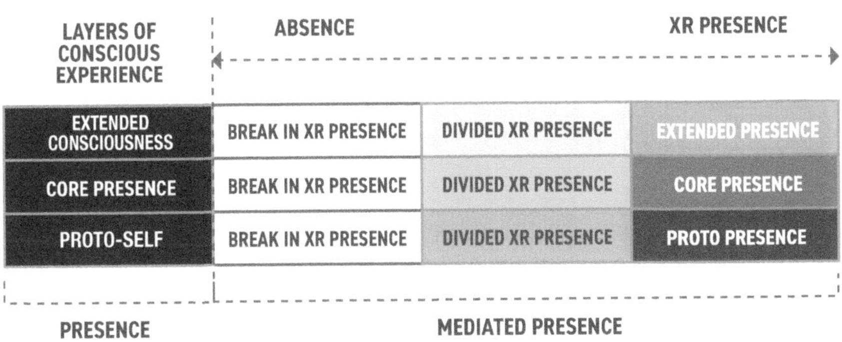

Figure VII.3 The focus aspect of mediated presence

Source: Illustration by © 2024 Peter (Zak) Zakrzewski, based on Waterworth & Waterworth, The Three Dimensions of Virtual Experience; Damasio, Theory of Consciousness.

in XR mediated environments. The same framework can be applied to iteration of a system after the audience enters into the play mode should the anticipated levels of presence not be attained by the participants. During the design phase, the team addressed the immersive substrate of the XR experience by supporting the alignment of the three layers of nonmediated presence that form the basis of our daily experience in the world: the proto-self, the core presence, and the extended presence. The initial design choices regarding the inscribed system are put to the test once the audience enters the XR environment.

At the iteration phase, the design team needs to revisit their choices to find out if there are any breaks in presence caused by the way in which the immersive system is capable of supporting the three layers of conscious experience. Along with the five senses, the three layers of presence must be integrated at each moment to maintain the illusion of reality in the particular situation the XR environment is presenting to the audience.

The design team can assess this alignment by paying close attention to the three key markers of conscious experience: the focus, the locus, and the sensus aspects of the participant's attention. While the focus aspect of the model helps the designers consider the degree of attention paid by the participant to events happening within the immersive experience, the locus addresses where the participant is focusing their attention, be it the physical world, the immersive world conveyed through media, or somewhere in between.

All of these options might be desirable depending on the type of experience the audience is seeking. Being able to attend to both the virtual and physical environments at the same time is highly desirable for AR applications but could be problematic for VR gaming. Iterating for presence means aligning the inscribed level of desired presence with the actual delivered immersive support.

In the context of XRXD, sensus refers to the level of attentional arousal engendered in the participant by the immersive experience. Arousal directly affects the

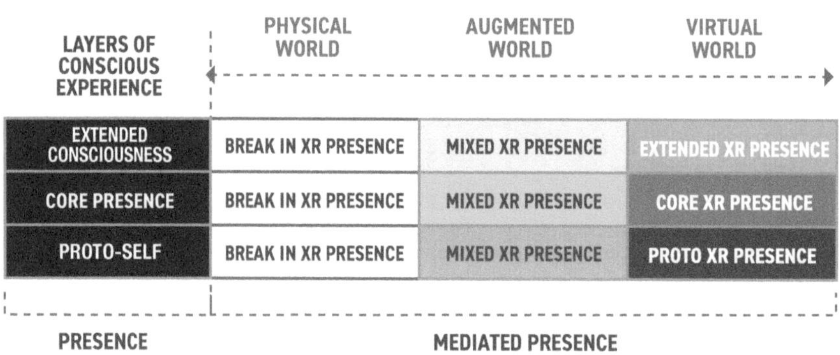

Figure VII.4 The locus aspect of mediated presence

Source: Concept and illustration by © 2024 Peter (Zak) Zakrzewski, based on Waterworth & Waterworth, The Three Dimensions of Virtual Experience; Damasio, Theory of Consciousness.

degree of engagement and activity exhibited by the participant during the experience. The sensus spectrum includes the state of consciousness known as flow originally investigated by Mihaly Csikszentmihalyi (2009). As a positive state of consciousness characterized by deep involvement, absorption, and enjoyment in challenging tasks, flow is well known to the game design community.

The challenge in supporting flow lies in ensuring that the levels of challenges faced by the participants are perfectly matched with their personal resources and skills. The multimodal usability studies we discussed in Chapter 9 will help the team assess the match between the audience expectations and skill level and the tasks presented by the XR system.

The assessment of the sensus associated with the XR experience can be performed based on the experience fluctuation model we originally presented in Chapter 11. As we discussed, low challenges matched by low skill levels elicit such states as worry, apathy, and boredom. High challenge levels not matched by similar skill levels can produce anxiety in the audience. Lower challenge levels matched

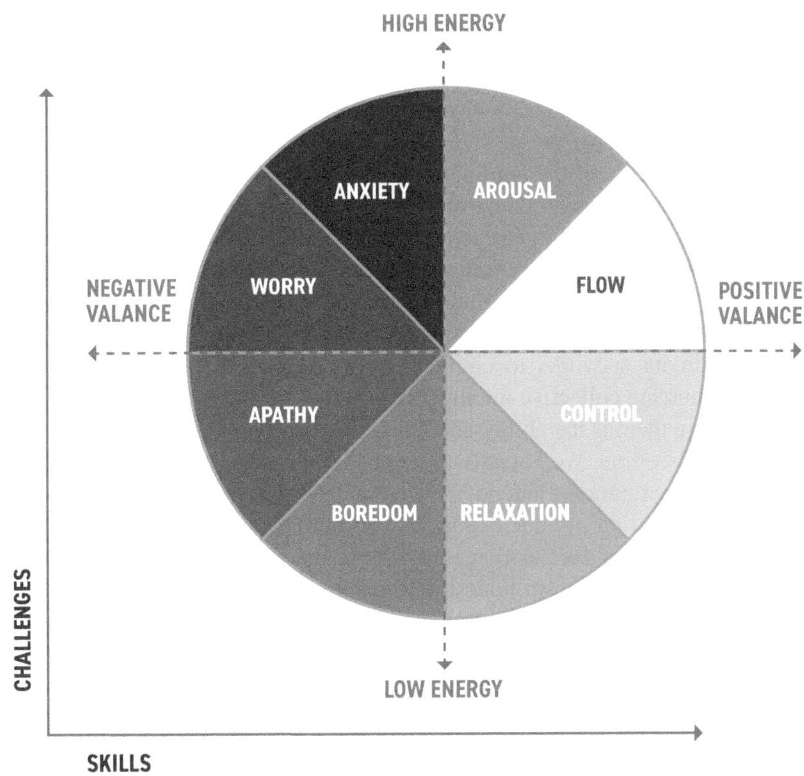

Figure VII.5 The Sensus aspect of mediated presence

Source: Illustration by © 2024 Peter (Zak) Zakrzewski, based on Riva & Waterworth, Presence and the Self. Delle Fave et al. Experience Fluctuation Model.

by high skill levels can produce the state of relaxation and control. The desirable state of flow emerges when the challenges presented by the immersive experience are matched by the equally high levels of participants' skills and resources. We augmented the original model with two additional continua of positive/negative valence and high and low energy to help the design teams assess the levels of affect felt by the participants during the experience.

Social XR: Optimizing XR Systems for Collective Intelligence

What used to be considered a hyper level of connections during the Web 2.0 era is about to grow in speed and strength in the Web 3.0 world of immersive media. The increased hyperconnectivity will affect cultural production in ways we cannot yet fully appreciate or anticipate. Christoph Riedl and colleagues argue that collective intelligence (CI) enabled by network society is one of the most promising paths toward solving complex problems in science and business.

One of their findings shows that for teams who excel in group task performance, the collaboration process is more important in predicting success than the skills of individual members (Riedl et al., 2021). In an interview with Peter, Henry Jenkins described his definition of collective intelligence as being based on the work of Pierre Levy, who argued that in a network society, nobody knows everything, but everybody knows something. By being connected to others in our network, we gain access to an enormous array of diverse kinds of expertise and knowledge we can rely on to make sense of the world around us. The more we broaden our access to different kinds of expertise, the stronger the position we are in to solve problems and create new opportunities (personal communication, 2009).

Jenkins suggests that the power of collective intelligence lies in making what any member knows available to the entire network on an ad-hoc basis at the right time when the information is needed. Participatory culture relies on critical information within its networks to circulate freely. Jenkins outlines key conditions needed for spurring collective intelligence.

First among them is the ability and willingness among the network members to engage in deliberation. We humans are very susceptible to committing logical fallacies regarding arguments that do not agree with our worldview. We are, however, very good at pointing out such fallacies committed by others. This odd feature of human cognition makes us potentially more rational collectively, provided we allow ourselves to engage in honest deliberation. A collectively intelligent community is predicated on pooling of knowledge, working through things together, weighing possible solutions, challenging each other's solutions, and showing proof for key assumptions and arguments.

Jenkins views open access to information as another critical condition of collective intelligence. A network optimized for collective intelligence must allow free circulation of information. It is also vitally important that the members of the network trust in the process behind the information exchange. A system optimized for generating collective intelligence must have built in mechanisms of reward and attribution for those making a meaningful contribution.

Scientists from such diverse disciplines as mathematics and therapeutics have successfully experimented with the ability of a group of connected individuals and computers to work together to solve a range of complex problems such as proving the Hales-Jewett theorem or simulating protein dynamics. Researchers at MIT's Center for Collective Intelligence have found that open and connected innovation contests generate effective solutions not only because potential solutions are offered by more people, but perhaps more importantly, because the ideas come from very different kinds of people (Duhaime et al., 2019). By analyzing a dataset on the activity of 12,000 scientists in 166 contests on the open-innovation site Innocentive, Jeppesen and Lakhani found that the most unique ideas came from people who were socially marginalized or from people whose technical expertise lay outside of the specialized field. Chemists, for instance, made highly creative contributions to solving biology problems (Jeppesen & Lakhani, 2010).

While collective intelligence addresses participation among human actors within traditional or Internet-based networks, the concept of Superminds adds synthetic AI-based agents to the augmented intelligence mix. Thomas Malone, the founder of the MIT Center for Collective Intelligence, highlights the creative potential of Superminds as "the surprising power of people and computers thinking together" (Malone, 2018, p. I). Mixing humans and advanced computational resources in human+computer networks is still a relatively nascent practice.

But we can already see the promise. Generative AI can augment human creativity by generating a seemingly limitless flow of new ideas exploring alternative possibilities. It does so by engaging in divergent explorations that make associations among remote concepts and produce innovative ideas drawn from them (Eapen et al., 2023).

To explore this potential, Tojin T. Eapen, Daniel J. Finkenstadt, Josh Folk, and Lokesh Venkataswamy at *Harvard Business Review* used Midjourney's text-to-image algorithm to create an image that combined an elephant and a butterfly. Midjourney responded by producing a chimera of a "phantafly" worthy of being included in the next Pixar film. Eapen and his colleagues argue that in addition to promoting divergent thinking, collaborating with advanced AI algorithms can also help teams to challenge the expertise bias, assist in idea evaluation, support idea refinement, and facilitate open collaboration (2023).

The notion of studying social phenomena such as idea flows in terms of laws and mathematics of motion goes back to the founder of political philosophy, Thomas Hobbes. The concept was further developed by generations of "social physicists," which included the founder of modern sociology, the French philosopher Auguste Comte, and Belgian statistician Adolphe Quetelet, who argued that society could be understood using mathematical modeling, probability, and statistics.

Helder Coelho defines social physics as quantitative social science aimed at mathematically describing the connections between the flow of ideas and information among individuals within social networks, which affects their habits, values, beliefs, decision-making, and behavior (Coelho, 2016).

In his book *Social Physics,* Alex Pentland focused on the potential of computational social science to study influence and emergent behavior within social networks and communities of peers in order to refine policy-level decision strategies (2015).

Current interest in quantifying human behavior was preceded by an often-misunderstood empirical behaviorist, B.F. Skinner, who was inspired by asking whether the science of human behavior could lead to a technology that could be applied for the betterment of humankind. Skinner's research was motivated by finding solutions to such grand challenges as war, overpopulation, pollution, and racial discrimination.

One of Pentland's contributions to social physics is the idea of leveraging the honest signals of behavior collected as digital footprints that can be analyzed using advanced data science algorithms. He refers to this process as reality mining. Pentland, along with his students and colleagues, has used this approach to predict such outcomes as which individuals might be likely to develop diabetes, which start-up team might secure investments, or who will pay back their loans.

Pentland rejects the idea of people as independent decision-makers. He argues for the importance of factoring in the dynamic social effects as the true driving forces behind such social phenomena as economic bubbles, political revolutions, social media memes, and the entire economy of Web 2.0. These social phenomena are outcomes of billions of microinteractions occurring between individuals trading not only goods and money but, maybe even more importantly, worldview-shaping ideas that affect their individual and collective behavior.

Pentland argues that decisions made by social actors are a blend of personal and socially acquired information. In the instances when personal information is weak, individuals tend to rely more on their social networks. Social learning has a confidence-enhancing effect in situations involving uncertainty (Pentland, 2015). Pentland suggests that the only way to understand the digitally hyperconnected world is to extend our definitions of key economic and political ideas to include the effects of millions of digital citizens and consumers engaging in communal voyages of discovery that influence their opinions, values, and behavior.

The important insight from social physics is that XR systems can be highly persuasive environments. Behavioral nudging can be best accomplished by providing users with types of social network incentives that affect idea flows within a given network. Having the algorithmically enhanced meta-level view of the XR world gives designers not only an understanding of how human+computer systems work but also the tools to take steps to iterate the systems and fix the problems inherent in their status quos. By creating the richest possible quantitative descriptions of social networks, social physics can build quantitative, predictive models of human behavior applicable to immersive interactions.

Pentland has identified three types of design-level strategies best suited to the social physics within sociotechnical systems.

1. Social mobilization enables designers to recruit people from target communities to help solve problems on an ad hoc basis. Pentland and his team used

this strategy to win the 2009 Defense Advanced Research Projects Agency (DARPA) Red Balloon Challenge in under 9 hours. The challenge posed to the competing teams was to locate and report ten red balloons placed randomly all over the United States. MIT's team chose to offer rewards to the people who correctly reported any red balloons and those who recruited the eventual balloon finders.

2. The second type of strategy involves tuning a given social network to provide sufficient idea diversity. Research showed that social networks of struggling communities suffer from two types of problems:

    a. Insufficient diversity of information sources and
    b. Echo-chamber effects. Pentland's research team was able to tune idea flows within the communities they studied by providing small nudges, or incentives that caused isolated people to engage more with others in their network. They also provide incentives to nudge people who were hyperconnected within the network to explore new contacts outside of their current network.

3. Pentland's third strategy involves leveraging social engagement to increase collaboration by connecting people around resonant problems within their communities. The strategy addresses the well-known tragedy-of-the-commons problem in which individuals actively acting in self-interest collectively diminish the communal resources (Pentland, 2015).

Tech ethnographer Tricia Wang suggests that many of the now-identified Web 2.0 problems came down to the issue of poor representation and participation between the people building and using the digital tools (Wang et al., 2024). Open participation is one of the critical building blocks of successful immersive systems.

Henry Jenkins's four conditions for broader participation in media circulation might serve as a useful guide for developing the next web platform. Any open circulation of media content is predicated on the existence of social networks that can link communities based on shared values through the exchange of meaningful bytes. The measure of the meaningfulness of the bytes is based on the attributes of a media text that resonate with a particular community's motivation for sharing the content. The resulting act of sharing is enabled by technical resources that make media content easy to circulate. Finally, the various economic structures, or business models, ensuring the economic viability of the underlying technology platforms must support the open flow of content.

Purposeful Emergence: Redesigning the Ecology of Sociotechnical Systems

Biologist E.O. Wilson argued that an ability to accurately describe and understand complex systems was the great 21st-century challenge of scientific holism and the greatest challenge in all of science today (Wilson, 1999, p. 93). As great as this challenge might be, the ability to create entire virtual worlds takes the XR design community beyond describing and understanding complexity and into the realm of creating and maintaining artificial ecologies.

System designers' structural and procedural control over all manner of human-made systems is part of a larger shift brought on by the discovery of software and genetic engineering that is disrupting the Darwinian logic of random mutation and natural selection with entirely new generative rules (Enriquez & Gullans, 2015). Today, system designers can leverage the power of two revolutionary rules of non-random mutation and unnatural selection to create synthetic ecosystems.

The emerging wide adoption of XR as a fully fledged communication medium brings us ever closer to a new domain—the world of artificially designed persuasive and purposeful sociotechnical systems. Outsourcing of computationally intensive cognitive tasks combined with purposeful intentionality of designing the artificial might offer us one of the most intriguing and promising models of human+computer intelligence augmentation. It is, however, important to point out that this promise is inextricably mixed with the specter of the reappearance of social engineering, as exemplified by the power exerted by such data mining companies as Meta Platforms and Google.

The concept of social engineering has its roots in the work of sociologist Ferdinand Tönnies, who argued that complex modern societies could not operate successfully without developing new methods of social management. To ensure such critical outcomes as safety and sustainability, critical executive-level decisions had to be made using advanced management techniques based on reliable statistical data. Tönnies suggested that social engineering was the best way to manage the Earth's finite resources and human capital while allowing the highest levels of freedom, prosperity, and well-being to be shared by all members of the population.

In the social sciences, social engineering is considered to be a top-down approach to influence, aimed at eliciting particular attitudes and behaviors on a system-level scale. Social engineers apply the scientific method to obtain and analyze social system data to design programs and apply appropriate methods to achieve desired results within social networks.

Karl Popper referred to social engineering as the application of the critical and rational methods of science to the specific problems facing open societies (Popper, 1979). He made a clear distinction between his proposal and the totalitarian social engineering he witnessed in Nazi Germany and the Soviet Union. Popper viewed democratic social engineering as a reasonable method of achieving positive social transformation and improving the individual and social wellbeing while also acknowledging that its coercive forms could lead to an increase in human suffering.

Discussion about the possibilities for coercive manipulation afforded by social engineering began seeping into the public discourse following the abuses perpetrated by various 20th-century authoritarian regimes. Critique of social engineering in such areas as advertising and bias-based journalism have gradually inspired deep mistrust toward such practices in Western capitalist societies.

The specter of misusing the transformative power of social engineering compels us to consider the role of system designers. As discussed, XR media environments give system designers the power to not only structure the initial conditions of sociotechnical systems but also iterate and recalibrate the system in response to data obtained from the system and user behavior.

System designers can only accomplish such tasks from the perspective of what family therapist and communication theorist Paul Watzlawick referred to as fourth-order knowledge. First-order knowledge is based on our sensory awareness of objects in the environment. Second-order knowledge concerns the knowledge we develop about objects we sense. We attain the third-order of knowledge when we create explanatory principles about our environment and the objects in it.

The mindset or a paradigm we create with the third order of knowledge can only be examined and iterated from the still-higher fourth order. Fourth-order knowledge is a type of understanding that gives family therapists and system designers a meta-level position—a superpower Donella Meadows referred to as mastery over paradigms. The fourth-order knowledge allows system designers to step outside of the frame of a particular sociotechnical system to communicate and think about the system itself (Watzlawick et al., 1967).

One of the original Macy Conference participants, Gregory Bateson, argued that interactive systems always demand thinking in terms of ecology (Bateson, 1972). The art of designing purposeful yet ecologically sound immersive systems lies in finding the right balance between the system's inherent ability to self-organize and the designers' task to keep the system on track towards its intended function. An attempt to exert too much control over the XR system can lead the design team to engage in a heavy-handed form of social engineering.

Like the best games, the most engaging immersive experiences allow participants to move freely within a system of well-thought-out constraints. No one is going to enjoy an immersive experience run like a Victorian boarding school. Hey! Designers! Leave them kids alone! As a class of system stakeholders, designers are but one element contributing to an overall emergence.

As mentioned before, in the context of systems design, emergence is said to occur when a system exhibits properties its components parts do not seem to possess on their own. System-level emergent behaviors occur only when the parts interact as a whole. Conway's Game of Life might be one of the best-known examples of emergent behavior arising out of a network of simple cellular automata arranged in grids of cells and evolving according to a set of simple rules guiding the interactions between the cells. Despite the rules being only defined locally, they allow the entire system to demonstrate an ability to produce very unexpected global-level behavior.

Clearly, human actors within immersive systems are a lot more complex than Conway's cellular automata. Within multiminded systems such as democratic societies and XR systems optimized for collective intelligence, self-organization is directly linked with the agency exercised by the human actors. Audience members' agency, along with the system's external environment, falls within the two system dimensions over which designers cannot exert direct control. This leaves the design team with the five system dimensions they can control. We covered the designer-controllable part of system iteration by outlining the five immersive system dimensions of (1) function, (2) structure, (3) process, (4) semantics, and (5) pragmatics.

Donella Meadows's list of 12 leverage points complements this checklist of iterative tools available to XR system designers with the hierarchy of leverage points in terms of their relative impact.

For the cautious Prometheus, mixing humans and nonhumans within immersive XR systems offers a glimpse at an intriguing possibility of co-designed or synthetic emergence. One of the most interesting examples of this approach is a method known as parametric design. Architectural practice based on parametric design does not yet have a robust theory but is mostly an area defined by the work of such well-known architects as Zaha Hadid, Frank Gehry, and Santiago Calatrava and promoted by such software companies as Autodesk.

The most circulated definition of parametric design comes from a somewhat disputed Wikipedia entry. It describes it as a design method in which various formal building elements and engineering components are shaped indirectly according to algorithmic processes rather than being affected by the designers directly (Wikipedia Authors, 2024). The name refers to the application of input parameters that guide the algorithmic search for the desired solution.

Within the parametric design paradigm, designers indirectly influence rather than dictate the form-finding process by factoring in the parameters as rules that, in turn, determine the relationship between design intent and design outcome. Parametric architectural form-finding optimizes the entire search process for achieving specific design goals chosen by the design team against a set of external and internal constraints. In the area of urban planning, parametric processes have allowed designers to optimize pedestrian and vehicle traffic by generating and comparing the performance of multiple traffic flow options.

The promise of applying genetic algorithms beyond generating unusual architectural forms was first expressed by Christopher Alexander, who, in his 1964 book titled *Notes on the Synthesis of Form*, discussed "the task of designing a complete environment for a million people" (Alexander, 1964, p. 2). He argued that an increasing number of design problems were reaching what he referred to as "insoluble levels of complexity" (p. 2). Applied to XR systems, a parametric design approach that mixes advanced computation with human judgment can bring us a step closer to humane purposeful emergence that benefits us all.

Summary

The iterative design process is depicted as a cycle in which hypotheses are formed, prototypes are developed, experiments are conducted, and insights are gathered, leading to refinements. This methodology is particularly suited to XR environments, which benefit from regular testing to align user experience with evolving expectations. Through iterative design, teams can catch potential issues early, improving usability and creating immersive, cost-effective experiences.

XR systems are complex sociotechnical entities, and theories like Galois's theory of groups and Russell's theory of logical types provide frameworks to understand and manage the interactions within these systems. These theories emphasize the interdependence and structure of parts within a system, encouraging designers to adopt a "meta-level" view. From this perspective, designers can oversee and modify systems, recognizing that some changes, such as rule adjustments or structure overhauls, are essential for achieving substantial transformations, akin to paradigm shifts.

Applying Donella Meadows's 12 leverage points, we underscore strategic areas for intervention in XR systems that range from modifying system parameters to adjusting the mindset or paradigm shaping the system's operation. By employing these leverage points thoughtfully, designers can guide immersive systems toward desired outcomes, such as optimized collective intelligence or enhanced user engagement. Embracing the XRXD framework allows designers to create purposeful immersive systems that align with user expectations and adapt dynamically to external influences.

Note

1 Antinomies are contradictions between two beliefs, each of which might in itself be reasonable. For example, in the statement "There is no absolute truth," the claim contradicts itself by attempting to state absolute truth about absolute truth.

References

Alexander, C. (1964). *Notes on the synthesis of form*. Harvard University Press.

Bateson, G. (1972). *Steps to an ecology of mind*. Ballantine Books.

Coelho, H. (2016). Social physics: How good ideas spread, the lessons from a new science. *JASSS*, 19(1). www.jasss.org/19/1/reviews/2.html

Csikszentmihalyi, M. (2009). *Flow: The psychology of optimal experience*. Harper and Row.

Duhaime, E. P., Bond, B. M., Yang, Q., Boer, P. de, & Malone, T. W. (2019). Recruiting hay to find needles: Recursive incentives and innovation in social networks. *SSRN Electronic Journal*.

Eapen, T. T., Finkenstadt, D. J., Folk, J., & Venkataswamy, L. (2023, August). How generative AI can augment human creativity. *Harvard Business Review*. https://hbr.org/2023/07/how-generative-ai-can-augment-human-creativity

Enriquez, J., & Gullans, S. (2015). *Evolving ourselves: Redesigning the future of humanity—one gene at a time*. Penguin Publishing Group.

Jeppesen, L. B., & Lakhani, K. R. (2010). Marginality and problem-solving effectiveness in broadcast search. *Organization Science*, 21(5), 1016–1033.

Lanier, J. (2017). *Dawn of the new everything*. Henry Holt and Co.

Malone, T. W. (2018). *Superminds: The surprising power of people and computers thinking together* (1st ed.). Little, Brown and Company.

Meadows, D. H. (1997). Leverage points: Places to intervene in a system. *Whole Earth Winter*. https://donellameadows.org/archives/leverage-points-places-to-intervene-in-a-system/

Meadows, D. H. (2009). *Thinking in systems: A primer*. Earthscan.

Parker, C. B. (2015, January 23). Michal Kosinski: Computers are better judges of your personality than friends. *Insights by Stanford Business*. www.gsb.stanford.edu/insights/michal-kosinski-computers-are-better-judges-your-personality-friends

Pentland, A. (2015). *Social physics: How social networks can make us smarter*. Penguin Press.

Popper, K. R. (1979). *Objective knowledge: An evolutionary approach*. Clarendon Press.

Riedl, C., Kim, Y. J., Gupta, P., Malone, T. W., & Woolley, A. W. (2021). Quantifying collective intelligence in human groups. *Proceedings of the National Academy of Sciences*, 118(21). https://doi.org/10.1073/pnas.2005737118

Wang, P., Miller, M. R., Han, E., DeVeaux, C., & Bailenson, J. N. (2024). Understanding virtual design behaviors: A large-scale analysis of the design process in Virtual Reality. *Design Studies*, 90, 101237. https://doi.org/10.1016/j.destud.2023.101237

Watzlawick, P., Bavelas, J. B., & Jackson, D. D. (1967). *Pragmatics of human communication: A study of interactional patterns, pathologies, and paradoxes*. Norton.

Wikipedia Authors. (2024). Parametric design. *Wikipedia*. https://en.wikipedia.org/w/index.php?title=Parametric_design&oldid=1249866860

Wilson, E. O. (1999). *Consilience: The unity of knowledge*. Vintage Books.

Zakrzewski, P. (2022). *Designing XR: A rhetorical design perspective for the ecology of human+computer systems*. Emerald Publishing Limited. http://doi.org/10.1108/9781802623659

Index

Note: Page numbers in *italics* indicate a figure, and page numbers in **bold** indicate a table on the corresponding page.